NONSMOOTH VECTOR FUNCTIONS AND CONTINUOUS OPTIMIZATION

Optimization and Its Applications

VOLUME 10

Managing Editor
Panos M. Pardalos (University of Florida)

Editor—Combinatorial Optimization
Ding-Zhu Du (University of Texas at Dallas)

Advisory Board
J. Birge (University of Chicago)
C.A. Floudas (Princeton University)
F. Giannessi (University of Pisa)
H.D. Sherali (Virginia Polytechnic and State University)
T. Terlaky (McMaster University)
Y. Ye (Stanford University)

Aims and Scope
Optimization has been expanding in all directions at an astonishing rate during the last few decades. New algorithmic and theoretical techniques have been developed, the diffusion into other disciplines has proceeded at a rapid pace, and our knowledge of all aspects of the field has grown even more profound. At the same time, one of the most striking trends in optimization is the constantly increasing emphasis on the interdisciplinary nature of the field. Optimization has been a basic tool in all areas of applied mathematics, engineering, medicine, economics and other sciences.

The series *Optimization and Its Applications* publishes undergraduate and graduate textbooks, monographs and state-of-the-art expository works that focus on algorithms for solving optimization problems and also study applications involving such problems. Some of the topics covered include nonlinear optimization (convex and nonconvex), network flow problems, stochastic optimization, optimal control, discrete optimization, multi-objective programming, description of software packages, approximation techniques and heuristic approaches.

NONSMOOTH VECTOR FUNCTIONS AND CONTINUOUS OPTIMIZATION

By

V. JEYAKUMAR
University of New South Wales, Sydney, NSW, Australia

D.T. LUC
University of Avignon, Avignon, France

 Springer

V. Jeyakumar
University of New South Wales
School of Mathematics and Statistics
Sydney
Australia

D.T. Luc
University of Avignon
Department of Mathematics
Avignon
France

ISBN-13: 978-1-4419-4472-6   e-ISBN-13: 978-0-387-73717-1

Dedicated to our families

Contents

Preface

Thinking in terms of choices is common in our cognitive culture. Searching for the best possible choice is a basic human desire, which can be satisfied, to some extent, by using the mathematical theory and methods for examining and solving optimization problems, provided that the situation and the objective are described quantitatively. An optimization problem is a mathematical problem of making the best choice from a set of possible choices and it has the form of optimizing (minimizing or maximizing) an objective function subject to constraints. Continuous optimization is the study of problems in which we wish to optimize a continuous (usually nonlinear) objective function of several variables often subject to a collection of restrictions on these variables. Thus, continuous optimization problems arise everyday as management and technical decisions in science, engineering, mathematics and commerce.

The mathematical studies of optimization are grounded in the development of calculus by Newton and Leibniz in the seventeenth century. The traditional differential calculus of vector functions is based on the very basic idea of gradient vectors or the Jacobian matrices, which have also played a fundamental role in many advances of mathematical and computational methods. These matrices do not always exist when a map or system is not differentiable (not smooth). A recent significant innovation in mathematical sciences has been the progressive use of nonsmooth calculus, an extension of the differential calculus, which is now a key tool of modern analysis in many areas of mathematics and engineering.

Several recent monographs have provided a systematic exposition and a state-of-the-art study of nonsmooth variational analysis. Focusing on the study of vector functions, this book presents a comprehensive account of the calculus of generalized Jacobian matrices and their applications to continuous optimization in finite dimensions. It was motivated by our desire to expose an elementary approach to nonsmooth calculus by using a set of matrices to replace the nonexistent Jacobian matrix of a continuous vector function. Such a set of matrices forms a new generalized Jacobian, called pseudo-Jacobian. It is a direct extension of the classical derivative and at the same time provides an axiomatic approach to nonsmooth calculus. It enjoys simple rules of calculus and gives a flexible tool for handling nonsmooth continuous optimization problems.

In Chapter 1, the notion of pseudo-Jacobian is introduced and illustrated by numerous examples from known generalized derivatives. The basic properties of pseudo-Jacobians and methods for constructing stable pseudo-Jacobians are also presented. In Chapter 2, a whole machinery of calculus is developed for pseudo-Jacobians including a mean value theorem and chain rules. Diversity and simplicity of calculus rules of pseudo-Jacobians empower us to combine different kinds of generalized derivatives in solving variational problems. In the remaining three chapters, applications to openness of continuous vector functions, nonsmooth mathematical programming, and to variational inequalities are given. They demonstrate that pseudo-Jacobians are amenable to the study of a number of important variational problems.

We hope that this book will be useful to graduate students and researchers in applied mathematics and related areas. We have attempted to present proofs of theorems that best represent the classical technique, so that readers with a modest background in undergraduate mathematical analysis can follow the material with minimum effort. Readers who are not very familiar with other notions of generalized derivatives of nonsmooth functions can skip Sections 1.3, 1.4, and 1.8 at their first reading.

Acknowledgment. We have been developing the material for the book for several years and it is a result of a long and fruitful collaboration between the authors, supported by the University of New South Wales. We are grateful to the University of New South Wales and the University of Avignon for their assistance during the preparation of the book. We have also benefited from feedback and suggestions from our colleagues. We wish to particularly thank Bruce Craven, Jean-Paul Penot, Alexander Rubinov, and Xiaoqi Yang. We are also grateful to Beata Wysocka for her suggestions and extensive comments that have contributed to the final preparation of the book. Finally, we wish to thank John Martindale and Robert Saley for their assistance in producing this book.

Sydney and Avignon   *V. Jeyakumar*
January 2007   *D.T. Luc*

1

Pseudo-Jacobian Matrices

In this chapter we introduce pseudo-Jacobian matrices for continuous vector functions. This concept, which has been termed as approximate Jacobian matrices in the earlier publications of the authors in [44–51] and [78-82] can be regarded as an axiomatic approach to generalized derivatives of nonsmooth vector functions. We then show that many well-known generalized derivatives are examples of pseudo-Jacobians.

1.1 Preliminaries

We begin by presenting some preliminary material on classical calculus.

Notations

Throughout the book $\mathbb{R}^n$ denotes the n-dimensional Euclidean space whose Euclidean norm for $x = (x_1, \ldots, x_n) \in \mathbb{R}^n$ is given by

$$\|x\| = [\sum_{i=1}^{n} (x_i)^2]^{1/2}.$$

The inner product between two vectors x and y in $\mathbb{R}^n$ is defined by

$$\langle x, y \rangle = \sum_{i=1}^{n} x_i y_i.$$

The closed unit ball of $\mathbb{R}^n$, denoted B_n, is defined by

$$B_n := \{x \in \mathbb{R}^n : \|x\| \leq 1\},$$

and the open unit ball of $\mathbb{R}^n$ is the interior of B_n, and is given by

$$\text{int}(B_n) := \{x \in \mathbb{R}^n : \|x\| < 1\}.$$

Given a nonempty set $A \subseteq \mathbb{R}^n$, the notation $\text{cl}(A)$ stands for the *closure* of A, and $\text{int}(A)$ stands for the *interior* of A. The *conic hull* and the *affine hull* of A are, respectively, defined by

$$\text{cone}(A) := \{ta : a \in A, t \in \mathbb{R}, t \geq 0\}$$

$$\text{aff}(A) := \{\sum_{i=1}^{k} t_i a_i : a_i \in A, t_i \in \mathbb{R}, i = 1, \ldots, k\}.$$

It is clear that $\text{cone}(A)$ is a cone; that is, it is invariant by multiplication with positive numbers, and $\text{aff}(A)$ is an affine subspace of $\mathbb{R}^n$.

Let $L(\mathbb{R}^n, \mathbb{R}^m)$ be the space of real $m \times n$-matrices. Each $m \times n$-matrix M can be regarded as a linear operator from $\mathbb{R}^n$ to $\mathbb{R}^m$; so for a vector $x \in \mathbb{R}^n$ one has $M(x) \in \mathbb{R}^m$. The transpose of M is denoted by M^{tr} and considered as a linear operator from $\mathbb{R}^m$ to $\mathbb{R}^n$. Sometimes the writing vM for $v \in \mathbb{R}^m$ is used instead of $M^{tr}(v)$. Let us endow $L(\mathbb{R}^n, \mathbb{R}^m)$ with the norm of linear operators

$$\|M\| = \sup_{\|x\| \leq 1} \|M(x)\|.$$

This norm is equivalent to the Euclidean norm defined by

$$|M| = (\|M_1\|^2 + \cdots + \|M_n\|^2)^{1/2},$$

where $M_1, \ldots, M_n \in \mathbb{R}^m$ are n columns of the matrix M. The closed unit ball in the space $L(\mathbb{R}^n, \mathbb{R}^m)$ is denoted $B_{m \times n}$.

Convex Sets

A set A in $\mathbb{R}^n$ is said to be *convex* if the segment joining any two points of A lies entirely in A, which means that for every $x, y \in A$ and for every real number $\lambda \in [0, 1]$, one has $\lambda x + (1 - \lambda)y \in A$. It follows directly from the definition that the intersection of convex sets, the Cartesian product of convex sets, the image and inverse image of a convex set under a linear transformation, and the interior and the closure of a convex set are convex. In particular, the sum $A_1 + A_2 := \{x + y : x \in A_1, y \in A_2\}$ of two convex sets A_1 and A_2 is convex; the conic hull of a convex set is convex.

The *convex hull* of A, denoted $\text{co}(A)$, consists of all convex combinations of elements of A; that is,

$$\text{co}(A) := \{\sum_{i=1}^{k} \lambda_i x_i : x_i \in A, \lambda_i \geq 0, i = 1, \ldots, k, \text{ and } \sum_{i=1}^{k} \lambda_i = 1\}.$$

It is the intersection of all convex sets containing A. The closure of the convex hull of A is denoted $\overline{\mathrm{co}}(A)$, which actually is the intersection of all closed convex sets containing A. The following result known as Caratheodory's theorem shows that the convex hull of a set in $\mathbb{R}^n$ can be obtained by convex combinations in which at most $n + 1$ elements take part.

Theorem 1.1.1 *Suppose that $A \subseteq \mathbb{R}^n$ is a nonempty set. Then each element of the convex hull of A can be expressed as a convex combination of at most $(n + 1)$ points of A.*

Proof. Let $x \in \mathrm{co}(A)$. By definition there are $x_1, \ldots, x_k \in A$ and positive numbers $\lambda_1, \ldots, \lambda_k$ with $\sum_{i=1}^k \lambda_i = 1$ such that

$$x = \sum_{i=1}^k \lambda_i x_i.$$

If $k \leq n + 1$, we are done. If not, the system of vectors $\{x_1 - x_k, \ldots, x_{k-1} - x_k\}$ is linearly dependent. Then, there exist real numbers, $\alpha_i, i = 1, \ldots, k - 1$, not all zero, such that

$$\sum_{i=1}^{k-1} \alpha_i (x_i - x_k) = 0.$$

Setting $\alpha_k = -\alpha_1 - \ldots - \alpha_{k-1}$, one deduces

$$\sum_{i=1}^k \alpha_i x_i = 0 \text{ and } \sum_{i=1}^k \alpha_i = 0.$$

Choose $\lambda = \max_{i=1,\ldots,k} \alpha_i / \lambda_i$ and set $\gamma_i = \lambda_i - \alpha_i / \lambda$. Then $\lambda > 0$ and $\gamma_i \geq 0$ with $\sum_{i=1}^k \gamma_i = 1$. Moreover, among γ_is there is at least one that equals zero and

$$x = x - 0 = \sum_{i=1}^k \lambda_i x_i - (1/\lambda) \sum_{i=1}^k \alpha_i x_i = \sum_{i=1}^k \gamma_i x_i$$

is a convex combination of less than k points of A. Continuing this process until $k = n + 1$ completes the proof.   □

Let A be a nonempty convex set in $\mathbb{R}^n$. The interior of the set A with respect to the affine hull $\mathrm{aff}(A)$ is called the *relative interior* of A and is defined by

$$\mathrm{ri}(A) := \{x \in \mathrm{aff}(A) : (x + \epsilon B_n) \cap \mathrm{aff}(A) \subseteq A \quad \text{for some } \epsilon > 0\}.$$

It is important to note that every nonempty convex set in $\mathbb{R}^n$ has a nonempty relative interior. The next theorem on separation of convex sets is one of the fundamental results of mathematical analysis.

Theorem 1.1.2 *Suppose that $A \subseteq \mathbb{R}^n$ is a nonempty convex set not containing the origin. Then there exists a nonzero vector ξ of $\mathbb{R}^n$ such that*

$$\langle \xi, x \rangle \geq 0 \quad \text{for every } x \in C.$$

If, in addition, C is closed, then the vector ξ can be chosen so that the above inequality is strict.

A simple proof of this theorem is obtained by the Hahn–Banach theorem which states that if A is an open convex set and L is a linear subspace of $\mathbb{R}^n$ with $A \cap L = \emptyset$, then there exists a vector ξ of $\mathbb{R}^n$ strictly separating A and L in the sense that $\langle \xi, x \rangle > \langle \xi, y \rangle = 0$ for all $x \in A$ and $y \in L$. A proof without referring to the Hahn–Banach theorem is given in Section 2.1.

Dini Directional Derivatives

Let $\phi : \mathbb{R}^n \to \mathbb{R}$ be a given function and let x and $u \in \mathbb{R}^n$. The *upper Dini directional derivative* of the function ϕ at x in the direction u, which is denoted $\phi^+(x; u)$, is defined by

$$\phi^+(x; u) := \limsup_{t \downarrow 0} \frac{\phi(x + tu) - \phi(x)}{t}.$$

Likewise, the *lower Dini directional derivative* of the function ϕ at x in the direction u, which is denoted $\phi^-(x; u)$, is defined by

$$\phi^-(x; u) := \liminf_{t \downarrow 0} \frac{\phi(x + tu) - \phi(x)}{t}.$$

The extended real values $+\infty$ and $-\infty$ are allowed in the above limits, which in fact is a peculiarity of nonsmooth functions. Note that if the upper and the lower Dini directional derivatives in a direction u are finite at a given point, then the function is continuous at that point along the direction u. The converse is not true in general. On the real line, the function $\phi(x) = \sqrt{|x|}$ is continuous, but its directional derivatives at $x = 0$ in directions $u = 1$ and $u = -1$ are infinite. When $\phi^-(x; u) = \phi^+(x; u)$ and is finite, the common value, denoted $\phi'(x; u)$, is called the *directional derivative* of ϕ in the direction u at x. When this is true for every direction u in $\mathbb{R}^n$, the function ϕ is said to be *directionally differentiable* at x.

One of the notable features of upper and lower Dini directional derivatives is that they always exist, even when the function is discontinuous. Although they are not necessarily finite, it is relatively easy to work with them, due to the following elementary properties and calculus rules.

Proposition 1.1.3 *Let ϕ and ψ be real functions on $\mathbb{R}^n$. Then the following assertions hold.*

(i)   Homogeneity: $\phi^+(x; u)$ is positively homogeneous in u; that is,

$$\phi^+(x; \lambda u) = \lambda \phi^+(x; u) \quad \text{for all } \lambda > 0.$$

(ii)   Scalar multiple: for $\lambda > 0$ one has $(\lambda \phi)^+(x; u) = \lambda \phi^+(x; u)$, and for $\lambda < 0$ one has $(\lambda \phi)^+(x; u) = \lambda \phi^-(x; u)$.

(iii)   Sum rule: $(\phi + \psi)^+(x; u) \leq \phi^+(x; u) + \psi^+(x; u)$ provided that the sum in the right–hand side exists.

(iv)   Product rule: $(\phi \psi)^+(x; u) \leq [\psi(x)\phi]^+(x; u) + [\phi(x)\psi]^+(x; u)$ provided that the sum in the right–hand side exists, the functions ϕ and ψ are continuous at x, and that either of the following conditions is satisfied: $\phi(x) \neq 0$; $\psi(x) \neq 0$; $\phi^+(x; u)$ is finite; and $\psi^+(x; u)$ is finite.

(v)   Quotient rule: $(\phi/\psi)^+(x; u) \leq ([\psi(x)\phi]^+(x; u) + [-\phi(x)\psi]^+(x; u))/[\psi(x)]^2$ provided that the expression in the right–hand side exists and the function ψ is continuous at x.

If, in addition, the functions ϕ and ψ are directionally differentiable at x, then the inequalities in the three last assertions become equalities.

Proof. This is immediate from the definition.   □

Properties and calculus rules of lower Dini directional derivatives can be obtained in a similar manner. The next result shows that upper and lower Dini directional derivatives are convenient tools for characterizing an extremum of a function.

Theorem 1.1.4 *Let $\phi : \mathbb{R}^n \to \mathbb{R}$. Then the following assertions hold.*

(i)   If $\phi(x) \leq \phi(x + tu)$ (respectively, $\phi(x) \geq \phi(x + tu)$) for all $t > 0$ sufficiently small, then $\phi^-(x; u) \geq 0$ (respectively, $\phi^+(x; u) \leq 0$). In particular, if ϕ is directionally differentiable at x, and $\phi(x) \leq \phi(y)$ (respectively, $\phi(x) \geq \phi(y)$) for every y in a small neighborhood of x, then its directional derivative at this point is positive (respectively, negative). Consequently, if $\phi'(x; u)$ is linear in u, it vanishes in all directions.

(ii)   If $\phi^+(x + tu; u) \geq 0$ for all $t \in (0, 1)$ and if the function $t \mapsto \phi(x + tu)$ is continuous on $[0, 1]$, then $\phi(x) \leq \phi(x + u)$.

Proof. The first assertion is clear. Let us prove the second one. Suppose, to the contrary, that $\phi(x) > \phi(x + u)$. Consider the function

$$h(t) := \phi(x + tu) - \phi(x) + t[\phi(x) - \phi(x + u)].$$

Clearly, h is continuous on the segment $[0, 1]$ and takes the value zero at the end points $t = 0$ and $t = 1$. Then, there exists some $t_0 \in [0, 1)$ at which h attains its maximum. Set $y := x + t_0 u$. Then $h(t_0) \geq h(t_0 + t)$ for $t \in [0, 1 - t_0]$, and hence

$$\phi(y + tu) - \phi(y) \leq t[\phi(x + u) - \phi(x)]$$

for $t > 0$ sufficiently small. By dividing both sides of the latter inequality by t and passing to the limit when t tends to 0 we deduce

$$\phi^+(y; u) \leq \phi(x + u) - \phi(x) < 0$$

which contradicts the hypothesis. The proof is complete. $\square$

We now derive a mean-value theorem for continuous functions.

Theorem 1.1.5 *Let $\phi : \mathbb{R}^n \to \mathbb{R}$ be continuous. Then for every two distinct points a and b in $\mathbb{R}^n$ one can find two points x and y in the interval $[a, b)$ such that*

$$\phi^+(x; b - a) \leq \phi(b) - \phi(a) \leq \phi^-(y; b - a).$$

In particular, if the upper Dini directional derivative $\phi^+(x; b - a)$ is continuous in the variable x on the interval $[a, b)$, then there is a point c between a and b such that

$$\phi(b) - \phi(a) = \phi'(c; b - a).$$

Proof. Consider the function

$$h(t) := \phi(a + t(b - a)) - \phi(a) + t[\phi(a) - \phi(b)].$$

Because h is continuous on the segment $[0, 1]$ and takes the value zero at the end points $t = 0$ and $t = 1$, there exist some points t_0 and t_1 in the interval $[0, 1)$ such that h attains its minimum at t_0 and maximum at t_1. Set $x := a + t_0(b - a)$ and $y := a + t_1(b - a)$. Now the first part of the theorem follows from Theorem 1.1.4. The second part is immediate from the first one and the classical intermediate value theorem. $\square$

The hypothesis on the continuity of the derivative $\phi^+(.; b - a)$ in the second part of Theorem 1.1.5 cannot be neglected. To see this, let us consider the function $\phi(x) = |x|$ on $\mathbb{R}$. It is directionally differentiable everywhere. For $a = -1$ and $b = 1$ we have

$$\phi'(x; b - a) = \begin{cases} -2 & \text{for } x < 0 \\ 2 & \text{for } x \geq 0, \end{cases}$$

which is discontinuous at $x = 0$. There exists no c between a and b such that $0 = \phi(b) - \phi(a) = \phi'(c; b - a)$. Notice, however, that $\phi(b) - \phi(a)$ does belong to the convex hull of the derivatives $\phi'(0; b - a)$ and $\phi'(0; a - b)$.

Let us denote by e_j the unit jth coordinate direction in $\mathbb{R}^n$. If ϕ is directionally differentiable at x in directions e_j and $-e_j$, and if $\phi'(x; e_j) = \phi'(x; -e_j)$ is finite, then this value, denoted $(\partial\phi(x)/\partial x_j)$, is called the partial derivative of ϕ at x in the jth variable. Thus, by definition

$$\frac{\partial\phi(x)}{\partial x_j} := \lim_{t \to 0} \frac{\phi(x + te_j) - \phi(x)}{t}.$$

The vector

$$\nabla\phi(x) := \left(\frac{\partial\phi(x)}{\partial x_1}, \ldots, \frac{\partial\phi(x)}{\partial x_n}\right)$$

is called the gradient of ϕ at x.

Lipschitz Functions

Let $\phi : \mathbb{R}^n \to \mathbb{R}$ be given and let U be an open set in $\mathbb{R}^n$. We say that ϕ is *Lipschitz* on U with a *Lipschitz constant* $k > 0$ if $|\phi(x) - \phi(y)| \leq k\|x - y\|$ for all x and y in U. We say that ϕ is Lipschitz near x, or locally Lipschitz at x, if, for some $t > 0$, ϕ is Lipschitz on the set $x + t\,\text{int}(B_n)$. The class of Lipschitz functions is quite large. It is invariant under usual operations of sum, product, and quotient. Lipschitz functions are continuous, but not always directionally differentiable. For instance, the function $\phi : \mathbb{R} \to \mathbb{R}$ with $\phi(x) = 0$ outside the interval $(0, 1)$, $\phi(x) = -2x + (2/3^i)$ on $[2/(3^{i+1}), 1/3^i)$, and $\phi(x) = 2x - 2/(3^{i+1})$ on $[1/(3^{i+1}), 2/(3^{i+1}))$, $i = 0, 1, 2, \ldots$, is Lipschitz on $\mathbb{R}$ with a Lipschitz constant $k = 2$. However, for $x = 0$ and $u = 1$ we have $\phi^+(x; u) = 1$ and $\phi^-(x; u) = 0$, which shows that ϕ is not directionally differentiable at x. Nevertheless, Lipschitz functions can be characterized by their upper and lower Dini directional derivatives as shown by the next result.

Proposition 1.1.6 *Let $\phi : \mathbb{R}^n \to \mathbb{R}$ be given and let U be an open set in $\mathbb{R}^n$. Then ϕ is Lipschitz on U with a Lipschitz constant $k > 0$ if and only if for every $x \in U$ and $u \in \mathbb{R}^n$ one has*

$$\max\{|\phi^-(x; u)|, |\phi^+(x; u)|\} \leq k\|u\|.$$

Proof. The conclusion follows from Theorem 1.1.5. □

Jacobian Matrices and Derivatives

For a vector function $f : \mathbb{R}^n \to \mathbb{R}^m$, the *directional derivative* of f at x in the direction u is defined by

$$f'(x; u) = \lim_{t \downarrow 0} \frac{f(x + tu) - f(x)}{t}.$$

When $f'(x; u)$ exists for every $u \in \mathbb{R}^n$, the function f is called *directionally differentiable* at x. Let $f_1, \ldots, f_m$ be the components of f. Then, f is directionally differentiable at x if and only if the component functions $f_1, \ldots, f_m$ are directionally differentiable at this point.

If the partial derivatives $(\partial f_i(x)/\partial x_j)$, $i = 1, \ldots, m$ and $j = 1, \ldots, n$ exist, then the $m \times n$-matrix $\nabla f(x)$, which is called the *Jacobian matrix* of f at x, is given by

$$\nabla f(x) = \begin{pmatrix} \frac{\partial f_1(x)}{\partial x_1} & \cdots & \frac{\partial f_1(x)}{\partial x_n} \\ & \cdots & \\ \frac{\partial f_m(x)}{\partial x_1} & \cdots & \frac{\partial f_m(x)}{\partial x_n} \end{pmatrix}.$$

Thus, the Jacobian matrix consists of m rows that are gradients of the component functions. We notice also that the Jacobian matrix uniquely depends upon the behavior of the function on the coordinate directions, so that its existence at a point does not imply that its component functions are directionally differentiable at that point. Moreover, the existence of a Jacobian matrix of a function does not ensure that the function is continuous. Below we present some properties of Jacobian matrices.

Proposition 1.1.7 *Let f and g be vector functions on $\mathbb{R}^n$ with values in $\mathbb{R}^m$ and let $\nabla f(x)$ and $\nabla g(x)$ be their Jacobian matrices. Then the following assertions hold.*

(i)   *The function f is directionally differentiable in every coordinate direction and the directional derivative $f'(x; e_j)$ is the transposed jth column vector of the Jacobian matrix $\nabla f(x)$.*

(ii)   *For every vector v in $\mathbb{R}^m$, the gradient of the real function $\phi(x) := v_1 f_1(x) + \cdots + v_m f_m(x)$ exists and $\nabla \phi(x) = v \nabla f(x)$.*

(iii)   *For every real number λ one has $\nabla(\lambda f)(x) = \lambda \nabla f(x)$.*

(iv)   *The Jacobian matrix at x of the sum function $f + g$ exists and $\nabla(f + g)(x) = \nabla f(x) + \nabla g(x)$.*

Proof.   This is immediate from the definition. $\square$

Jacobian matrices are very useful in expressing classical derivatives of smooth functions. We say that $f : \mathbb{R}^n \to \mathbb{R}^m$ is *Gâteaux differentiable* at x if there is an $m \times n$-matrix M such that for each $u \in \mathbb{R}^n$ one has

$$\lim_{t \downarrow 0} \frac{f(x + tu) - f(x)}{t} = M(u).$$

In this case M is called the *Gâteaux derivative* of f at x. It follows that if f is Gâteaux differentiable at x, then it is directionally differentiable at this point and $f'(x; u) = \nabla f(x)(u)$, so that M coincides with the Jacobian matrix of f at x. The converse is also true, namely, if f is directionally differentiable at x and the function $f'(x; u)$ is linear in u, then f is Gâteaux differentiable at this point provided that $\nabla f(x)(u) = f'(x; u)$ for every $u \in \mathbb{R}^n$.

When the matrix M satisfies

$$\lim_{u \to 0} \frac{f(x + u) - f(x) - M(u)}{\|u\|} = 0,$$

it is called the *Fréchet derivative* of f at x and f is said to be *Fréchet differentiable* at x. Moreover, if

$$\lim_{y \to x, u \to 0} \frac{f(y + u) - f(y) - M(u)}{\|u\|} = 0,$$

then f is said to be *strictly (Hadamard) differentiable* and M is its *strict (Hadamard) derivative* at x. It follows that a strictly differentiable function is Fréchet differentiable, which is also Gâteaux differentiable. The converse is in general not true. For instance, the real-valued function $\phi(x) = x^2 \cos(1/x)$ for $x \neq 0$ and $\phi(0) = 0$ is Fréchet differentiable, but not strictly differentiable at $x = 0$. We end this preliminary section with a sufficient condition for strict differentiability of a vector function in terms of Jacobian matrices.

Proposition 1.1.8 *Let $f : \mathbb{R}^n \to \mathbb{R}^m$ be a continuous vector function, and let $x \in \mathbb{R}^n$. Assume that the Jacobian matrix $\nabla f(y)$ of f at every point y in a neighborhood of x exists and that the map $y \mapsto \nabla f(y)$ is continuous on line segments in a neighborhood of x and continuous at x. Then f is strictly differentiable at x.*

Proof. By considering the components separately we may restrict ourselves to the case where f is a scalar function. Set $u^i = \sum_{j=i}^{n} u_j e_j, i = 1, \ldots, n$ and $u^{n+1} = 0$ for a vector $u = (u_1, \ldots, u_n)$ in $\mathbb{R}^n$. Then

$$f(y + u) - f(y) = \sum_{i=1}^{n} [f(y + u^i) - f(y + u^{i+1})].$$

Because the segment $[y+u^i, y+u^{i+1}]$ is parallel to the ith axis, we apply the mean value theorem (Theorem 1.1.5) to find a point y^i from that segment such that $f(y + u^i) - f(y + u^{i+1}) = \nabla f(y^i)(u^i - u^{i+1})$. We notice that y^i converges to x as y tends to x and u tends to 0. It follows that

$$f(y + u) - f(y) - \nabla f(x)(u) = \sum_{i=1}^{n}[\nabla f(y^i) - \nabla f(x)](u^i - u^{i+1}).$$

Dividing both sides of this equality by $\|u\|$ and passing to the limit as u tends to 0 and y tends to x, we obtain that $\nabla f(x)$ is the strict derivative of f at x. ☐

1.2 Pseudo-Jacobian Matrices

Although the concept of pseudo-Jacobian is available for functions defined on a neighborhood of the point under consideration, we describe it for continuous functions so as not to blur the presentation of the concept.

Definition

Let $f : \mathbb{R}^n \to \mathbb{R}^m$ be a continuous vector function. We say that a nonempty closed set of $m \times n$-matrices $\partial f(x) \subset L(\mathbb{R}^n, \mathbb{R}^m)$ is a *pseudo-Jacobian* of f at x if for every $u \in \mathbb{R}^n$ and $v \in \mathbb{R}^m$ one has

$$(vf)^+(x; u) \leq \sup_{M \in \partial f(x)} \langle v, M(u) \rangle, \tag{1.1}$$

where vf is the real function $(vf)(x) = \sum_{i=1}^{m} v_i f_i(x)$ for every $x \in \mathbb{R}^n$. Each element of $\partial f(x)$ is called a *pseudo-Jacobian matrix* of f at x. If equality holds in (1.1), we say that $\partial f(x)$ is a *regular pseudo-Jacobian* of f at x.

Note that this definition encompasses three known procedures of vector analysis: scalarization of the vector function f through all directions v in $\mathbb{R}^m$; approximation of the scalarized functions vf by means of upper Dini directional derivatives; and sublinearization of the approximations by a set of matrices. To illustrate this, let us consider the vector function $f : \mathbb{R}^2 \to \mathbb{R}^2$ defined by

$$f(x, y) = (\sqrt{|x|}, \sqrt{|y|}).$$

For each direction $v = (v_1, v_2)$ in $\mathbb{R}^2$ the scalarized function vf is given by

$$(vf)(x, y) = v_1\sqrt{|x|} + v_2\sqrt{|y|}.$$

The upper Dini directional derivative of vf at $(0,0)$ in direction $u = (u_1, u_2)$ is calculated as

$$(vf)^+((0,0);(u_1,u_2)) = \limsup_{t \downarrow 0} \frac{v_1\sqrt{|u_1|} + v_2\sqrt{|u_2|}}{\sqrt{t}}$$

$$= \text{sign}(v_1\sqrt{|u_1|} + v_2\sqrt{|u_2|}) \times \infty,$$

where $0 \times \infty$ is understood to be 0. Let M be a 2×2-matrix whose entries are real numbers a_{ij}, $i, j = 1, 2$. Then

$$\langle v, M(u) \rangle = \sum_{i,j=1}^{2} a_{ij} v_i u_j.$$

Because the variables x and y in the function vf are separable, it suffices to use matrices M with $a_{12} = a_{21} = 0$ in determining a pseudo-Jacobian. It is now easy to prove that for any positive numbers α and β, the set of matrices M with $|a_{11}| \geq \alpha$, $|a_{22}| \geq \beta$ and $a_{12} = a_{21} = 0$ is a pseudo-Jacobian of f at $(0,0)$.

It is worth observing that the set of matrices M with $|a_{11}| \geq 1$, $a_{11} = a_{22}$ and $a_{12} = a_{21} = 0$ is not a pseudo-Jacobian of f at $(0,0)$, although it satisfies (1.1) whenever v belongs to the set of coordinate directions $\{(1,0), (-1,0), (0,1), (0,-1)\}$.

We notice also that $\partial f(x)$ is not unique and that we do not assume that it is a convex or bounded subset of $L(\mathbb{R}^n, \mathbb{R}^m)$. This makes the concept rather flexible and covers a number of nonsmooth generalized derivatives (see Section 1.3). The use of matrices in the sub-linearization in (1.1) greatly facilitates the development of the pseudo-Jacobian based calculus as we show throughout the book. A pseudo-Jacobian produces upper estimates for the upper Dini derivatives $(vf)^+(x; u)$ via (1.1) for all $v \in \mathbb{R}^m$ and $u \in \mathbb{R}^n$. Therefore, like outer approximation of a set, it may be arbitrarily large, but can gradually be narrowed by imposing additional restrictions so that it suits a problem at hand. Our interest, often, is to obtain a pseudo-Jacobian, which is as small as possible (in the sense of set inclusion). However, for a given nonsmooth function the smallest pseudo-Jacobian does not necessarily exist. For the function $f(x) = x^{1/3}$ on the real line, one has $(vf)^+(0; u) = +\infty$ if $vu > 0$, and $(vf)^+(0; u) = -\infty$ if $vu < 0$. Any set of the form $[\alpha, \infty)$ is a pseudo-Jacobian of f at 0. Conversely, a pseudo-Jacobian of f at 0 must contain at least a sequence of positive numbers converging to ∞. Hence, the smallest pseudo-Jacobian for this function does not exist.

Basic Properties

Proposition 1.2.1 *The following properties of pseudo-Jacobians hold:*

(i) A closed set $\partial f(x) \subset L(\mathbb{R}^n, \mathbb{R}^m)$ is a pseudo-Jacobian of f at x if and only if for every $u \in \mathbb{R}^n$ and $v \in \mathbb{R}^m$ one has

$$(vf)^-(x; u) \geq \inf_{M \in \partial f(x)} \langle v, M(u) \rangle. \tag{1.2}$$

(ii) If $\partial f(x) \subseteq L(\mathbb{R}^n, \mathbb{R}^m)$ is a pseudo-Jacobian of f at x, then every closed subset $A \subseteq L(\mathbb{R}^n, \mathbb{R}^m)$ containing $\partial f(x)$ is a pseudo-Jacobian of f at x.

(iii) If $\{\partial_i f(x)\}_{i=1}^{\infty} \subseteq L(\mathbb{R}^n, \mathbb{R}^m)$ is a decreasing (by inclusion) sequence of bounded pseudo-Jacobians of f at x, then $\bigcap_{i=1}^{\infty} \partial_i f(x)$ is a pseudo-Jacobian of f at x.

Proof. Let $u \in \mathbb{R}^n$ and $v \in \mathbb{R}^m$ be arbitrarily given. Then we have

$$(-vf)^+(x; u) = \limsup_{t \downarrow 0} \frac{(-vf)(x + tu) - (-vf)(x)}{t}$$

$$= -\liminf_{t \downarrow 0} \frac{(vf)(x + tu) - (vf)(x)}{t}$$

$$= -(vf)^-(x; u).$$

This and the equality

$$\sup_{M \in \partial f(x)} \langle -v, M(u) \rangle = -\inf_{M \in \partial f(x)} \langle v, M(u) \rangle$$

show the equivalence between (1.1) and (1.2).

The property in (ii) is evident from the definition. For the property (iii), we notice that each set $\partial_i f(x)$ is compact, hence the intersection of the family $\{\partial_i f(x) : i = 1, 2, \ldots\}$ is nonempty and compact. Moreover, for each $u \in \mathbb{R}^n$ and $v \in \mathbb{R}^m$ it follows from the definition of pseudo-Jacobian that

$$(vf)^+(x; u) \leq \langle v, M_i(u) \rangle$$

for some $M_i \in \partial_i f(x)$, $i = 1, 2, \ldots$. Because $\{M_i\}_{i=1}^{\infty}$ is bounded, we may assume that it has a limit $M_0 \in \bigcap_{i=1}^{\infty} \partial_i f(x)$. Letting i go to infinity in the above inequality we obtain

$$(vf)^+(x; u) \leq \langle v, M_0(u) \rangle \leq \sup_{M \in \bigcap_{i=1}^{\infty} \partial_i f(x)} \langle v, M(u) \rangle,$$

which completes the proof. $\square$

In the third property of Proposition 1.2.1, if the sets $\partial_i f(x)$, $i = 1, 2, \ldots$, are unbounded, then the conclusion is no longer true. An example of this can be obtained when the intersection of these sets is empty. Indeed, as we have already seen, on the real line, the sets $\partial_k f(0) := [k, \infty)$, $k = 1, 2, \ldots,$. are pseudo-Jacobians of the function $f(x) = x^{1/3}$ at 0. Their intersection is an empty set, so that it cannot be a pseudo-Jacobian of f at that point.

Classical Derivatives

Now we show that all classical derivatives are examples of pseudo-Jacobians.

Proposition 1.2.2 *Let* $f : \mathbb{R}^n \to \mathbb{R}^m$ *be continuous and Gâteaux differentiable at* x. *Then* $\{\nabla f(x)\}$ *is a pseudo-Jacobian of* f *at* x. *Conversely, if* f *admits a singleton pseudo-Jacobian at* x, *then it is Gâteaux differentiable at this point and its derivative coincides with the pseudo-Jacobian matrix.*

Proof. If f is Gâteaux differentiable at x, then for each $u \in \mathbb{R}^n$ and $v \in \mathbb{R}^m$ one has

$$(vf)^+(x; u) = \langle v, \nabla f(x)(u) \rangle,$$

which shows that the singleton set $\{\nabla f(x)\}$ is a pseudo-Jacobian of f at x. Conversely, assume that f admits a singleton pseudo-Jacobian at x, say $\partial f(x) = \{M\}$. Then by Proposition 1.2.1,

$$(vf)^+(x; u) = (vf)^-(x; u) = \langle v, M(u) \rangle$$

for every $u \in \mathbb{R}^n$ and $v \in \mathbb{R}^m$. Hence for each $u \in \mathbb{R}^n$, the directional derivative of f at x in the direction u :

$$f'(x; u) = \lim_{t \downarrow 0} \frac{f(x + tu) - f(x)}{t}$$

exists and equals $M(u)$. This means that f is Gâteaux differentiable and $\nabla f(x) = M$. $\square$

Proposition 1.2.3 *Let* $f : \mathbb{R}^n \to \mathbb{R}^m$ *be continuous, Gâteaux differentiable at* x, *and let* $\partial f(x)$ *be a bounded pseudo-Jacobian of* f *at* x. *Then for every* $v \in \mathbb{R}^m$ *there is some matrix* M *of the convex hull* $\mathrm{co}(\partial f(x))$ *such that* $[\nabla f(x)]^{tr}(v) = M^{tr}(v)$. *In particular,* $\nabla f(x) \in \mathrm{co}(\partial f(x))$ *whenever* $m = 1$.

Proof. It follows from the hypothesis that, for each $u \in \mathbb{R}^n$ and $v \in \mathbb{R}^n$,

$$\inf_{M \in \partial f(x)} \langle v, M(u) \rangle \leq \langle v, \nabla f(x)(u) \rangle = (vf)^+(x; u) \leq \sup_{M \in \partial f(x)} \langle v, M(u) \rangle,$$

which implies that

$$\langle v, \nabla f(x)(u) \rangle \in \{\langle v, M(u) \rangle : M \in \mathrm{co}(\partial f(x))\}.$$

The set $\{vM : M \in \mathrm{co}(\partial f(x))\} \subset \mathbb{R}^n$ is convex and compact, therefore there exists some $M \in \mathrm{co}(\partial f(x))$ such that $v\nabla f(x) = vM$. When $m = 1$, by choosing $v = 1$, we get $\nabla f(x) = M$. $\qquad\square$

1.3 Nonsmooth Derivatives

In this section we show that many generalized derivatives of modern nonsmooth analysis are examples of pseudo-Jacobians. Readers who are not familiar with these generalized derivatives may skip this section at the first reading.

Clarke's Generalized Jacobians

Suppose that $\phi : \mathbb{R}^n \to \mathbb{R}$ is a locally Lipschitz function at x. Let $u \in \mathbb{R}^n$ be given. The Clarke directional derivative of the function ϕ at x in the direction u, which is denoted $\phi^\circ(x; u)$, is defined by

$$\phi^\circ(x; u) := \limsup_{t\downarrow 0, x' \to x} \frac{\phi(x' + tu) - \phi(x')}{t}.$$

Because ϕ is locally Lipschitz, this upper limit is finite, and actually as the function of u, $\phi^0(x; u)$ is a convex, positively homogeneous function, that is,

$$\phi^0(x; su) = s\phi^0(x; u), \quad \text{for } s > 0$$
$$\phi^0(x; u + v) \leq \phi^0(x; u) + \phi^0(x; v).$$

The Clarke subdifferential of ϕ at x is defined by

$$\partial^C \phi(x) := \{\xi \in \mathbb{R}^n : \langle \xi, u \rangle \leq \phi^0(x; u) \quad \text{for } u \in \mathbb{R}^n\}.$$

One of the notable properties of this subdifferential is that it is a nonempty convex and compact set in $\mathbb{R}^n$ and $\phi^0(x; \cdot)$ satisfies the relation

$$\phi^0(x; u) = \max_{\xi \in \partial^C \phi(x)} \langle \xi, u \rangle.$$

Moreover, $\partial^C \phi(x)$ is a singleton if and only if ϕ is strictly differentiable at x.

Now, suppose that $f : \mathbb{R}^n \to \mathbb{R}^m$ is a vector function that is locally Lipschitz at x, that is, as in the scalar case, there exists a neighborhood U of x and a positive k such that

$$\|f(x_1) - f(x_2)\| \leq k\|x_1 - x_2\| \quad \text{for all} \quad x_1, x_2 \in U.$$

Using a theorem due to Rademacher, a locally Lipschitz function is differentiable almost everywhere (in the sense of Lebesgue measure) on U, we define the *Clarke generalized Jacobian* of f at x, denoted $\partial^C f(x)$, by

$$\partial^C f(x) := \mathrm{co} \left\{ \lim_{i \to \infty} \nabla f(x_i) : x_i \in \Omega, \; x_i \to x \right\},$$

where Ω is the set of points in U at which f is differentiable. The set of all limits in the right–hand side without the convex hull is called the *B-subdifferential* of f at x and is denoted $\partial^B f(x)$. The following summarize some basic properties of the Clarke generalized Jacobian.

(i) $\partial^C f(x)$ is a nonempty convex and compact subset of $L(\mathbb{R}^n, \mathbb{R}^m)$, and
$\partial^C(-f)(x) = -\partial^C f(x)$.

(ii) $\partial^C f(x)$ is a singleton if and only if f is strictly differentiable at x.

(iii)   (Robustness) $\partial^C f(x) = \{\lim_{i \to \infty} v_i : v_i \in \partial^C f(x_i), x_i \to x\}$.

(iv)   For locally Lipschitz functions $f : \mathbb{R}^n \to \mathbb{R}^m$, $g : \mathbb{R}^n \to \mathbb{R}^k$,

$$\partial^C(f, g)(x) \subseteq \left\{ \begin{pmatrix} M \\ N \end{pmatrix} : M \in \partial^C f(x), \; N \in \partial^C g(x) \right\}.$$

(v) $\partial^C(f_1 + f_2)(x) \subseteq \partial^C f_1(x) + \partial^C f_2(x)$, where $f_1, f_2 : \mathbb{R}^n \to \mathbb{R}^m$ are locally Lipschitz.

(vi)   (Lebourg's mean value theorem) For $a, b \in \mathbb{R}^n$,

$$f(b) - f(a) \in \mathrm{co}\left(\partial^C f([a, b])(b - a)\right)$$

and when $m = 1$, there is some $c \in (a, b)$ such that

$$f(b) - f(a) \in \partial^C f(c)(b - a).$$

The link between the Clarke generalized Jacobian of the vector function f and the Clarke directional derivative of the real function vf, $v \in \mathbb{R}^m$, at x in the direction $u \in \mathbb{R}^n$ is given by

$$(vf)^\circ(x; u) = \max_{M \in \partial^C f(x)} \langle v, M(u) \rangle.$$

Proposition 1.3.1 *Let $f : \mathbb{R}^n \to \mathbb{R}^m$ be locally Lipschitz at x. Then the Clarke generalized Jacobian $\partial^C f(x)$ of f at x is a pseudo-Jacobian of f at this point.*

Proof. For each $u \in \mathbb{R}^n$ and $v \in \mathbb{R}^m$, one has

$$(vf)^+(x; u) \leq (vf)^\circ(x; u).$$

Now the assertion follows from the fact that $(vf)^\circ(x; u)$

$$= \max_{M \in \partial^C f(x)} \langle v, M(u) \rangle.$$

$\square$

We note that the inequality in the proof of the preceding proposition may be strict, so that in general the Clarke generalized Jacobian is not a regular pseudo-Jacobian. Let us look at a numerical example of a locally Lipschitz function where the Clarke generalized Jacobian strictly contains a pseudo-Jacobian.

Example 1.3.2 Consider the function $f : \mathbb{R}^2 \to \mathbb{R}^2$, defined by

$$f(x, y) = (|x|, |y|).$$

It is easy to verify that the set

$$\partial f(0) = \left\{ \begin{pmatrix} 1 & 0 \\ 0 & 1 \end{pmatrix}, \begin{pmatrix} 1 & 0 \\ 0 & -1 \end{pmatrix}, \begin{pmatrix} -1 & 0 \\ 0 & 1 \end{pmatrix}, \begin{pmatrix} -1 & 0 \\ 0 & -1 \end{pmatrix} \right\}$$

is a pseudo-Jacobian of f at 0. On the other hand, the Clarke generalized Jacobian is given by

$$\partial^C f(0) = \left\{ \begin{pmatrix} \alpha & 0 \\ 0 & \beta \end{pmatrix} : \alpha, \beta \in [-1, 1] \right\}$$

which is also a pseudo-Jacobian of f at 0 and contains $\partial f(0)$.

Observe in this example that $\partial^C f(0)$ is the convex hull of $\partial f(0)$. However, this is not always the case. The following example illustrates that even for the case where $m = 1$, the convex hull of a pseudo-Jacobian of a locally Lipschitz function may be strictly contained in the Clarke generalized Jacobian.

Example 1.3.3 Consider the function $f : \mathbb{R}^2 \to \mathbb{R}$, defined by

$$f(x, y) = |x| - |y|.$$

Then it can easily be verified that

$$\partial_1 f(0) = \{(1, 1), (-1, -1)\} \qquad \text{and} \qquad \partial_2 f(0) = \{(1, -1), (-1, 1)\}$$

are pseudo-Jacobians of f at 0; whereas

$$\partial^C f(0) = \text{co}\{(1,1), (-1,1), (1,-1), (-1,-1)\}.$$

Observe that the convex hull of the pseudo-Jacobian $\partial_1 f(0)$ is a proper subset of the Clarke generalized Jacobian $\partial^C f(0)$ and that the two pseudo-Jacobians $\partial_1 f(0)$ and $\partial_2 f(0)$ are not included in each other.

Mordukhovich's Coderivatives

Let C be a nonempty subset of $\mathbb{R}^n$. The distance function $d(\cdot, C)$ to the set C is given by

$$d(x, C) := \inf_{c \in C} \|x - c\|$$

and the set of best approximations of x in $\text{cl}(C)$, denoted $P(x, C)$, is given by

$$P(x, C) := \{c \in C : \|x - c\| = d(x, C)\}.$$

The limiting normal cone to C at $\overline{x} \in \text{cl}(C)$ is the closed cone

$$N(C, \overline{x}) := \{\lim v_i : v_i \in \text{cone}(x_i - P(x_i, C)), x_i \to \overline{x}\}$$

where $\text{cone}(x - P(x, C))$ is the cone generated by the set $\{x - P(x, C)\}$, that is,

$$\text{cone}(x - P(x, C)) := \{t(x - y) : t \geq 0, y \in P(x, C)\}.$$

In other words, $N(C, \overline{x})$ consists of all limits $\lim t_i a_i$, where $t_i \geq 0$ and $a_i \in x_i - P(x_i, C)$, $x_i \to \overline{x}$.

Now suppose that $f : \mathbb{R}^n \to \mathbb{R}^m$. Then, the graph of f is the set

$$\text{graph}(f) := \{(x, f(x)) \in \mathbb{R}^n \times \mathbb{R}^m : x \in \mathbb{R}^n\}.$$

The *Mordukhovich coderivative* of f at x_0 is the set-valued map $D^M f(x_0) : \mathbb{R}^m \rightrightarrows \mathbb{R}^n$ defined by

$$D^M f(x_0)(v) := \{u \in \mathbb{R}^n : (u, -v) \in N(\text{graph}(f), (x_0, f(x_0)))\}.$$

The normal cone $N(C, x_0)$ can also be written in the form

$$N(C, x_0) = \{\lim v_i : v_i \in \hat{N}(C, x_i), x_i \in C, x_i \to x_0\},$$

where $\hat{N}(C, x)$ is the cone consisting of all vectors $\xi \in \mathbb{R}^n$ satisfying

$$\limsup_{x' \in C, \ x' \to x} \frac{\langle \xi, x' - x \rangle}{\|x' - x\|} \leq 0,$$

which is the dual to the Bouligand contingent cone

$$T(C, x) := \{\lim t_i(x_i - x) : t_i > 0, x_i \in C, x_i \to x\}.$$

When the two cones $N(C, x_0)$ and $\hat{N}(C, x_0)$ coincide, the set C is said to be regular at x_0. Note that in general, the set $D^M f(x_0)(v)$ is neither convex nor bounded. Here are some basic properties of $D^M f$:

(i)   (Robustness) $D^M f(x)(v) = \{\lim \xi_i : \xi_i \in D^M f(x_i)(v_i), v_i \to v, x_i \to x$ with $f(x_i) \to f(x)\}$.

(ii)   When f is strictly differentiable at x_0, one has

$$D^M f(x_0)(v) = (\nabla f(x_0))^{tr}(v) \text{ for every } v \in \mathbb{R}^m.$$

(iii) For $f_1, f_2 : \mathbb{R}^n \to \mathbb{R}^m$, if the following qualification condition holds

$$D^M f_1(x_0)(0) \cap (-D^M f_2(x_0)(0)) = \{0\},$$

then $D^M(f_1 + f_2)(x_0) \subseteq D^M f_1(x_0) + D^M f_2(x_0)$.

(iv)   When f is locally Lipschitz at x_0, $D^M f(x_0)$ consists of $n \times m$-matrices and satisfies the following set equality

$$[\partial^C f(x_0)]^{tr}(v) = [\text{co}\, (D^M f(x_0))](v)$$

for all $v \in \mathbb{R}^m$. Moreover, if there is some subset $\Gamma \subseteq L(\mathbb{R}^n, \mathbb{R}^m)$ such that

$$[\overline{\text{co}}\, (D^M f(x_0))](v) = \overline{\text{co}}\{A^{tr}(v) : A \in \Gamma\},$$

or equivalently

$$\sup_{\xi \in D^M f(x_0)(v)} \langle \xi, u \rangle = \sup_{A \in \Gamma} \langle v, A(u) \rangle,$$

then f is locally Lipschitz at x_0.

We write $[D^M f(x_0)]^{tr}$ to indicate the set of transposed matrices of $D^M f(x_0)$.

Proposition 1.3.4 *Let* $f : \mathbb{R}^n \to \mathbb{R}^m$ *be locally Lipschitz at* x. *Then* $[D^M f(x)]^{tr}$ *is a pseudo-Jacobian of* f *at this point.*

Proof. This follows immediately from the above observation and Proposition 1.3.1.   □

As it was shown by Example 1.3.3, a locally Lipschitz function may have a pseudo-Jacobian strictly smaller than the Mordukhovich coderivative. When f is not locally Lipschitz, the set $D^M f(x)(v)$ may be empty. This may happen, for instance, when f is strictly differentiable except for a point x and $\|\nabla f(x')(v)\|$ goes to ∞ as x' tends to x.

Warga's Unbounded Derivative Containers

Let $f : \mathbb{R}^n \to \mathbb{R}^m$ be a continuous function and V an open set in $\mathbb{R}^n$. A collection $\{\Lambda^\varepsilon f(x) \subseteq L(\mathbb{R}^n, \mathbb{R}^m) : \varepsilon > 0, x \in V\}$ is said to be an *unbounded derivative container* for f if

(i) $\Lambda^\varepsilon f(x) \subset \Lambda^{\varepsilon'} f(x)$ for $\varepsilon < \varepsilon'$.

(ii) For every compact set $C \subseteq V$, there is a sequence $\{f_i\}_{i \geq 1}$ of continuously differentiable functions defined in a neighborhood of C, an integer $i_C \geq 1$, and a positive number δ_C such that $\{f_i\}$ uniformly converges to f on C and $\Lambda^\varepsilon f(x)$ contains $\nabla f_i(y)$ for all $i \geq i_C$ and for all $y \in V$ with $\|y - x\| < \delta_C$.

When $\Lambda^\varepsilon f(x), \varepsilon > 0, x \in V$ are all closed and uniformly bounded, the unbounded derivative container $\Lambda^\varepsilon f$ is called a derivative container. Here are some properties of unbounded derivative containers:

(i)   If $\Lambda^\varepsilon f(x)$ is an unbounded derivative container of f, then any family $\Omega^\varepsilon f(x) \subseteq L(\mathbb{R}^n, \mathbb{R}^m)$ with $\Omega^\varepsilon f(x) \subseteq \Omega^{\varepsilon'} f(x)$ for $\varepsilon' > \varepsilon, x \in V$ and $\Lambda^\varepsilon f(x) \subseteq \Omega^\varepsilon f(x)$, is also an unbounded derivative container of f.

(ii) The function f is locally Lipschitz if and only if it has a derivative container, in which case

$$\partial^C f(x) \subseteq \mathrm{co}\Big(\bigcap_{\varepsilon > 0} \Lambda^\varepsilon f(x) \Big).$$

The next proposition shows that unbounded derivative containers are instances of pseudo-Jacobians.

Proposition 1.3.5 *Let $f : \mathbb{R}^n \to \mathbb{R}^m$ be a continuous function. Let $\{\Lambda^\varepsilon f(x) \subseteq L(\mathbb{R}^n, \mathbb{R}^m) : \varepsilon > 0, x \in V\}$ be an unbounded derivative container for f. Then for every $\varepsilon > 0$, the closure of $\Lambda^\varepsilon f(x)$, is a pseudo-Jacobian of f at x.*

Proof. Let $\{t_i\}$ be a sequence of positive numbers converging to 0 such that

$$(vf)^+(x; u) = \lim_{i \to \infty} \frac{(vf)(x + t_i u) - (vf)(x)}{t_i}.$$

Here we allow the limit to take $+\infty$ and $-\infty$. Let us take C to be a closed neighborhood of x in V. Then, there exists a smaller neighborhood C_0 such that $\|y - x\| < \delta_C$ for all $y \in C_0$. For $i \geq i_C$ sufficiently large, $x + t_i u \in C_0$ and as the sequence $\{vf_i\}$ converges uniformly on C_0 to vf, one finds $k_i \geq i_C$ such that

$$\|vf(y) - vf_{k_i}(y)\| < t_i/i,$$

for every $y \in C_0$. Then, for every $u \in \mathbb{R}^n$ and $v \in \mathbb{R}^m$, we obtain

$$\lim_{i \to \infty} \frac{(vf)(x + t_i u) - (vf)(x)}{t_i}$$

$$= \lim_{i \to \infty} \frac{1}{t_i} [(vf)(x + t_i u) - (vf_{k_i})(x + t_i u) + (vf_{k_i})(x + t_i u) - (vf_{k_i})(x)$$

$$+ (vf_{k_i})(x) - (vf)(x)]$$

$$= \lim_{i \to \infty} \frac{1}{t_i} [(vf_{k_i})(x + t_i u) - (vf_{k_i})(x)]. \tag{1.3}$$

Because f_{k_i} is continuously differentiable, we apply the classical mean value theorem to find $y_i \in (x, x + t_i u)$ such that

$$(vf_{k_i})(x + t_i u) - (vf_{k_i})(x) = v \nabla f_{k_i}(y_i)(t_i u).$$

Substituting this expression into (1.3) and noting $\nabla f_{k_i}(y_i) \in \Lambda^\varepsilon f(x)$, we obtain

$$(vf)^+(x; u) \leq \sup_{M \in \Lambda^\varepsilon f(x)} \langle v, M(u) \rangle.$$

This shows that the closure of $\Lambda^\varepsilon f(x)$ is a pseudo-Jacobian of f at x. $\square$

Ioffe's Prederivatives

We pause to recall the notion of support functions that characterize closed convex sets. Given a nonempty subset C of $\mathbb{R}^n$, its support function, denoted σ_C, is defined by

$$\sigma_C(u) := \sup_{x \in C} \langle u, x \rangle.$$

The support function σ_C is sublinear, that is,

$$\sigma_C(u_1 + u_2) \leq \sigma_C(u_1) + \sigma_C(u_2),$$
$$\sigma_C(tu) = t\sigma_C(u), \ t > 0.$$

Moreover, the support function of C coincides with the support function of the closed convex hull $\overline{\text{co}}(C)$ of C. When C is closed, $\sigma_C(\cdot)$ is finite valued if and only if C is compact. It is also known that a given function $\sigma : \mathbb{R}^n \to \mathbb{R}$ is sublinear and continuous if and only if there is a nonempty convex and compact set $C \subseteq \mathbb{R}^n$ such that $\sigma = \sigma_C$. Any such C is unique.

Let $\Omega : \mathbb{R}^n \rightrightarrows \mathbb{R}^m$ be a set-valued map. It is called a *fan* if the following properties hold.

(a) $\Omega(u)$ is nonempty, convex, and compact for each $u \in \mathbb{R}^n$.
(b) $\Omega(u_1 + u_2) \subseteq \Omega(u_1) + \Omega(u_2)$ for each $u_1, u_2 \in \mathbb{R}^n$.

(c) $\Omega(tu) = t\Omega(u)$ for each $u \in \mathbb{R}^n$ and $t \in \mathbb{R}$.

(d) $\|\Omega\| := \sup_{\|u\| \leq 1, v \in \Omega(u)} \|v\| < \infty$.

It turns out that a fan can be characterized by a bi-sublinear function. Namely, given a fan $\Omega : \mathbb{R}^n \rightrightarrows \mathbb{R}^m$, we define a function $\sigma : \mathbb{R}^n \times \mathbb{R}^m \to \mathbb{R}$ by

$$\sigma(u, v) := \sup_{y \in \Omega(u)} \langle y, v \rangle \text{ for } (u, v) \in \mathbb{R}^n \times \mathbb{R}^m.$$

It follows that σ is sublinear and finite–valued in each variable. For every fixed $u \in \mathbb{R}^n$, $\sigma(u, \cdot)$ is the support function of the convex and compact set $\Omega(u)$. For each fixed $v \in \mathbb{R}^m$, $\sigma(\cdot, v)$ is the support function of a certain convex and compact set that is unique and is denoted by $\Omega^*(v) \subseteq \mathbb{R}^n$. It is not hard to see that the set-valued map $v \mapsto \Omega^*(v)$ from $\mathbb{R}^m$ to $\mathbb{R}^n$ is a fan that we call conjugate to Ω. Conversely, given a continuous and bisublinear function $\sigma : \mathbb{R}^n \times \mathbb{R}^m \to \mathbb{R}$, let $\Omega(u)$ be the convex and compact set in $\mathbb{R}^m$ whose support function is $\sigma(u, \cdot)$ and let $\Omega^*(v)$ be the convex and compact set in $\mathbb{R}^n$ whose support function is $\sigma(\cdot, v)$. Then the set-valued maps $u \mapsto \Omega(u)$ and $v \mapsto \Omega^*(v)$ are both fans and conjugate to each other.

Let $f : \mathbb{R}^n \to \mathbb{R}^m$ be a continuous function and let $\Omega : \mathbb{R}^n \rightrightarrows \mathbb{R}^m$ be a fan. We say that Ω is a *prederivative* of f at x if

$$f(x + u) - f(x) \in \Omega(u) + r(u)\|u\|B_m,$$

where $r(u) \to 0$ as $u \to 0$. We say that Ω is a *strict prederivative* of f at x if

$$f(x' + u) - f(x') \in \Omega(u) + r(x', u)\|u\|B_m,$$

where $r(x'; u) \to 0$ as $x' \to x$ and $u \to 0$.

Proposition 1.3.6 *Assume that a fan Ω is generated by a set of $m \times n$-matrices. If it is a prederivative of f at x, then it is a pseudo-Jacobian of f at x.*

Proof. Let $u \in \mathbb{R}^n$ and $v \in \mathbb{R}^m$. Because Ω is a prederivative of f at x, for each $t > 0$,

$$(vf)(x + tu) - (vf)(x) \in t\langle v, \Omega(u) \rangle + t\|u\|r(u)\langle v, B_m \rangle.$$

Consequently,

$$\frac{(vf)(x + tu) - (vf)(x)}{t} \leq \sup_{M \in \Omega, b \in B_m} (\langle v, M(u) \rangle + \|u\|r(tu)\langle v, b \rangle).$$

By passing to the limit as $t \to 0$, one obtains

$$(vf)^+(x; u) \leq \sup_{M \in \Omega} \langle v, M(u) \rangle$$

which shows that Ω is a pseudo-Jacobian of f at x. $\square$

It follows directly from the definition that a strict prederivative is also a prederivative. Hence when being defined by $m \times n$-matrices, it is also a pseudo-Jacobian. When f is locally Lipschitz, Ioffe showed that the fan defined by the Clarke generalized Jacobian is the smallest strict prederivative of f, hence any other fan containing this fan is also a strict prederivative and f may have a pseudo-Jacobian strictly smaller than its strict prederivative.

The Gowda and Ravindran H-Differentials

Suppose that $f : \mathbb{R}^n \to \mathbb{R}^m$ is continuous. We say that a nonempty set $T(x) \subset L(\mathbb{R}^n, \mathbb{R}^m)$ is an *H-differential* of f at x if for every sequence $\{x_i\}$ converging to x, there exists a subsequence $\{x_{i_k}\}$ and a matrix $A \in T(x)$ such that

$$f(x_{i_k}) - f(x) - A(x_{i_k} - x) = o(\|x_{i_k} - x\|),$$

where

$$\lim_{k \to \infty} \frac{o(\|x_{i_k} - x\|)}{\|x_{i_k} - x\|} = 0.$$

If f has an H-differential at x, then it is said to be *H-differentiable* at x.

When f is Fréchet differentiable at x, the set $\{\nabla f(x)\}$ is evidently an H-differential of f at x. This is not necessarily the case when f is merely Gâteaux differentiable. Moreover, when f is locally Lipschitz, the Clarke generalized Jacobian is an H-differential of f.

Proposition 1.3.7 *Let* $f : \mathbb{R}^n \to \mathbb{R}^m$ *be H-differentiable with an H-differential* $T(x)$. *Then the closure of the set* $T(x)$ *is a pseudo-Jacobian of* f *at* x.

Proof. Let $u \in \mathbb{R}^n$ and $v \in \mathbb{R}^m$. Let $\{t_i\}$ be a sequence of positive numbers converging to 0 such that

$$(vf)^+(x; u) = \lim_{i \to \infty} \frac{(vf)(x + t_i u) - (vf)(x)}{t_i}.$$

Because $T(x)$ is an H-differential of f at x, there exists a subsequence $\{t_{i_k}\}$ and a matrix $A \in T(x)$ such that

$$f(x + t_{i_k} u) - f(x) - A(t_{i_k} u) = o(\|t_{i_k} u\|).$$

This implies that

$$(vf)^+(x; u) = \langle v, Au \rangle \leq \sup_{M \in T(x)} \langle v, Mu \rangle,$$

which shows that $\mathrm{cl}(T(x))$ is a pseudo-Jacobian of f at x. $\square$

The following example illustrates that a pseudo-Jacobian of f at x is not necessarily an H-differential.

Example 1.3.8 Let $f : \mathbb{R} \to \mathbb{R}$ be defined by

$$f(x) = \sqrt{|x|}.$$

Trivially, the set $\mathbb{R}$ is a pseudo-Jacobian of f at $x = 0$. However, it is not an H-differential of f at $x = 0$. Indeed, no real numbers $\alpha \in \mathbb{R}$ satisfy

$$f(x_i) - f(0) - \alpha(x_i - 0) = o(|x_i|),$$

where $\{x_i\}_1^\infty$ is a subsequence of the sequence $\{1/i\}_1^\infty$. Actually, the function is not H-differentiable at this point.

1.4 Pseudo-Differentials and Pseudo-Hessians of Scalar Functions

We specialize in this section the concept of pseudo-Jacobians to scalar functions. This leads to a new concept of pseudo-differential of continuous functions and pseudo-Hessian matrices of continuously differentiable functions.

Pseudo-differentials

Let $f : \mathbb{R}^n \to \mathbb{R}$ be continuous. We say that a closed subset $\partial f(x) \subseteq \mathbb{R}^n$ is a *pseudo-differential* of f at x if considered as a subset of $L(\mathbb{R}^n, \mathbb{R})$ it is a pseudo-Jacobian of f at x.

Because there are only two directions in $\mathbb{R}$ (the positive direction and the negative direction), the definition of pseudo-differential is reduced to the two following inequalities: for each $u \in \mathbb{R}^n$,

$$f^+(x; u) \leq \sup_{x^* \in \partial f(x)} \langle x^*, u \rangle \tag{1.4}$$

$$f^-(x; u) \geq \inf_{x^* \in \partial f(x)} \langle x^*, u \rangle. \tag{1.5}$$

By definition, as a function of variable u, the function in the right–hand side of (1.4) is the support function of the set $\partial f(x)$ and is convex and positively homogeneous. The function in the right hand side of (1.5) is concave and positively homogeneous. Thus, the lower Dini directional derivative $f^-(x; \cdot)$

and the upper Dini directional derivative of $f^+(x; \cdot)$ at x are sandwiched between these two positively homogeneous functions.

As we have seen in the previous section, if f is Lipschitz near x, then the Clarke subdifferential $\partial^C f(x)$ and the Mordukhovich coderivative $D^M f(x)$ are examples of pseudo-differentials. Some more examples of pseudo-differentials are given below.

The Clarke–Rockafellar Subdifferential

Suppose that $f : \mathbb{R}^n \to \mathbb{R}$ is continuous. The Clarke–Rockafellar directional derivative of f at x in the direction u is given by

$$f^\uparrow (x; u) := \sup_{\delta > 0} \limsup_{y \to x, t \downarrow 0} \inf_{\|u' - u\| \leq \delta} \frac{f(y + tu') - f(y)}{t}.$$

The *Clarke–Rockafellar subdifferential* of f at x is defined by

$$\partial^{CR} f(x) := \{\xi \in \mathbb{R}^n : \langle \xi, u \rangle \leq f^\uparrow(x; u) \text{ for all } u \in \mathbb{R}^n\}.$$

The original definition of the Clarke–Rockafellar subdifferential is given for lower semicontinuous functions, in which case one assumes that $f(x)$ is finite and the upper limit is taken over $y \to x$ with $f(y) \to f(x)$ only. When f is locally Lipschitz, the Clarke–Rockafellar subdifferential is exactly the Clarke subdifferential. We need the following approximate mean value theorem of Zagrodny: Let $f : \mathbb{R}^n \to \mathbb{R}$ be continuous and let $a, b \in \mathbb{R}^n$ be distinct points. Then there exist a sequence $\{x_i\}$ converging to $c \in [a, b]$ and $\xi_i \in \partial^{CR} f(x_i)$ such that

$$\lim_{i \to \infty} \langle \xi_i, b - a \rangle \geq f(b) - f(a).$$

Proposition 1.4.1 *Assume that* $f : \mathbb{R}^n \to \mathbb{R}$ *is continuous. Then* $\partial^{CR} f(x)$ *is a pseudo-differential of* f *at* x *provided the set-valued map* $y \mapsto \partial^{CR} f(y)$ *is upper semicontinuous at* x.

Proof. Let $\{t_i\}_{i=1}^\infty$ be a sequence of positive numbers converging to 0 such that

$$f^+(x; u) = \lim_{i \to \infty} \frac{f(x + t_i u) - f(x)}{t_i}.$$

For each $i = 1, 2, \ldots$, using Zagrodny's mean value theorem, we can find a sequence $\{c_{ij}\}_j$ converging to some $c_i \in [x, x + t_i u]$ and $\xi_{ij} \in \partial^{CR} f(c_{ij})$ such that

$$f(x + t_i u) - f(x) \leq \lim_{j \to \infty} \langle \xi_{ij}, t_i u \rangle.$$

We notice that $c_i \to x$ as i tends to ∞. Let $\varepsilon > 0$ be arbitrary. By the upper semicontinuity assumption of the Clarke–Rockafellar subdifferential, we may assume that there is some $i_0 > 0$ such that

$$\partial^{CR} f(c_{ij}) \subset \partial^{CR} f(x) + \varepsilon B_n, \text{ for } i, j > i_0.$$

It follows that

$$f^+(x; u) \leq \sup_{\xi \in \partial^{CR} f(x), \beta \in B_n} \langle \xi + \varepsilon \beta, u \rangle.$$

As ε is arbitrary, we obtain

$$f^+(x; u) \leq \sup_{\xi \in \partial^{CR} f(x)} \langle \xi, u \rangle.$$

Similarly, by applying Zagrodny's mean value theorem to $f(x) - f(x + s_i u)$, where $\{s_i\}$ is a sequence of positive numbers converging to 0 such that

$$f^-(x; u) = \lim_{i \to \infty} \frac{f(x + s_i u) - f(x)}{s_i},$$

we deduce

$$f^-(x; u) \geq \inf_{\xi \in \partial^{CR} f(x)} \langle \xi, u \rangle.$$

Thus $\partial^{CR} f(x)$ is a pseudo-differential of f at x.   □

Notice that the Clarke–Rockafellar subdifferential of a continuous function may be empty at a point, so that in general without any further hypotheses, it is not a pseudo-differential.

Subdifferentials of Convex Functions

Let $f : \mathbb{R}^n \to \mathbb{R} \cup \{\infty\}$ be a function whose values are either real numbers or ∞. The effective domain of f is the set

$$\text{dom}(f) := \{x \in \mathbb{R}^n : f(x) < \infty\}$$

and its epigraph is the set

$$\text{epi}(f) := \{(x, t) \in \mathbb{R}^n \times \mathbb{R} : f(x) \leq t\}.$$

We say that f is *convex* if its epigraph is a convex set, which means that for every two points $w_1, w_2 \in \text{epi}(f)$ and for every positive $\lambda \in [0, 1]$ the convex combination $\lambda w_1 + (1 - \lambda) w_2$ belongs to $\text{epi}(f)$, or equivalently for every two points $x_1, x_2 \in \text{dom}(f)$ and for every positive $\lambda \in [0, 1]$ one has

$$f(\lambda x_1 + (1 - \lambda) x_2) \leq \lambda f(x_1) + (1 - \lambda) f(x_2).$$

Convex functions enjoy many interesting properties. Some of them are exposed in the next lemma.

Lemma 1.4.2 *Let x_0 be an interior point of the effective domain of a convex function f. Then the following properties hold.*

(i) f is locally Lipschitz at x_0.
(ii)   The directional derivative of f at x_0 in any direction $u \in \mathbb{R}^n$ exists and is given by

$$f'(x;u) = \lim_{t \downarrow 0} \frac{f(x_0 + tu) - f(x_0)}{t} = \inf_{t>0} \frac{f(x_0 + tu) - f(x_0)}{t}.$$

Proof. Without loss of generality we may suppose that $x_0 = 0$. The proof is divided into four steps.

(a) f is bounded above on a neighborhood of $x = 0$.
Indeed, choose a system of $(n+1)$ affinely independent vectors $a_1, \ldots, a_{n+1} \in \mathbb{R}^n$ so small that the set $U := \text{int}(\text{co}\{a_1, \ldots, a_{n+1}\})$ contains 0 and is contained in the effective domain of f. Set $\alpha := \max\{f(a_1), \ldots, f(a_{n+1})\}$. Then for every $x \in U$, one expresses it as a convex combination of $a_1, \ldots, a_{n+1}$ by $x = \sum_{i=1}^{n+1} \lambda_i a_i$ with $\lambda_i \geq 0, i = 1, \ldots, n+1$ and $\sum_{i=1}^{n+1} \lambda_i = 1$, so that the convexity of f gives

$$f(x) \leq \sum_{i=1}^{n+1} \lambda_i f(x_i) \leq \alpha.$$

(b) f is bounded in a neighborhood of $x_0 = 0$.
Choose a positive δ so small that $2\delta B_n \subseteq U$. For each $x \in 2\delta B_n$, one has $-x \in 2\delta B_n$ as well; hence $0 = (x + (-x))/2$ and by convexity

$$f(0) \leq \frac{1}{2}f(x) + \frac{1}{2}f(-x) \leq \frac{1}{2}f(x) + \frac{1}{2}\alpha.$$

By this, f is bounded below by $2f(0) - \alpha$ on the set $2\delta B_n$ and hence, in view of (a), it is bounded near $x_0 = 0$.
(c) f is Lipschitz on δB_n.
Denote by β a bound of $|f(x)|$ on $2\delta B_n$. Let x_1, x_2 be two arbitrary distinct points of the set δB_n. Then the point

$$x_3 := x_2 + \frac{\delta}{\|x_2 - x_1\|}(x_2 - x_1)$$

belongs to $2\delta B_n$. Solving for x_2 yields

$$x_2 = \frac{\delta}{\|x_2 - x_1\| + \delta}x_1 + \frac{\|x_2 - x_1\|}{\|x_2 - x_1\| + \delta}x_3.$$

Because f is convex, one deduces

$$f(x_2) \le \frac{\delta}{\|x_2 - x_1\| + \delta} f(x_1) + \frac{\|x_2 - x_1\|}{\|x_2 - x_1\| + \delta} f(x_3),$$

which implies

$$f(x_2) - f(x_1) \le \frac{\|x_2 - x_1\|}{\|x_2 - x_1\| + \delta} (f(x_3) - f(x_1)) \le \gamma \|x_2 - x_1\|,$$

where $\gamma = (2\beta)/\delta$ is a constant independent of x_1 and x_2. Interchanging the roles of x_1 and x_2 will give the Lipschitz property of f on δB_n.
(d) The function $t \mapsto (f(x_0 + tu) - f(x_0))/t$ is nondecreasing for $t > 0$. Indeed, let $0 < t_1 < t_2$ such that $x_0 + t_2 u \in \text{dom}(f)$. Then

$$x + t_1 u = \frac{t_2 - t_1}{t_2} x + \frac{t_1}{t_2} (x + t_2 u).$$

Since f is convex, one has

$$\frac{f(x_0 + t_1 u) - f(x_0)}{t_1} \le \frac{f(x_0 + t_2 u) - f(x_0)}{t_2}$$

as requested. By this, the second assertion of the lemma follows.   □

Assume that $f : \mathbb{R}^n \to \mathbb{R} \cup \{\infty\}$ is a convex function. Let x be an interior point of the effective domain of f. The subdifferential of f at x in the sense of convex analysis (or convex analysis subdifferential) is the set

$$\partial^{ca} f(x) := \{\xi \in \mathbb{R}^n : \langle \xi, u \rangle \le f'(x; u) \quad \text{for every } u \in \mathbb{R}^n\}.$$

Direct verification shows that this set is convex. Moreover, it is a compact set when x is an interior point of the effective domain of f, because in view of Lemma 1.4.2 the function is locally Lipschitz at this point.

Proposition 1.4.3 *Suppose that $f : \mathbb{R}^n \to \mathbb{R} \cup \{\infty\}$ is a convex function and x is an interior point of the effective domain of f. Then the subdifferential $\partial^{ca} f(x)$ of f at x coincides with the set of vectors $\xi \in \mathbb{R}^n$ satisfying*

$$\langle \xi, u \rangle \le f(x + u) - f(x), \quad \text{for every } u \in \mathbb{R}^n.$$

Moreover, this subdifferential also coincides with the Clarke subdifferential. Consequently, when f is real-valued, the subdifferential $\partial^{ca} f(x)$ is a pseudo-differential of f at x.

Proof. Denote by J the set of all vectors ξ such that $\langle \xi, u \rangle \le f(x + u) - f(x)$, for every $u \in \mathbb{R}^n$. The conclusion $\partial^{ca} f(x) \subseteq J$ is evident in view of Lemma 1.4.2. For the converse inclusion, let $\xi \in J$ and let $u \in \mathbb{R}^n \setminus \{0\}$, then for $t > 0$ we have

$$\langle \xi, tu \rangle \le f(x + tu) - f(x).$$

By dividing both sides of this inequality by t and letting t tend to 0, we obtain, in view of Lemma 1.4.2, that

$$\langle \xi, u \rangle \le f'(x; u).$$

Hence $\xi \in \partial^{ca} f(x)$ and the equality $\partial^{ca} f(x) = J$ holds.

To complete the proof it suffices now to show that

$$f'(x; u) = f^\circ(x; u)$$

for every $u \in \mathbb{R}^n$. It follows easily from the definition of the Clarke directional derivative that

$$f'(x; u) \le f^\circ(x; u).$$

To prove the opposite inequality, we express the Clarke directional derivative in the form

$$f^\circ(x; u) = \lim_{\epsilon \downarrow 0} \sup_{x' \in x + \epsilon \delta B_n} \sup_{0 < t < \epsilon} \frac{f(x' + tu) - f(x')}{t},$$

where δ is a fixed, but arbitrary positive number. Using Lemma 1.4.2 we derive the following expression,

$$f^\circ(x; u) = \lim_{\epsilon \to 0} \sup_{x' \in x + \epsilon \delta B_n} \frac{f(x' + \epsilon u) - f(x')}{\epsilon}.$$

For $x' \in x + \epsilon \delta B_n$, the Lipschitz continuity of f, say with a Lipschitz constant β, yields

$$\left| \frac{f(x' + \epsilon u) - f(x')}{\epsilon} - \frac{f(x + \epsilon u) - f(x)}{\epsilon} \right| \le 2\delta\beta$$

which implies

$$f^\circ(x; u) \le \lim_{\epsilon \to 0} \frac{f(x + \epsilon u) - f(x)}{\epsilon} + 2\delta\beta$$
$$\le f'(x; u) + 2\delta\beta.$$

Letting δ tend to 0 in the above inequality, we derive

$$f^\circ(x; u) \le f'(x; u),$$

and the required equality follows. $\square$

Mordukhovich's Subdifferentials

When $f : \mathbb{R}^n \to \mathbb{R}$ is merely continuous, the *Mordukhovich basic subdifferential* of f at x is defined by

$$\partial^M f(x) = \limsup_{x' \to x, \varepsilon \downarrow 0} \partial_\varepsilon^F f(x'),$$

where $\partial_\varepsilon^F f(x')$ is the Fréchet ε-subdifferential of f at x' given by

$$\partial_\varepsilon^F f(x') := \left\{ x^* \in \mathbb{R}^n : \liminf_{\|h\| \to 0} \frac{f(x' + h) - f(x') - \langle x^*, h \rangle}{\|h\|} \geq \varepsilon \right\}.$$

It can be seen that the basic subdifferential consists of all vectors $u \in \mathbb{R}^n$ such that

$$(u, -1) \in N(\mathrm{epi}(f), (x, f(x))).$$

The set

$$\partial_s^M f(x) := \{ u \in \mathbb{R}^n : (u, 0) \in N(\mathrm{epi}(f), (x, f(x))) \}$$

is called the *Mordukhovich singular subdifferential* of f at x.

Corollary 1.4.4 *Assume that $f : \mathbb{R}^n \to \mathbb{R}$ is locally Lipschitz at x. Then the Mordukhovich basic subdifferential $\partial^M f(x)$ is a pseudo-differential of f at x.*

Proof. This follows immediately from Proposition 1.3.4. $\qquad\qquad\square$

Notice that when f is not locally Lipschitz at x, the Mordukhovich basic subdifferential may be empty. For instance, the function $f(x) = \sqrt{|x|}$ for $x \in \mathbb{R}$ has $\partial^M f(0) = \emptyset$. Its singular subdifferential at 0 is the whole space $\mathbb{R}$. The following example shows that even when the basic subdifferential of f is nonempty, it is not necessarily a pseudo-differential.

Example 1.4.5 Let $f : \mathbb{R} \to \mathbb{R}$ be defined by

$$f(x) = \begin{cases} x^2 \sin(1/x) & \text{if } x < 0, \\ -x^{1/3} & \text{else.} \end{cases}$$

Direct calculation shows that $\partial^M f(0) = [-1, 1]$ which cannot be a pseudo-differential of f at 0 because $f^-(0; 1) = -\infty$ and (1.5) is not verified. Note, however, that the singular subdifferential of f at 0 is the set $(-\infty, 0]$ and its union with the basic subdifferential forms a pseudo-differential of f at that point.

A locally Lipschitz function may have a pseudo-Jacobian that is strictly contained in the basic subdifferential. To see this let us consider the function f given in Example 1.3.3. The basic subdifferential of this function at $(0,0)$ is the set

$$\{(t,1) \in \mathbb{R}^2 : -1 \le t \le 1\} \cup \{(t,-1) \in \mathbb{R}^2 : -1 \le t \le 1\}$$

which contains the pseudo-differential $\partial f(0,0) = \{(1,-1),(-1,1)\}$ as a proper subset.

Ioffe's Approximate Subdifferentials

Suppose that $f : \mathbb{R}^n \to \mathbb{R}$ is continuous. The *Ioffe approximate subdifferential* of f at x, denoted $\partial^{IA} f(x)$, is defined by

$$\partial^{IA} f(x) = \limsup_{x' \to x, \varepsilon \downarrow 0} \partial_\varepsilon^- f(x'),$$

where

$$\partial_\varepsilon^- f(x) := \left\{ \xi \in \mathbb{R}^n : \langle \xi, u \rangle \le \liminf_{u' \to u, t \downarrow 0} \frac{f(\xi + tu') - f(x)}{t} + \varepsilon \|u\| \text{ for all } u \right\}.$$

Corollary 1.4.6 *Assume that $f : \mathbb{R}^n \to \mathbb{R}$ is locally Lipschitz at x. Then the Ioffe approximate subdifferential $\partial^{IA} f(x)$ is a pseudo-differential of f at x.*

Proof. It suffices to observe that the Ioffe approximate subdifferential coincides with the Mordukhovich basic subdifferential and apply Corollary 1.4.4. $\qquad\square$

The definition of the approximate subdifferential above is adapted to the finite-dimensional case. In general spaces the Ioffe approximate subdifferential and the Mordukhovich basic subdifferential are distinct.

The Michel–Penot Subdifferential

Suppose that $f : \mathbb{R}^n \to \mathbb{R}$ is continuous. The Michel–Penot upper and lower directional derivatives of f at x are, respectively, given by

$$f^\diamond(x; u) = \sup_{z \in \mathbb{R}^n} \limsup_{t \downarrow 0} t^{-1}[f(x + tz + tu) - f(x + tz)]$$

and

$$f_\diamond(x; u) = \inf_{z \in \mathbb{R}^n} \liminf_{t \downarrow 0} t^{-1}[f(x + tz + tu) - f(x + tz)].$$

The corresponding *Michel–Penot subdifferential* is defined by

$$\partial^{MP} f(x) := \{x^* \in \mathbb{R}^n : f^\diamond(x; u) \ge \langle x^*, u \rangle \text{ for all } u\}.$$

Principal properties of $\partial^{MP} f$ are listed below.

(i) $\partial^{MP} f(x)$ is a convex set, and it is compact when f is locally Lipschitz near x.

(ii)   The function f is Gâteaux differentiable at x if and only if $\partial^{MP} f(x)$ is a singleton in which case $\partial^{MP} f(x) = \{\nabla f(x)\}$.

(iii)   When f is convex, $\partial^{MP} f(x)$ coincides with the subdifferential of f at x in the sense of convex analysis, that is, $x^* \in \partial^{MP} f(x)$ if and only if $\langle x^*, u \rangle \leq f(x + u) - f(x)$ for all u.

It is shown in the next proposition that the Michel–Penot subdifferential of a locally Lipschitz function is also a pseudo-differential. Example 2.1.15 gives a function that is not locally Lipschitz and admits a pseudo-differential strictly smaller than the Michel–Penot subdifferential.

Proposition 1.4.7 *Assume that $f : \mathbb{R}^n \to \mathbb{R}$ is locally Lipschitz. Then the set $\partial^{MP} f(x)$ is a pseudo-differential of f at x.*

Proof. Because f is locally Lipschitz, it follows that the Michel–Penot upper and lower directional derivatives $f^\diamond(x, \cdot)$ and $f_\diamond(x, \cdot)$ are finite and sublinear, and $\partial^{MP} f(x)$ is convex and compact. Moreover,

$$f^\diamond(x; u) = \max_{x^* \in \partial^{MP} f(x)} \langle x^*, u \rangle,$$

$$f_\diamond(x; u) = \min_{x^* \in \partial^{MP} f(x)} \langle x^*, u \rangle.$$

Because $f^+(x; u) \leq f^\diamond(x; u)$ and $f^-(x; u) \geq f_\diamond(x; u)$ for each $u \in \mathbb{R}^n$, we conclude that $\partial^{MP} f(x)$ is a pseudo-differential of f at x. $\square$

Treiman's Linear Generalized Gradients

Suppose that $f : \mathbb{R}^n \to \mathbb{R}$ is continuous. A vector $v \in \mathbb{R}^n$ is said to be a *proximal subgradient* to f at x if there is some $\mu > 0$ such that

$$f(x') - f(x) \geq \langle v, x' - x \rangle - \mu \|x' - x\|^2$$

for x' in some neighborhood of x.

A sequence of proximal subgradients $\{v_k\} \to v$ to f at x is said to be linear if either v_k is a proximal subgradient to f at x for every k or there exists a sequence $\{x_k\}$ converging to x with $x_k \neq 0$, and $\mu, \delta > 0$ such that

$$f(x_k + h) - f(x_k) \geq \langle v_k, h \rangle - (\mu / \|x_k - x\|_f) \|h\|^2$$

for every h with $\|h\| \leq \delta \|x_k - x\|_f$, where

$$\|x_k - x\|_f = \|x_k - x\| + |f(x_k) - f(x)|.$$

Treiman's linear generalized gradient, denoted $\partial^l f(x)$, of f at x is the closure of the set of all limits of linear sequences of proximal subgradients to f at x.

We list some basic properties of linear generalized gradients.

(i)   If x is a local minimizer of f, then $0 \in \partial^l f(x)$.

(ii)   If $f : \mathbb{R}^n \to \mathbb{R}$ is continuous and $g : \mathbb{R}^n \to \mathbb{R}$ is locally Lipschitz, then

$$\partial^l (f + g)(x) \subseteq \partial^l f(x) + \partial^l g(x),$$

$$\partial^l (\alpha f)(x) = \alpha \partial^l f(x) \text{ for } \alpha > 0.$$

(iii) If f is locally Lipschitz, then $\mathrm{co}(\partial^l f(x)) = \partial^{MP} f(x)$.

Proposition 1.4.8 *Assume that $f : \mathbb{R}^n \to \mathbb{R}$ is locally Lipschitz. Then the set $\partial^l f(x)$ is a pseudo-differential of f at x.*

Proof. Invoke Proposition 1.4.7 and property (iii) above.   □

The Demyanov-Rubinov Quasidifferentials

Suppose that $f : \mathbb{R}^n \to \mathbb{R}$ is directionally differentiable at x. We say that f is *quasidifferentiable* at x if the directional derivative $f'(x; u)$ can be represented in the form

$$f'(x; u) = \max_{a \in A}\langle a, u \rangle + \min_{b \in B}\langle b, u \rangle,$$

where A and B are some convex and compact sets in $\mathbb{R}^n$. The pair $[A, B]$ is called the *quasidifferential* of f at x.

Here are some basic properties of quasidifferentials.

(i)   If f is differentiable at x, then it is quasidifferentiable at this point with a quasidifferential $[\nabla f(x), \{0\}]$.

(ii)   If f is convex and $\partial^{ca} f(x)$ is its subdifferential, then f is quasidifferentiable with a quasidifferential $[\partial^{ca} f(x), \{0\}]$.

(iii) If f_1 and f_2 are quasidifferentiable at x with quasidifferentials $[A_1, B_1]$ and $[A_2, B_2]$, respectively, then $f_1 + f_2$ and λf_1 with $\lambda \in \mathbb{R}$ are quasidifferentiable at this point with quasidifferentials $[A_1 + A_2, B_1 + B_2]$ and $[\lambda A_1, \lambda B_1]$.

It is clear that every pair of convex and compact sets $[A', B']$ satisfying

$$A - B' = A' - B$$

is also a quasidifferential of f at x.

Proposition 1.4.9 *Let* $f : \mathbb{R}^n \to \mathbb{R}$ *be continuous. Assume that* $f :$ $\mathbb{R}^n \to \mathbb{R}$ *is quasidifferentiable at* x *and that the pair of sets* $[A, B]$ *is a quasidifferential of* f *at* x. *Then the set* $A + B$ *is a pseudo-differential of* f *at* x.

Proof. Clearly, from the quasidifferentiability of f at x, we obtain that, for every $u \in \mathbb{R}^n$,

$$
\begin{aligned}
f^+(x; u) &= \max_{a \in A}\langle a, u \rangle + \min_{b \in B}\langle b, u \rangle \\
&\leq \max_{a \in A}\langle a, u \rangle + \max_{b \in B}\langle b, u \rangle \\
&\leq \max_{c \in A+B}\langle c, u \rangle
\end{aligned}
$$

and

$$
\begin{aligned}
f^-(x; u) &= \max_{a \in A}\langle a, u \rangle + \min_{b \in B}\langle b, u \rangle \\
&\geq \min_{a \in A}\langle a, u \rangle + \min_{b \in B}\langle b, u \rangle \\
&\geq \min_{c \in A+B}\langle c, u \rangle.
\end{aligned}
$$

This shows that $A + B$ is a pseudo-differential of f at x. $\square$

When f is positively homogeneous, the Demyanov–Rubinov convexificator is defined as a convex set $C \subset \mathbb{R}^n$ that satisfies the following relation

$$
\min_{c \in C}\langle c, x \rangle \leq f(x) \leq \max_{c \in C}\langle c, x \rangle \quad \text{for every } x.
$$

Because f is positively homogeneous, $f'(0; u) = f(u)$ for every u. By the relation above, this convexificator is a pseudo-differential of f at 0.

Pseudo-Hessian Matrices

In the rest of this section we apply the concept of pseudo-Jacobians to introduce generalized Hessian matrices for continuously differentiable scalar functions.

Let $f : \mathbb{R}^n \to \mathbb{R}$ be continuously differentiable. The derivative map ∇f is a continuous vector function from $\mathbb{R}^n$ to $\mathbb{R}^n$. We say that a closed subset of $n \times n$-matrices $\partial^2 f(x) \subseteq L(\mathbb{R}^n, \mathbb{R}^n)$ is a *pseudo-Hessian* of f at x if it is a pseudo-Jacobian of ∇f at x.

Pseudo-Hessians share all properties of pseudo-Jacobians. We list some of them in the next proposition.

Proposition 1.4.10 *Let* $f : \mathbb{R}^n \to R$ *be continuously differentiable. The following assertions hold.*

(i) If $\partial^2 f(x) \subseteq L(\mathbb{R}^n, \mathbb{R}^m)$ is a pseudo-Hessian of f at x, then every closed subset $A \subseteq L(\mathbb{R}^n, \mathbb{R}^m)$ containing $\partial^2 f(x)$ is a pseudo-Hessian of f at x.

(ii) If f is twice Gâteaux differentiable at x, then the Hessian $\{\nabla^2 f(x)\}$ is a pseudo-Hessian of f at x. Moreover, f is twice Gâteaux differentiable at x if and only if it admits a singleton pseudo-Hessian at this point.

Proof. Invoke Propositions 1.2.1 and 1.2.2.   □

Now we give some instances of pseudo-Hessians of continuously differentiable functions.

The Hiriart-Urruty, Strodiot, and Hien Nguyen Generalized Hessians

Suppose that $f : \mathbb{R}^n \to \mathbb{R}$ is differentiable whose derivative is locally Lipschitz. Such a function is called a $C^{1,1}$-function. Because ∇f is locally Lipschitz, it is differentiable almost everywhere.

The *generalized Hessian* of f at x in the sense of Hiriart-Urruty, Strodiot, and Hien Nguyen is given by

$$\partial_H^2 f(x) = \mathrm{co}\{\lim \nabla^2 f(x_i) : x_i \in \Omega,\ x_i \to x\},$$

where Ω is the set of points at which f is twice differentiable. In other words, it is the Clarke generalized Jacobian of the gradient vector function ∇f at x.

Proposition 1.4.11 *Assume that $f : \mathbb{R}^n \to \mathbb{R}$ is a $C^{1,1}$-function. Then the set $\partial_H^2 f(x)$ is a pseudo-Hessian of f at x.*

Proof. The conclusion follows from Proposition 1.3.1.   □

We note that a $C^{1,1}$-function may have a pseudo-Hessian that is strictly smaller than the generalized Hessian above. Such examples can easily be constructed by integrating the functions of Examples 1.3.2 and 1.3.3.

Another concept of a generalized Hessian, introduced by Cominetti and Correa for $C^{1,1}$-functions, is given as follows. Suppose that $f : \mathbb{R}^n \to \mathbb{R}$ is differentiable whose derivative is locally Lipschitz. The second order directional derivative of f at x in the directions $(u, v) \in \mathbb{R}^n \times \mathbb{R}^n$ is defined by

$$f^{00}(x; u, v) = \limsup_{y \to x, t \to 0} \frac{\langle \nabla f(y + tu), v \rangle - \langle \nabla f(y), v \rangle}{t}.$$

The *generalized Hessian* in the sense of Cominetti and Correa is defined as a set-valued map $\partial^{00} f(x) : \mathbb{R}^n \rightrightarrows \mathbb{R}^n$, which is given by

$$\partial^{00} f(x)(u) = \{x^* \in \mathbb{R}^n : f^{00}(x; u, v) \geq \langle x^*, v \rangle \text{ for all } v \in \mathbb{R}^n\}.$$

Corollary 1.4.12 *Let* $f : \mathbb{R}^n \rightarrow \mathbb{R}$ *be a* $C^{1,1}$*-function and let* $A \subset L(\mathbb{R}^n, \mathbb{R}^m)$ *be a closed set such that* $A(u) \supseteq \partial^{00} f(x)(u)$ *for all* $u \in \mathbb{R}^n$. *Then* A *is a pseudo-Hessian of* f *at* x.

Proof. It is known that for each $u \in \mathbb{R}^n$,

$$\partial^{00} f(x)(u) = \partial_H^2 f(x)(u).$$

The conclusion is derived from Proposition 1.4.11   □

Mordukhovich's Second-Order Subdifferentials

Suppose that $f : \mathbb{R}^n \rightarrow \mathbb{R}$ is a C^1-function. The Mordukhovich coderivative $D^M \nabla f(x)$ of the vector function ∇f at x is called the *Mordukhovich second-order subdifferential* of f at x.

Proposition 1.4.13 *Let* $f : \mathbb{R}^n \rightarrow \mathbb{R}$ *be a* $C^{1,1}$*-function. Then* $[D^M \nabla f(x)]^{tr}$ *is a pseudo-Hessian of* f *at* x.

Proof. Invoke Proposition 1.3.4.   □

Note that the original construction of the Mordukhovich second order subdifferential was given for set-valued maps without smoothness assumption. When ∇f is not locally Lipschitz, the set-valued map $D^M \nabla f(x):$ $\mathbb{R}^n \rightrightarrows \mathbb{R}^n$ is not necessarily defined by matrices, and so it cannot be a pseudo-Hessian of f.

1.5 Recession Matrices and Partial Pseudo-Jacobians

When dealing with non-Lipschitz functions, we have unwillingly to face unbounded pseudo-Jacobians. In such situations recession directions serve as a useful tool to describe the global picture of pseudo-Jacobians.

Recession Pseudo-Jacobian Matrices

Let $A \subseteq \mathbb{R}^n$ be a nonempty set. The *recession cone* or asymptotic cone of the set A, denoted A_∞, is defined by

$$A_\infty := \{\lim t_i a_i : a_i \in A, t_i \downarrow 0\}.$$

Elements of A_∞ are called recession directions of A. We say that A is *asymptotable* if for every $v \in A_\infty \setminus \{0\}$, and for every sequence $\{t_i\}_{i \geq 1}$ of positive numbers converging to ∞, there is a sequence $\{v_i\}_{i \geq 1}$ converging to v such that $t_i v_i \in A$ for all i.

Lemma 1.5.1 *Let $A, B \subseteq \mathbb{R}^n$ and $C \subseteq \mathbb{R}^m$ be nonempty. Then the following assertions hold.*

(i) A_∞ *is a closed cone.*

(ii) A *is bounded if and only if* $A_\infty = \{0\}$.

(iii) *If A is convex and closed, then* $A = A + A_\infty$.

(iv) $\mathrm{co}(A_\infty) \subseteq (\mathrm{co}A)_\infty$. *Equality holds provided* $\mathrm{co}(A_\infty)$ *contains no nontrivial linear subspaces;*

(v) $(A \cup B)_\infty = A_\infty \cup B_\infty$.

(vi) $(A \cap B)_\infty \subseteq A_\infty \cap B_\infty$. *Equality holds provided A and B are closed, convex, and* $A \cap B \neq \emptyset$.

(vii) $(A + B)_\infty \subseteq A_\infty + B_\infty$ *provided* $A_\infty \cap -B_\infty = \{0\}$; *and* $A_\infty + B_\infty \subseteq (A+B)_\infty$ *provided A is asymptotable. Equality holds when B is bounded.*

(viii) $(A \times C)_\infty \subseteq A_\infty \times C_\infty$. *Equality holds provided A is asymptotable.*

Proof. The first assertion is immediate from the definition. For the second assertion, if A is bounded, every sequence $\{t_i a_i\}_{i \geq 1}$ with $a_i \in A$ and $t_i \downarrow 0$ converges to 0. Hence $A_\infty = \{0\}$. Conversely, if A is unbounded, then there is a sequence $\{a_i\}$ in A with $\lim_{i \to \infty} \|a_i\| = \infty$. The sequence $\{a_i/\|a_i\|\}_{i \geq 1}$ is bounded and so we may assume that it converges to some vector $v \neq 0$. We have $v \in A_\infty$ and therefore A_∞ is not trivial.

Let A be convex and closed. To show (iii), it suffices to establish $A + A_\infty \subseteq A$ because the inclusion $A \subseteq A + A_\infty$ is always true. Let $u \in A_\infty$ and $a \in A$. By definition $u = \lim_{i \to \infty} t_i a_i$ for some $a_i \in A$ and $t_i \downarrow 0$. As A is convex, we have $(1 - t_i)a + t_i a_i \in A$, and by the closeness of A, we have $a + u = \lim_{i \to \infty}[(1 - t_i)a + t_i a_i] \in A$.

For assertion (iv), let $u, v \in A_\infty$, say $u = \lim_{i \to \infty} t_i a_i$ and $v = \lim_{i \to \infty} s_i b_i$ for some $a_i, b_i \in A$, $t_i \downarrow 0$, and $s_i \downarrow 0$. By taking $\alpha_i = t_i + s_i$ and $c_i = (t_i/\alpha_i)a_i + (s_i/\alpha_i)b_i \in \mathrm{co}(A)$ we obtain $u + v = \lim_{i \to \infty} \alpha_i c_i \in (\mathrm{co}(A))_\infty$. Suppose that $\mathrm{co}(A_\infty)$ contains no nontrivial linear subspaces and let $v \in (\mathrm{co}(A))_\infty$, say $v = \lim_{i \to \infty} t_i b_i$ for some $b_i \in \mathrm{co}(A)$ and $t_i \downarrow 0$. We apply Caratheodory's theorem (Theorem 1.1.1) to find $\lambda_{ij} \geq 0$, $a_{ij} \in A, j = 1, \ldots, n+1$ such that

$$b_i = \sum_{j=1}^{n+1} \lambda_{ij} a_{ij} \text{ and } \sum_{j=1}^{n+1} \lambda_{ij} = 1.$$

Consider the sequences $\{t_i\lambda_{ij}a_{ij}\}_{i=1}^{\infty}, j = 1, \ldots, n+1$. We claim that they are bounded. Indeed, if not, without loss of generality one may assume that $\lim_{i\to\infty}\|t_i\lambda_{i1}a_{i1}\| = \infty$, $\|\lambda_{i1}a_{i1}\| \geq \|\lambda_{ij}a_{ij}\|$ and $\lim_{i\to\infty}(\lambda_{ij}a_{ij})/\|\lambda_{i1}a_{i1}\| = a_j \in A_{\infty}, j = 1, \ldots, n+1$. We derive

$$0 = \lim_{i\to\infty}\frac{v}{\|\lambda_{i1}a_{i1}\|} = \lim_{i\to\infty}\sum_{j=1}^{n+1}\frac{\lambda_{ij}a_{ij}}{\|\lambda_{i1}a_{i1}\|} = \sum_{j=1}^{n+1}a_j.$$

This implies $-a_1 = \sum_{j=2}^{n+1}a_j \neq 0$, which contradicts the hypothesis. In this way, the sequences $\{t_i\lambda_{ij}a_{ij}\}_{i=1}^{\infty}, j = 1, \ldots, n+1$ are bounded and we may assume that they converge respectively to $v_j \in A_{\infty}, j = 1, \ldots, n+1$. Then $v = \sum_{j=1}^{n+1}v_j \in \mathrm{co}(A_{\infty})$, as requested.

The fifth assertion and the first part of the sixth assertion are immediate from the definition. Let us consider the case when A and B are closed and convex with $A \cap B \neq \emptyset$. Let $u \in A_{\infty} \cap B_{\infty}$ and let $a \in A \cap B$. By the assumption, we have $a+tu \in A \cap B$ for every $t \geq 0$. This gives $u \in (A \cap B)_{\infty}$. We take up assertion (vii). Let $u \in (A+B)_{\infty}$, say $u = \lim_{i\to\infty}t_i(a_i+b_i)$ for some $a_i \in A, b_i \in B$, and $t_i \downarrow 0$. If the sequence $\{t_ia_i\}_{i\geq 1}$ is bounded, then so is $\{t_ib_i\}_{i\geq 1}$. We may assume that these sequences converge to $v \in A_{\infty}$ and $w \in B_{\infty}$, respectively. Then $u = v + w \in A_{\infty} + B_{\infty}$. In the other case, both of them are unbounded and we may assume further that $\|a_i\| \geq \|b_i\|$ for all i, with $\lim_{i\to\infty}a_i/\|a_i\| = u_0 \in A_{\infty}$. We derive that $\lim_{i\to\infty}b_i/\|a_i\| = \lim_{i\to\infty}(a_i/\|a_i\| + u/\|a_i\|) = -u_0 \in B_{\infty}$, which contradicts the hypothesis. Now, let $u \in A_{\infty}$ and $v \in B_{\infty}$, say $v = \lim_{i\to\infty}s_ib_i$ with $b_i \in B$ and $s_i \downarrow 0$. Because A is asymptotable, there is $a_i \in A$ such that the sequence $\{s_ia_i\}_{i\geq 1}$ converges to u. Hence $u + v = \lim_{i\to\infty}s_i(a_i + b_i) \in (A + B)_{\infty}$. When B is bounded, one has $B_{\infty} = \{0\}$ by (ii), and $(A + B)_{\infty} = A_{\infty} = A_{\infty} + B_{\infty}$.

The inclusion of the last assertion is obtained directly from the definition. When A is asymptotable, equality is obtained by an argument similar to the previous assertion. □

Recall that a map is open if the image of every open set is open.

Lemma 1.5.2 *Let $A \subseteq \mathbb{R}^n$ be a nonempty set and let L be a linear map from $\mathbb{R}^n$ to $\mathbb{R}^m$. Then one has*

$$L(A_{\infty}) \subseteq (L(A))_{\infty}.$$

Equality holds under each of the following conditions:

(i) L is open and $L^{-1}(L(A)) = A$.
(ii) $\mathrm{Ker}L \cap A_{\infty} = \{0\}$.

Proof.   Let $v \in L(A_\infty)$. Then, there exist $u \in A_\infty$ with $L(u) = v$, a sequence $\{x_i\}_{i=1}^\infty \subseteq A$, and a sequence of positive numbers $\{t_i\}_{i=1}^\infty$ converging to 0 such that $\lim_{i\to\infty} t_i x_i = u$. By the continuity of L, one has $v = \lim_{i\to\infty} L(t_i x_i) \in (L(A))_\infty$.

Under condition (i), let $v \in (L(A))_\infty$; that is, $v = \lim_{i\to\infty} t_i y_i$ for $y_i \in L(A)$ and $t_i > 0$ with $\lim_{i\to\infty} t_i = 0$. Because L is open, given $u \in L^{-1}(v)$ we can find a sequence $\{u_i\}_{i=1}^\infty$ in $\mathbb{R}^n$ with $\lim_{i\to\infty} u_i = u$ and $L(u_i) = t_i y_i$ for all $i = 1, 2, \dots$. Setting $x_i = u_i/t_i$, we have $x_i \in L^{-1}(L(A)) = A$ so that $u \in A_\infty$. Consequently $v \in L(A_\infty)$. Assume that (ii) holds. Let $v \in (L(A))_\infty$, that is, $v = \lim_{i\to\infty} t_i y_i$ for $y \in L(A)$ and $t_i \downarrow 0$. Let $x_i \in A$ be such that $y_i = L(x_i)$. If $\{||x_i||\}_{i=1}^\infty$ is bounded, $\lim_{i\to\infty} t_i x_i = 0$. Consequently $v = \lim_{i\to\infty} t_i L(x_i) = \lim_{i\to\infty} L(t_i x_i) = 0 \in L(A_\infty)$. If $\{||x_i||\}_{i=1}^\infty$ is unbounded, one may assume that $\{x_i/||x_i||\}_{i=1}^\infty$ converges to some $u \in A_\infty$. The sequence $\{t_i||x_i||\}_{i=1}^\infty$ is bounded, otherwise one should have

$$L(u) = \lim_{i\to\infty} \frac{v}{t_i||x_i||} = 0 \text{ with } ||u|| = 1 \, ,$$

contradicting the condition $\mathrm{Ker}L \cap A_\infty = \{0\}$. Therefore, we may assume that $\{t_i||x_i||\}_{i=1}^\infty$ converges to some $\alpha \geq 0$. By this,

$$v = \lim_{i\to\infty} L(t_i||x_i||\frac{x_i}{||x_i||}) = \alpha L(u) \in L(A_\infty)$$

and the inclusion becomes an equality. $\square$

Suppose now that $f : \mathbb{R}^n \to \mathbb{R}^m$ is continuous and that $\partial f(x)$ is a pseudo-Jacobian of f at x. The set $(\partial f(x))_\infty$ denotes the recession cone of $\partial f(x)$. Elements of $(\partial f(x))_\infty$ are called *recession matrices* of $\partial f(x)$.

Proposition 1.5.3 *Assume that $\partial f(x)$ is a pseudo-Jacobian of f at x. Then the following assertions hold.*

(i) $\partial f(x)$ *is bounded if and only if* $(\partial f(x))_\infty = \{0\}$.
(ii)   *If $\partial f(x)$ is convex, then* $\partial f(x) = \partial f(x) + (\partial f(x))_\infty$.
(iii)   *If $\partial f(x)$ is convex and $0 \in \partial f(x)$, then* $(\partial f(x))_\infty \subset \partial f(x)$.

Proof. Invoke Lemma 1.5.1. $\square$

Example 1.5.4 Define $f : \mathbb{R}^2 \to \mathbb{R}^2$ by

$$f(x, y) = (\sqrt{|x|}\, \mathrm{sign}(x) + |y|, \ \sqrt{|y|}\, \mathrm{sign}(y) + |y|).$$

Then f is not locally Lipschitz at $(0, 0)$ and so the Clarke generalized Jacobian does not exist. However, for each $c \in R$, the set

$$\partial f(0,0) = \left\{ \begin{pmatrix} \alpha & 1 \\ 0 & \beta \end{pmatrix}, \begin{pmatrix} \alpha & -1 \\ 0 & \beta \end{pmatrix} \; : \; \alpha, \beta \geq c \right\}$$

is a pseudo-Jacobian of f at $(0,0)$. The recession cone of $\partial f(0,0)$ is given by

$$\partial^\infty f(0,0) = \left\{ \begin{pmatrix} \alpha & 0 \\ 0 & \beta \end{pmatrix} : \alpha \geq 0, \; \beta \geq 0 \right\}.$$

We observe that $\partial f(0,0)$ is not convex. It does not contain the zero matrix and the inclusion (iii) of Proposition 1.5.3 does not hold.

Partial Pseudo-Jacobians

Suppose that $f : \mathbb{R}^{n_1} \times \mathbb{R}^{n_2} \to \mathbb{R}^m$ is continuous in both variables $(x, y) \in \mathbb{R}^{n_1} \times \mathbb{R}^{n_2}$. A pseudo-Jacobian $\partial_x f(x, y) \subset L(\mathbb{R}^{n_1}, \mathbb{R}^m)$ of the function $x \mapsto f(x, y)$ with $y \in \mathbb{R}^{n_2}$ being fixed, is called a *partial pseudo-Jacobian* of f at (x, y) with respect to x. Similarly, $\partial_y f(x, y) \subset L(\mathbb{R}^{n_2}, \mathbb{R}^m)$ is called a partial pseudo-Jacobian of f at (x, y) with respect to y.

For a subset $Q \subset L(\mathbb{R}^{n_1} \times \mathbb{R}^{n_2}, \mathbb{R}^m)$ we denote

$$\text{Proj}_x Q := \{M \in L(\mathbb{R}^{n_1}, \mathbb{R}^m) : \text{ for some} N \in L(\mathbb{R}^{n_2}, \mathbb{R}^m), (MN) \in Q\},$$

$$\text{Proj}_y Q := \{N \in L(\mathbb{R}^{n_2}, \mathbb{R}^m) : \text{ for some} M \in L(\mathbb{R}^{n_1}, \mathbb{R}^m), (MN) \in Q\}.$$

Proposition 1.5.5 *Let $f : \mathbb{R}^{n_1} \times \mathbb{R}^{n_2} \to \mathbb{R}^m$ be continuous. If $\partial f(x, y) \subset L(\mathbb{R}^{n_1} \times \mathbb{R}^{n_2}, \mathbb{R}^m)$ is a pseudo-Jacobian of f at (x, y), then $\text{Proj}_x \partial f(x, y)$ is a partial pseudo-Jacobian of f at (x, y) with respect to x, and $\text{Proj}_y \partial f(x, y)$ is a partial pseudo-Jacobian of f at (x, y) with respect to y.*

Proof. Let $u \in \mathbb{R}^{n_1}$ and $w \in \mathbb{R}^m$. Consider $(u, 0) \in \mathbb{R}^{n_1} \times \mathbb{R}^{n_2}$. We have

$$(wf(\cdot, y))^+(x; u) = \limsup_{t \downarrow 0} \frac{(wf)(x + tu, y) - (wf)(x, y)}{t}$$

$$= \limsup_{t \downarrow 0} \frac{(wf)((x, y) + t(u, 0)) - (wf)(x, y)}{t}$$

$$\leq \sup_{(MN) \in \partial f(x,y)} \langle w, (MN)(u, 0) \rangle$$

$$\leq \sup_{(MN) \in \partial f(x,y)} \langle w, M(u) \rangle = \sup_{M \in \text{Proj}_x \partial f(x,y)} \langle w, M(u) \rangle.$$

This shows that $\text{Proj}_x f(x, y)$ is a pseudo-Jacobian of the function $f(\cdot, y)$ at x. A similar proof is available for $\text{Proj}_y f(x, y)$. □

Notice that if $\partial_x f(x, y)$ and $\partial_y f(x, y)$ are partial pseudo-Jacobians of f at (x, y) with respect to x and y, respectively, then it is not necessary that the set $(\partial_x f(x, y), \partial_y f(x, y))$ is a pseudo-Jacobian of f at (x, y). For instance, let f be a function that is not differentiable at (x, y), but admits partial derivatives $(\partial/\partial x)f(x,y)$ and $(\partial/\partial y)f(x,y)$. Then $\{(\partial/\partial x)f(x,y)\}$ and $\{(\partial/\partial y)f(x,y)\}$ are partial pseudo-Jacobians of f at (x, y). However, $\{((\partial/\partial x)f(x,y),\ (\partial/\partial y)f(x,y))\}$ is not a pseudo-Jacobian of f at (x, y), since if it were then, by Proposition 1.1.2, f would be Gâteaux differentiable at (x, y). We show later that some continuity of partial pseudo-Jacobians is needed in order to obtain a pseudo-Jacobian.

Proposition 1.5.6 *Let* $f\ :\ \mathbb{R}^{n_1} \times \mathbb{R}^{n_2} \to \mathbb{R}^m$ *be continuous and let* $\partial f(x,y) \subset L(\mathbb{R}^{n_1} \times \mathbb{R}^{n_2}, \mathbb{R}^m)$ *be a pseudo-Jacobian of* f *at* (x, y). *Then we have*

$$\mathrm{Proj}_x(\partial f(x,y))_\infty \subset (\mathrm{Proj}_x \partial f(x,y))_\infty$$

$$\mathrm{Proj}_y(\partial f(x,y))_\infty \subset (\mathrm{Proj}_y \partial f(x,y))_\infty.$$

Proof. This follows from Lemma 1.5.1 by considering the projections as linear maps from $L(\mathbb{R}^{n_1} \times \mathbb{R}^{n_2}, \mathbb{R}^m)$ onto $L(\mathbb{R}^{n_1}, \mathbb{R}^m)$ and $L(\mathbb{R}^{n_2}, \mathbb{R}^m)$. □

We note that in general equality does not hold in the conclusion of the above proposition as the following example demonstrates.

Example 1.5.7 Let $f : \mathbb{R} \times \mathbb{R} \to \mathbb{R}$ be defined by

$$f(x,y) = y^{1/3}.$$

Then the set

$$\partial f(0,0) = \{(\alpha, \alpha^2) : \alpha \in \mathbb{R}\}$$

is a pseudo-Jacobian of f at $(0,0)$. We have

$$(\partial f(0,0))_\infty = \{(0,\alpha) : \alpha \geq 0\} \quad \text{and} \quad \mathrm{Proj}_x(\partial f(0,0))_\infty = \{0\}$$

and

$$\mathrm{Proj}_x \partial f(0,0) = \mathbb{R} \quad \text{and} \quad (\mathrm{Proj}_x \partial f(0,0))_\infty = \mathbb{R}.$$

1.6 Constructing Stable Pseudo-Jacobians

A pseudo-Jacobian sometimes produces sharp conditions, but tends to be unstable as it is based on estimates of the function along line directions. When dealing with parametric models, normally generalized derivatives that share a certain degree of robustness (stability) are preferred. Our aim in this section is to explain how we construct a stable (upper-semicontinuous) pseudo-Jacobian from a given collection of pseudo-Jacobians around a point.

Upper Semicontinuous Set-Valued Maps

Let $F : \mathbb{R}^n \rightrightarrows \mathbb{R}^m$ be a set-valued map.

The *Kuratowski–Painlevé upper limit* of F at x is defined by

$$\limsup_{x' \to x} F(x') = \{\lim y_i : y_i \in F(x_i), x_i \to x \text{ as } i \to \infty\}$$

allowing $x' = x$ when taking limits. This upper limit is denoted $\widehat{F}(x)$.

The *recession upper limit* (or outer horizon limit) of F at x, which is denoted $F^\infty(x)$, is defined by

$$F^\infty(x) := \limsup_{x' \to x, t \downarrow 0} tF(x').$$

In other words, $F^\infty(x)$ is a closed cone consisting of all limits: $\lim t_i a_i$ where $a_i \in F(x_i)$, $x_i \to x$, and $t_i \downarrow 0$.

The cosmic upper limit of F consists of the pair of maps $(\widehat{F}, F^\infty)$. It follows from the definitions above that $\widehat{F}(x)$ is a closed set and $F^\infty(x)$ is a closed cone.

From now on we use the following weak version of upper semicontinuity of set-valued maps. We say that F is *upper semicontinuous* at x if for every $\varepsilon > 0$, there exists some $\delta > 0$ such that

$$F(x + \delta B_n) \subseteq F(x) + \varepsilon B_m.$$

When F is single-valued, upper semicontinuity reduces to continuity of a function in the usual sense. When F is compact-valued, F is upper semicontinuous at x if and only if for every open set $V \subset \mathbb{R}^m$ containing $F(x)$, there is a neighborhood U of x such that $F(U) \subset V$, which is the original definition of upper semicontinuity of set-valued maps.

Below we collect some elementary properties of upper semicontinuous set-valued maps for future use.

Lemma 1.6.1 *Let $F : \mathbb{R}^n \rightrightarrows \mathbb{R}^m$ be a set-valued map and let $x \in \mathbb{R}^n$. Then the following assertions hold.*

(i)   *If $F(U)$ is compact for some closed neighborhood U of x, then F is upper semicontinuous at x if and only if F is closed in the sense that $x_i \to x$, $y_i \to y$ and $y_i \in F(x_i)$ imply $y \in F(x)$.*

(ii)   *If F is upper semicontinuous at x, then*

$$F^\infty(x) \subseteq (F(x))_\infty.$$

(iii)   *If F is compact-valued and upper semicontinuous, then the set-valued map $\overline{\mathrm{co}}(F)$ is compact-valued and upper semicontinuous too.*

Proof. The first assertion is obvious. To prove the second assertion, let $v \in F^\infty(x)$; that is, $v = \lim t_i a_i$ where $a_i \in F(x_i)$, $x_i \to x$, and $t_i \downarrow 0$. By the upper semicontinuity of F, there is $i_0 > 0$ such that

$$F(x_i) \subset F(x) + B_m \quad \text{for } i > i_0.$$

It follows that $v \in (F(x) + B_m)_\infty$. In view of Lemma 1.5.1, $v \in (F(x))_\infty$. Assume now that F is compact-valued and semicontinuous. It is evident that $\overline{co}(F)$ is compact-valued too. By the first assertion, it suffices to show that $\overline{co}(F)$ is closed. Let $x_i \to x$, $y_i \to y$, and $y_i \in \overline{co}(F(x_i))$. Note that $F(x_i)$ being compact, one has $\overline{co}F(x_i) = coF(x_i)$. We apply Caratheodory's theorem to find $\lambda_{ij} \geq 0$, $a_{ij} \in F(x_i)$, $j = 1, \dots, m+1$ such that

$$y_i = \sum_{j=1}^{m+1} \lambda_{ij} a_{ij} \quad \text{and} \quad \sum_{j=1}^{m+1} \lambda_{ij} = 1.$$

Without loss of generality we may assume that $\lambda_{ij} \to \lambda_{0j} \geq 0$, $a_{ij} \to a_{0j} \in F(x), j = 1, \dots, m+1$, and $\sum_{j=1}^{m+1} \lambda_{0j} = 1$ when i tends to ∞. Thus we derive

$$y = \sum_{j=1}^{m+1} \lambda_{0j} a_{0j} \in \overline{co}(F(x)),$$

as required. $\qquad\square$

Given a sequence of pseudo-Jacobians $\{\partial_i f(x)\}_{i=1}^\infty$ of f at x, its *recession upper limit* is by definition

$$\lim_{i \to \infty}^{\infty} \partial_i f(x) = \limsup_{i \to \infty, t_i \downarrow 0} t_i \partial_i f(x).$$

This limit is a closed cone. It is trivial if and only if for some i_0, the union of all $\partial_i f(x), i \geq i_0$ is bounded.

For a convex cone $K \subseteq \mathbb{R}^n$ and $\delta > 0$, the conic δ-neighborhood of K, denoted K^δ, is defined by

$$K^\delta := \{x + \delta \|x\| B_n : x \in K\}.$$

It can be seen that when K is convex, closed, and pointed (i.e., $K \cap (-K) = \{0\}$), the cone K^δ is also convex, closed, and pointed for δ sufficiently small.

The next result is a generalization of Proposition 1.2.1 (iii) to a sequence of unbounded pseudo-Jacobians.

Proposition 1.6.2 *Let $\{\partial_i f(x)\}_{i=1}^\infty$ be a decreasing sequence of pseudo-Jacobians of f at x. Then for every $\delta > 0$ the set*

$$(\bigcap_{i=1}^{\infty} \partial_i f(x)) \cup ((\lim_{i \to \infty} \partial_i f(x))^\delta \setminus \text{int}(B_{m \times n}))$$

is a pseudo-Jacobian of f at x.

Proof. Let $u \in \mathbb{R}^n, u \neq 0$, and $v \in \mathbb{R}^m$ with $v \neq 0$. For each $i = 1, 2, \ldots$ there is some $M_i \in \partial_i f(x)$ such that

$$(vf)^+(x; u) \leq \langle v, M_i(u) \rangle + \frac{1}{i}.$$

If the sequence $\{M_i\}_{i=1}^{\infty}$ is bounded, then we may assume that it converge to some element M of the intersection $\bigcap_{i=1}^{\infty} \partial_i f(x)$. The above inequality produces

$$(vf)^+(x; u) \leq \langle v, M(u) \rangle.$$

If that sequence is unbounded, then we may assume that $\lim_{i \to \infty} \|M_i\| = \infty$ and $\lim_{i \to \infty} M_i/\|M_i\| = M$ for some $M \in (\lim_{i \to \infty}^{\infty} \partial_i f(x)) \setminus \text{int}(B_{m \times n})$. For a given $\delta > 0$, when i is sufficiently large, we have

$$M_i/\|M_i\| \in (\lim_{i \to \infty}^{\infty} \partial_i f(x))^\delta \setminus \text{int}(B_{m \times n})$$

and $\|M_i\| \geq 1$. Consequently,

$$(vf)^+(x; u) \leq \sup_{M \in (\lim_{i \to \infty}^{\infty} \partial_i f(x))^\delta \setminus \text{int}(B_{m \times n})} \langle v, M(u) \rangle.$$

This completes the proof. $\qquad\qquad\qquad\qquad\qquad\qquad\qquad\qquad\square$

Notice that the conclusion of Proposition 1.6.2 is in general not true with $\delta = 0$ when all the terms of the sequence $\{\partial_i f(x)\}$ are unbounded.

Upper Semicontinuous Hulls

Given a set-valued map $F : \mathbb{R}^n \rightrightarrows \mathbb{R}^m$, it is always possible to construct an upper semicontinuous map T so that $F(x) \subseteq T(x)$ for every x and has certain minimality properties.

We say that F is *locally bounded* at x if there exists a neighborhood U of x such that the set $F(U)$ is bounded. When F is locally bounded at any point, it is called locally bounded. From now on in this section, it is assumed that the values of F are nonempty sets around a point under consideration.

Lemma 1.6.3 *Assume that F is locally bounded at x. Then the set-valued map G defined by*

$$G(x') = \begin{cases} F(x') & \text{if } x' \neq x, \\ \widehat{F}(x) & \text{if } x' = x, \end{cases}$$

where $\widehat{F}(x)$ is the Kuratowski–Painlevé upper limit of F at x, is upper semi-continuous at x. Moreover, if F is locally bounded, then $\widehat{F}$ is the smallest by inclusion among upper semicontinuous, closed-valued maps that contain F.

Proof. Suppose, to the contrary, that G is not upper semicontinuous at x. Then there exist $\delta > 0$ and $x_i \to x, y_i \in F(x_i)$ as $i \to \infty$ such that $y_i \notin \widehat{F}(x) + \delta B_m$. Because F is locally bounded at x, the sequence $\{y_i\}$ is bounded and we may assume that it converges to some y. We have $y \notin \widehat{F}(x) + (\delta/2)B_m$ because $y_i \notin \widehat{F}(x) + \delta B_m$. On the other hand, by the definition of $\widehat{F}$, one has $y \in \widehat{F}(x)$ which is a contradiction.

For the second part, as we have already noticed that $\widehat{F}(x)$ is a closed set, we need only to show the upper semicontinuity of $\widehat{F}$. Indeed, for every $\varepsilon > 0$, by the first part, there is $\delta > 0$ such that

$$F(x') \subseteq \widehat{F}(x) + \varepsilon B_m \quad \text{for } x' \in x + \delta B_n.$$

Consequently,

$$\widehat{F}(x') \subseteq \widehat{F}(x) + \varepsilon B_m \quad \text{for } x' \in x + \tfrac{\delta}{2} B_n$$

and by this, $\widehat{F}$ is upper semicontinuous. Furthermore, if H is an upper semicontinuous, closed-valued map with $H(x') \supseteq F(x')$ for every x', then we have

$$H(x) \supseteq \limsup_{x' \to x} H(x') \supseteq \limsup_{x' \to x} F(x') = \widehat{F}(x).$$

Thus $\widehat{F}$ is the smallest one. □

The map $\widehat{F}$ is sometimes called the *upper semicontinuous hull* of F. We notice that the above result is no longer true when F is not locally bounded. For instance, the set-valued map $F : \mathbb{R} \rightrightarrows \mathbb{R}$ given by

$$F(x) = \begin{cases} \{\tfrac{1}{x}, 0\} & \text{if } x \neq 0, \\ \{0\} & \text{if } x \neq 0 \end{cases}$$

has $\widehat{F} = F$ which is evidently not upper semicontinuous at $x = 0$.

Lemma 1.6.4 *The set-valued map $F^\infty \cap B_m$ defined by*

$$(F^\infty \cap B_m)(x) = F^\infty(x) \cap B_m$$

is upper semicontinuous.

Proof. Because $F^\infty(x) \cap B_m$ is compact, by virtue of Lemma 1.6.1, it suffices to show that $y \in F^\infty(x) \cap B_m$ when $y = \lim_{i\to\infty} y_i$, where $y_i \in F^\infty(x_i) \cap B_m, x_i \to x$ as $i \to \infty$. If $y = 0$, then it is obvious that $y \in F^\infty(x) \cap B_m$. If $y \neq 0$, then we may assume $\|y\| = 1$ and $\|y_i\| = 1$. By the definition of F^∞, for each i, there exists a sequence $\{x_{i_j}\}_{j=1}^\infty$ converging to x_i and $y_{i_j} \in F(x_{i_j})$ such that $\|y_{i_j}\| \to \infty$ and $y_{i_j}/\|y_{i_j}\| \to y_i$ as $j \to \infty$. By a diagonal process we find a sequence $\{x_{i_k i_k}\}_{k=1}^\infty$ converging to x and $y_{i_k i_k} \in F(x_{i_k i_k})$ such that $\|y_{i_k i_k}\| \to \infty$ and $y_{i_k i_k}/\|y_{i_k i_k}\| \to y$ as $k \to \infty$. This shows that $y \in F^\infty(x) \cap B_m$ and the proof is complete.   □

Lemma 1.6.5 *Let $0 < \alpha < 1$ be given and let $x \in \mathbb{R}^n$ be fixed. The following assertions hold.*

(i)   The set-valued map $F_1 : \mathbb{R}^n \rightrightarrows \mathbb{R}^m$ defined by

$$F_1(x') = \begin{cases} F(x') & \text{if } x' \notin x + \alpha\text{int}(B_n), \\ \text{cl}(F(x + \alpha B_n)) & \text{otherwise} \end{cases}$$

is upper semicontinuous at every point $x' \in x + \alpha\text{int}(B_n)$.
(ii)   The set-valued maps F_2, F_3, and $F_4 : \mathbb{R}^n \rightrightarrows \mathbb{R}^m$ defined by

$$F_2(x') = \begin{cases} F(x') & \text{if } x' \neq x, \\ \widehat{F}(x) + (F^\infty(x))^\alpha & \text{if } x' = x, \end{cases}$$

$$F_3(x') = \begin{cases} F(x') & \text{if } x' \neq x, \\ \widehat{F}(x) \cup [(F^\infty(x))^\alpha \setminus \text{int}(B_m)] & \text{if } x' = x, \end{cases}$$

$$F_4(x') = \begin{cases} \widehat{F}(x') \cup [(F^\infty(x'))^{\alpha/2} \setminus \text{int}(B_m)] & \text{if } x' \neq x, \\ \widehat{F}(x) \cup [(F^\infty(x))^\alpha \setminus \text{int}(B_m)] & \text{if } x' = x \end{cases}$$

are upper semicontinuous at x.

Proof. For the first assertion let $x_0 \in x + \alpha\text{int}(B_n)$. Put $\varepsilon = \alpha - \|x - x_0\| > 0$. Then for every $x' \in x_0 + \varepsilon\text{int}(B_n)$, one has $x' \in x + \alpha\text{int}(B_n)$. By definition, F_1 is constant on $x_0 + \varepsilon\text{int}(B_n)$, hence it is upper semicontinuous at x_0.

For the map F_2, suppose to the contrary that it is not upper semicontinuous at x. Then one can find a sequence $\{x_i\}$ converging to x, a positive constant $\varepsilon > 0$ and $y_i \in G(x_i)$ such that

$$y_i \notin \widehat{F}(x) + (F^\infty(x))^\alpha + \varepsilon B_m, \quad i \geq 1. \tag{1.6}$$

Consider the sequence $\{y_i\}$. If it is bounded, then we may assume it converges to some y_0. By definition we derive $y_0 \in \widehat{F}(x)$ which contradicts (1.6). If the sequence $\{y_i\}$ is unbounded, we may assume $\lim_{i\to\infty} \|y_i\| = \infty$

and $\lim_{i\to\infty} y_i/\|y_i\| = u$ for some $u \in F^\infty(x)$, $\|u\| = 1$. Pick any $y_0 \in F(x)$ and consider the sequence $\{(y_i - y_0)/\|y_i\|\}$. This sequence has the same limit u. Moreover, as $u \in \operatorname{int}((F^\infty(x))^\alpha)$, we have $(y_i - y_0)/\|y_i - y_0\| \in (F^\infty(x))^\alpha$ for i sufficiently large. Because the set $(F^\infty(x))^\alpha$ is a cone, thus we conclude

$$y_i \in y_0 + (F^\infty(x))^\alpha \subseteq \widehat{F}(x) + (F^\infty(x))^\alpha$$

for i large. This contradicts (1.6) and shows that F_2 is upper semicontinuous at x.

For the map F_3 the proof is similar. Let us consider the map F_4. If it is not upper semicontinuous at x, then there exist some $\varepsilon > 0$, $x_i \to x$, and $y_i \in F_4(x_i) \setminus (F_4(x) + \varepsilon B_m)$. We need to consider two cases: either $y_i \in \widehat{F}(x_i)$ or $y_i \in (F^\infty(x'))^{\alpha/2} \setminus \operatorname{int}(B_m)$. In the first case, if the sequence $\{y_i\}_{i=1}^\infty$ is bounded, then it can be assumed to converge to some y_0. It is clear that $y_0 \in \widehat{F}(x)$ and hence, when i is sufficiently large, $y_i \in \widehat{F}(x) + \varepsilon B_m$, a contradiction. If that sequence is unbounded, we may assume that $\lim_{i\to\infty} \|y_i\| = \infty$ and $\lim_{i\to\infty} y_i/\|y_i\| = u$ for some $u \neq 0$. For each i, choose x_i' with $\|x_i' - x_i\| < 1/i$ and $y_i' \in F(x_i')$ with $\|y_i' - y_i\| < 1/i$. Then $\lim_{i\to\infty} y_i'/\|y_i'\| = u \in F^\infty(x)$. By this, when i is large, one has $y_i' \in (F^\infty(x))^{\alpha/2}$, which again contradicts the hypothesis. For the second case, we may assume that $\|y_i\| = 1$ and $\lim_{i\to\infty} y_i = u$ for some $u \neq 0$. Then $u \in F^\infty(x) \setminus \operatorname{int}(B_m)$. Thus, for i sufficiently large, $y_i \in F_4(x) + \varepsilon B_m$ and a contradiction occurs as well. The proof is complete. $\qquad\square$

Pseudo-Jacobian Maps

Now we turn to pseudo-Jacobian matrices. Suppose that $f : \mathbb{R}^n \to \mathbb{R}^m$ is continuous and that a pseudo-Jacobian $\partial f(x)$ of f at x is given for every x. The set-valued map $\partial f : x \mapsto \partial f(x)$ is called a *pseudo-Jacobian map* of f.

Theorem 1.6.6 *Let ∂f be a pseudo-Jacobian map of f. Then the following assertions hold.*

(i) *If ∂f is locally bounded at x, then the pseudo-Jacobian map $\mathcal{J}f$ defined by*

$$\mathcal{J}f(x') = \begin{cases} \partial f(x') & \text{if } x' \neq x, \\ \widehat{\partial f}(x) & \text{if } x' = x \end{cases}$$

is upper semicontinuous at x.

(ii) *If ∂f is locally bounded, then $\widehat{\partial f}$ is the smallest among upper semicontinuous pseudo-Jacobian maps that contain ∂f.*

(iii) *For every $\alpha > 0$, the pseudo-Jacobian maps defined as in Lemma 1.6.5 are upper semicontinuous at x. Moreover, if G is any pseudo-Jacobian map that is upper semicontinuous at x and contains ∂f, then*

$$G(x) \supseteq \widehat{\partial f}(x) \quad and \quad (G(x))_\infty \supseteq (\partial f)^\infty(x).$$

Proof. The first two assertions are immediate from Lemma 1.6.3. The first part of the third assertion is obtained from Lemma 1.6.5. For the second part of (iii), it is clear that $\widehat{\partial f}(x) \subseteq G(x)$ by the upper semicontinuity of G. By the same reason and by Lemma 1.6.1, we have $G^\infty(x) \subseteq (G(x))_\infty$. Moreover, the inclusion $\partial f(x') \subseteq G(x')$ for every x' implies $(\partial f)^\infty(x) \subseteq G^\infty(x)$. It follows that $(\partial f)^\infty(x) \subseteq (G(x))_\infty$ and the proof is complete.   □

Proposition 1.6.7 *Let $f : \mathbb{R}^n \to \mathbb{R}^m$ be locally Lipschitz. If f admits an upper semicontinuous pseudo-Jacobian map ∂f such that $\nabla f(x) \in \partial f(x)$ whenever $\nabla f(x)$ exists, then $\partial^B f(x) \subseteq \partial f(x)$.*

Proof. Let $M \in \partial^B f(x)$. By definition, there is a sequence $\{x_i\}$ converging to x such that $\nabla f(x_i)$ exists and M is the limit of $\{\nabla f(x_i)\}$. Because $\nabla f(x_i) \in \partial f(x_i)$ by hypothesis, and as ∂f is upper semicontinuous, we conclude $M \in \partial f(x)$.   □

Now we obtain the minimality of the B-subdifferential and the Clarke generalized Jacobian.

Corollary 1.6.8 *For a locally Lipschitz function, the B-subdifferential is the smallest with respect to inclusion among upper semicontinuous pseudo-Jacobian maps that contain the Jacobian matrices when they exist, and when $m = 1$ the Clarke generalized subdifferential map is the smallest among upper semicontinuous, convex-valued pseudo-Jacobian maps.*

Proof. This is immediate from Lemma 1.6.3 and Proposition 1.6.7.   □

Notice that the B-subdifferential map of a locally Lipschitz function need not be the smallest by inclusion among upper semicontinuous pseudo-Jacobian maps as illustrated in the example below.

Example 1.6.9 Define $f : \mathbb{R} \to \mathbb{R}$ by the formula

$$f(x) = \begin{cases} 0 & \text{if } x \in (-\infty, 0] \cup [1, \infty) \cup \{\bigcup_{k=1}^\infty [4^{-k}, 4^{1-k}/3]\}; \\ 2x - \frac{2}{3}4^{1-k} & \text{if } x \in \bigcup_{k=1}^\infty [4^{1-k}/3, (\frac{2}{3})4^{1-k}]; \\ 2(4)^{k-1} - 2x & \text{if } x \in \bigcup_{k=1}^\infty [(\frac{2}{3})4^{1-k}, 4^{1-k}]. \end{cases}$$

The B-subdifferential of f is given by

$$\partial^B f(x) = \begin{cases} \{0\} & \text{if } x \in (-\infty, 0) \cup (1, \infty) \cup \left\{ \bigcup_{k=1}^{\infty} \left(4^{-k}, 4^{1-k}/3\right) \right\}; \\[2mm] \{0; 2\} & \text{if } x = (\tfrac{1}{3})4^{1-k}, \ k = 1, 2, \ldots \\[2mm] \{0; -2\} & \text{if } x = 4^{-k}, \ k = 1, 2, \ldots; \\[2mm] \{2\} & \text{if } x \in \bigcup_{k=1}^{\infty} \left((\tfrac{1}{3})4^{1-k}, (\tfrac{2}{3})4^{1-k}\right); \\[2mm] \{-2\} & \text{if } x \in \bigcup_{k=1}^{\infty} \left((\tfrac{2}{3})4^{1-k}, 4^{1-k}\right); \\[2mm] \{0; -2; 2\} & \text{if } x = 0. \end{cases}$$

Now define $\partial f(x) = \{-2, 2\}$ for every $x \in \mathbb{R}$. It is an upper semicontinuous pseudo-Jacobian map of f. At $x = 0$ we have $\partial f(0) \subseteq \partial^B f(0)$ and these two maps are not comparable.

It is known that when $f : \mathbb{R}^n \to \mathbb{R}^m$ is locally Lipschitz, the Clarke generalized Jacobian map is bounded and upper semicontinuous. For $m = 1$, the Michel–Penot subdifferential is bounded, but not upper semicontinuous in general.

Example 1.6.10 Let $f : \mathbb{R}^2 \to \mathbb{R}^2$ be defined by

$$f(x, y) = (|x| - |y|, \ |x|).$$

Define

$$\partial f(x, y) = \left\{ \begin{pmatrix} \text{sign}(x) & -\text{sign}(y) \\ \text{sign}(x) & 0 \end{pmatrix} \right\} \quad \text{for } x \neq 0, \ y \neq 0,$$

$$\partial f(0, y) = \left\{ \begin{pmatrix} 1 & -\text{sign}(y) \\ 1 & 0 \end{pmatrix}, \begin{pmatrix} -1 & -\text{sign}(y) \\ -1 & 0 \end{pmatrix} \right\} \quad \text{for } y \neq 0,$$

$$\partial f(x, 0) = \left\{ \begin{pmatrix} \text{sign}(x) & 1 \\ \text{sign}(x) & 0 \end{pmatrix}, \begin{pmatrix} \text{sign}(x) & -1 \\ \text{sign}(x) & 0 \end{pmatrix} \right\} \quad \text{for } x \neq 0;$$

$$\partial f(0, 0) = \left\{ \begin{pmatrix} 1 & -1 \\ 1 & 0 \end{pmatrix}, \begin{pmatrix} 1 & 1 \\ 1 & 0 \end{pmatrix}, \begin{pmatrix} -1 & 1 \\ -1 & 0 \end{pmatrix}, \begin{pmatrix} -1 & -1 \\ -1 & 0 \end{pmatrix} \right\}.$$

It is easy to see that ∂f above is a bounded and upper semicontinuous pseudo-Jacobian map of f, which is smaller than the Clarke generalized Jacobian.

Example 1.6.11 Let $f : \mathbb{R}^2 \to \mathbb{R}^2$ be defined by

$$f(x, y) = (\sqrt{|x|}\,\text{sign}(x), \ \sqrt{|y|}\,\text{sign}(y) + x).$$

This function is not locally Lipschitz. Define

$$\partial f(x,y) = \left\{ \begin{pmatrix} \frac{1}{(2\sqrt{|x|})} & 0 \\ 1 & \frac{1}{(2\sqrt{|y|})} \end{pmatrix} \right\} \quad \text{for} \ \ x \neq 0, \ y \neq 0,$$

$$\partial f(0,y) = \left\{ \begin{pmatrix} \alpha & 0 \\ 1 & \frac{1}{(2\sqrt{|y|})} \end{pmatrix} : \alpha \geq 0 \right\} \quad \text{for} \ y \neq 0,$$

$$\partial f(x,0) = \left\{ \begin{pmatrix} \frac{1}{(2\sqrt{|x|})} & 0 \\ 1 & \beta \end{pmatrix} : \beta \geq 0 \right\} \quad \text{for} \ x \neq 0,$$

$$\partial f(0,0) = \left\{ \begin{pmatrix} \alpha & 0 \\ 1 & \beta \end{pmatrix} : \alpha, \beta \geq c \right\},$$

where c is any real number. It is easy to see that ∂f defined above is a pseudo-Jacobian map of f, that is unbounded at either $x = 0$ or $y = 0$, and is upper semicontinuous provided $c \leq 0$.

1.7 Gâteaux and Fréchet Pseudo-Jacobians

Let $f : \mathbb{R}^n \to \mathbb{R}^m$ be continuous and let $\partial f(x) \subset L(\mathbb{R}^n, \mathbb{R}^m)$ be a closed set of $m \times n$-matrices. We say that $\partial f(x)$ is a *Gâteaux pseudo-Jacobian* of f at x if for every $u \in \mathbb{R}^n$ and for every $t > 0$, there is some $M_t \in \partial f(x)$ such that

$$f(x + tu) - f(x) = M_t(tu) + o(t),$$

where $o(t)/t \to 0$ as $t \to 0$, and it is a *Fréchet pseudo-Jacobian* of f at x if for each y in a neighborhood of x, there exists a matrix $M_y \in \partial f(x)$ such that

$$f(y) - f(x) = M_y(y - x) + o(\|y - x\|),$$

where $o(\|y - x\|) / \|y - x\| \to 0$ as $y \to x$.

It follows immediately from the definition that any Fréchet pseudo-Jacobian is a Gâteaux pseudo-Jacobian. The converse is not always true, which can be seen in the next example.

Example 1.7.1 Define $f : \mathbb{R}^2 \to \mathbb{R}$ by

$$f(x_1, x_2) = \begin{cases} x_1 e^{-x_2/((x_1 - \sqrt{x_2})^2 - x_2/4))} & \text{if } x_2 > 0, \ \sqrt{x_2}/2 < x_1 < (3\sqrt{x_2})/2, \\ 0 & \text{otherwise.} \end{cases}$$

Then $\{(0,0)\}$ is a Gâteaux pseudo-Jacobian, but not a Fréchet pseudo-Jacobian of f at $(0,0)$. Indeed, for each $u \in \mathbb{R}^2, u \neq 0$, for t sufficiently small, one has $f(tu) = 0$. Hence $f(tu) - f(0) = 0$. On the other hand, by taking $y = (x_1, x_1^2)$, we have $f(y) - f(0) = x_1 e^4$, which shows that the set $\{(0,0)\}$ cannot be a Fréchet pseudo-Jacobian of f at $(0,0)$. Actually, the function f is Gâteaux differentiable and not Fréchet differentiable at 0.

The next result justifies the terminology of Gâteaux pseudo-Jacobian.

Proposition 1.7.2 *We have the following properties of Gâteaux pseudo-Jacobians.*

(i)   Every Gâteaux pseudo-Jacobian is a pseudo-Jacobian.

(ii)   If f is Gâteaux differentiable at x, then $\{\nabla f(x)\}$ is a Gâteaux pseudo-Jacobian of f at x. Conversely, if f admits a singleton Gâteaux pseudo-Jacobian $\{A\}$ at x, then f is Gâteaux differentiable at x and $A = \nabla f(x)$.

Proof. For the first assertion, let $u \in \mathbb{R}^n$ and $v \in \mathbb{R}^m$. Let $\{t_i\}$ be a sequence of positive numbers converging to 0 such that

$$(vf)^+(x; u) = \lim_{i \to \infty} \frac{(vf)(x + t_i u) - (vF)(x)}{t_i}.$$

Because $\partial f(x)$ is a Gâteaux pseudo-Jacobian of f at x, for each i, there exists $M_{t_i} \in \partial f(x)$ such that

$$\frac{\langle v, f(x + t_i u) \rangle - \langle v, f(x) \rangle}{t_i} = \langle v, M_{t_i}(u) \rangle + \langle v, o(t_i) \rangle.$$

Passing to the limit, we get that $\lim_{i \to \infty} (\langle v, o(t_i) \rangle / t_i) = 0$ and

$$(vf)^+(x; u) = \lim_{i \to \infty} \frac{(vf)(x+t_i u)-(vF)(x)}{t_i} \leq \sup_{N \in \partial F(x)} (\langle v, N(u) \rangle + \frac{\langle v, o(t_i) \rangle}{t_i})$$

$$= \sup_{N \in \partial F(x)} \langle v, N(u) \rangle,$$

which shows that $\partial f(x)$ is a pseudo-Jacobian of F at x. The second assertion follows directly from the definition. $\qquad\square$

A similar result is true for Fréchet pseudo-Jacobians.

Proposition 1.7.3 *We have the following properties of Fréchet pseudo-Jacobians.*

(i)   Every Fréchet pseudo-Jacobian is a pseudo-Jacobian.

(ii)   If f is Fréchet differentiable, then $\{\nabla f(x)\}$ is a Fréchet pseudo-Jacobian of f at x. Conversely, if f admits a singleton Fréchet pseudo-Jacobian $\{A\}$ at x, then f is Fréchet differentiable at x and $A = \nabla f(x)$.

Proof. Because every Fréchet pseudo-Jacobian is Gâteaux pseudo-Jacobian, thus the first property follows from Proposition 1.7.2. Now if f is Fréchet differentiable at x_0, then, in a neighborhood of x_0,

$$f(x) - f(x_0) = \nabla f(x_0)(x - x_0) + o(\|x - x_0\|).$$

It is obvious that the singleton $\{\nabla f(x_0)\}$ is a Fréchet pseudo-Jacobian of f at x_0. Furthermore, let $\{M\}$ be a singleton Fréchet pseudo-Jacobian of f at x_0; then for each x in a neighborhood of x_0 we have

$$f(x) - f(x_0) - M(x - x_0) = o(\|x - x_0\|)\,,$$

which shows that f is Fréchet differentiable and $\nabla f(x_0) = M$. $\square$

We note that if f is Fréchet differentiable and $\partial f(x)$ is a Fréchet pseudo-Jacobian of f at x, then $\nabla f(x)$ is not necessarily an element of $\partial f(x)$. For instance, the constant function $f : \mathbb{R}^2 \to \mathbb{R}$ defined by $f(x) = 0$ admits a Fréchet pseudo-Jacobian $\partial f(0) = \{(\alpha, \beta) : \alpha^2 + \beta^2 = 1\}$ at $x = 0$, which evidently does not contain $\nabla f(0) = (0, 0)$. Furthermore, not every pseudo-Jacobian, even when being a singleton, is a Fréchet pseudo-Jacobian, as we have seen in Example 1.7.1.

Proposition 1.7.4 *Suppose that $f : \mathbb{R}^n \to \mathbb{R}^m$ is locally Lipschitz and $\partial f(x)$ is a bounded pseudo-Jacobian of f at x. Then $\mathrm{co}(\partial f(x))$ is a Fréchet pseudo-Jacobian of f at x. In particular, the Clarke generalized Jacobian $\partial^C f(x)$, and, when $m = 1$, the Michel–Penot subdifferential $\partial^{MP} f(x)$ are Fréchet pseudo-Jacobians of f at x.*

Proof. Suppose, to the contrary, that $\mathrm{co}(\partial f(x))$ is not a Fréchet pseudo-Jacobian of f at x. Then there exist a sequence $\{x_k\}_{k=1}^{\infty}$ converging to x and a positive ε such that

$$f(x_k) - f(x) \notin \mathrm{co}\,(\partial f(x))\,(x_k - x) + \varepsilon\|x_k - x\|B_m$$

for $k \geq 1$. The set on the right hand side is convex, therefore there exists a vector $v_k \in R^m$ with $\|v_k\| = 1$ such that

$$\langle v_k, f(x_k) - f(x)\rangle \geq \sup_{M \in \partial f(x), b \in B_m} \langle v_k, M(x_k - x) + \varepsilon\|x_k - x\|b\rangle.$$

Set $t_k = \|x_k - x\|$ and $u_k = (x_k - x)/t_k$. Without loss of generality one may assume that $\{u_k\}_{k=1}^{\infty}$ converges to some $u \neq 0$ and $\{v_k\}_{k=1}^{\infty}$ converges to some $v \neq 0$. Then we deduce

$$\langle v_k, f(x + t_k u) - f(x)\rangle = \langle v_k, f(x + t_k u) - f(x_k)\rangle + \langle v_k, f(x_k) - f(x)\rangle$$
$$\geq -\lambda\|t_k(u_k - u)\| + \sup_{M \in \partial f(x)} \langle v_k, t_k M(u_k)\rangle + \varepsilon t_k,$$

where λ is a Lipschitz continuity constant of f near x. By dividing both sides of the above inequality by t_k and passing to the limit when $k \to \infty$, we obtain

$$(v \circ f)^+(x; u) \geq \sup_{M \in \partial f(x)} \langle v, M(u) \rangle + \varepsilon,$$

which contradicts the fact that $\partial f(x)$ is a pseudo-Jacobian of f and x.

The second part of the proposition is immediate by observing that the Clarke generalized Jacobian and the Michel–Penot subdifferential are convex and bounded pseudo-Jacobians (see Proposition 1.3.1 and Proposition 1.4.7). □

Note that a locally Lipschitz function may have a Fréchet pseudo-Jacobian smaller than the Clarke generalized Jacobian. For instance, the function $f(x) = |x|$ admits a Fréchet pseudo-Jacobian $\{1, -1\}$ at 0, while $\partial^C f(0) = [-1, 1]$. In this example $\partial^C f(0)$ is the convex hull of the Fréchet pseudo-Jacobian $\{1, -1\}$. The next example shows that a locally Lipschitz function may have a Fréchet pseudo-Jacobian whose convex hull is smaller than the Clarke generalized Jacobian.

Example 1.7.5 Suppose that $f : \mathbb{R}^2 \to \mathbb{R}$ is defined by

$$f(x) = \begin{cases} x^2 \sin \frac{1}{x} + |y| & x \neq 0, \\ |y| & \text{else.} \end{cases}$$

It is easy to check that this function is locally Lipschitz. A simple calculation confirms that the set

$$\partial f(0,0) := \big\{ (0, \beta) : \beta \in [-1, 1] \big\}$$

is a Fréchet pseudo-Jacobian of f at $(0,0)$, whereas its Clarke generalized Jacobian is the set

$$\partial^C f(0,0) := \big\{ (\alpha, \beta) : \alpha, \beta \in [-1, 1] \big\}.$$

Hence $\overline{\text{co}}(\partial f(0,0))$ is a proper subset of $\partial^C f(0,0)$.

Next we give an example of a continuous function that is not locally Lipschitz and has an unbounded Fréchet pseudo-Jacobian.

Example 1.7.6 Suppose that $f : \mathbb{R}^2 \to \mathbb{R}^2$ is defined by

$$f(x, y) = \big(|x|^{1/2} \, \text{sign}(x), y^{1/3} + |x| \big) .$$

This function is not locally Lipschitz. It is easy to see that the set

$$\partial f(0,0) := \left\{ \begin{pmatrix} \alpha & 0 \\ \beta & \gamma \end{pmatrix} : \alpha \geq 0, \, -1 \leq \beta \leq 1, \, \gamma \in \mathbb{R} \right\}$$

is a Fréchet pseudo-Jacobian of f at $(0,0)$.

Note also that a non-Lipschitz function may have a bounded Fréchet pseudo-Jacobian as shown in the next example.

Example 1.7.7 Let $f : \mathbb{R} \to \mathbb{R}$ be defined by

$$f(x) = \begin{cases} x^2 \sin \frac{1}{x^2} & x \neq 0, \\ 0 & x = 0. \end{cases}$$

Then $\{0\}$ is a Fréchet pseudo-Jacobian of f at 0, and f is not locally Lipschitz at 0. For real functions on $\mathbb{R}$, the notions of Fréchet differentiability and Gâteaux differentiablity coincide.

Besides the Clarke generalized Jacobian, several known generalized derivatives are instances of Fréchet pseudo-Jacobians. Some of them are presented below.

Proposition 1.7.8 (the Gowda–Ravindran H-differential) *Suppose that $f : \mathbb{R}^n \to \mathbb{R}^m$ is continuous. Let $T(x_0)$ be an H-differential of f at x_0, then its closure $\mathrm{cl}(T(x_0))$ is a Fréchet pseudo-Jacobian of f at x_0.*

Proof. In fact, suppose to the contrary that $\mathrm{cl}(T(x_0))$ is not a Fréchet pseudo-Jacobian of f at x_0. Then there exists a sequence $\{x_k\}$ converging to x_0 such that

$$\lim_{k \to \infty} \frac{d\big(f(x_k) - f(x_0), T(x_0)(x - x_0)\big)}{\|x_k - x_0\|} \geq \varepsilon$$

for some $\varepsilon > 0$, where $d(f(x_k) - f(x_0), T(x_0)(x_k - x_0))$ denotes the distance from $f(x_k) - f(x_0)$ to $T(x_0)(x_k - x_0)$. This contradicts the assumption that $T(x_0)$ is an H-differential of f at x_0. $\qquad \square$

It is clear that Proposition 1.3.7 is a direct consequence of Proposition 1.7.8. We notice also that the converse statement of Proposition 1.7.8 is not true in general, that is, a Fréchet pseudo-Jacobian is not necessarily an H-differential. The next simple example shows that a continuous function that admits a Fréchet pseudo-Jacobian may not be H-differentiable.

Example 1.7.9 Consider the function $f : \mathbb{R}^2 \to \mathbb{R}^2$ defined by

$$f(x, y) = (-x + y^{1/3}, -x^3 + y).$$

A direct calculation shows that

$$\partial f(0, 0) := \left\{ \begin{pmatrix} -1 & \alpha \\ 0 & 1 \end{pmatrix} : \alpha \geq 1 \right\}$$

is a Fréchet pseudo-Jacobian of f at $(0,0)$. However, it is easy to see that the function is not H-differentiable at $(0,0)$.

Proposition 1.7.10 (Ioffe's prederivative) *Let Ω_Q be a fan generated by a closed set $Q \subseteq L(\mathbb{R}^n, \mathbb{R}^m)$ by the rule*

$$\Omega_Q(u) = Q(u) \quad \text{for} \quad u \in \mathbb{R}^n.$$

Assume that f admits a prederivative of this form, then Q is a Fréchet pseudo-Jacobian of f at x. Conversely, if $\partial f(x)$ is a Fréchet pseudo-Jacobian of f at x that is convex and compact, then the fan generated by $\partial f(x)$ is a prederivative of f at x.

Proof. This follows easily from the definition of the prederivative. $\qquad\square$

Proposition 1.7.11 (Warga's unbounded derivative container) *Let $f : \mathbb{R}^n \to \mathbb{R}^m$ be continuous and let $\{\Lambda^\varepsilon f(x)\}$ be an unbounded derivative container of f on V. Then for each $x_0 \in V$ and $\varepsilon > 0$, the set $\overline{\mathrm{co}}(\Lambda^\varepsilon f(x_0))$ is a Fréchet pseudo-Jacobian of f at x_0 .*

Proof. Suppose, to the contrary, that $\overline{\mathrm{co}}(\Lambda^\varepsilon f(x_0))$ is not a Fréchet pseudo-Jacobian of f at x_0. Then there exists a sequence $\{x_k\}$ converging to x_0 such that
$$\frac{d(f(x_k) - f(x_0), \overline{\mathrm{co}}(\Lambda^\varepsilon f(x_0))(x_k - x_0))}{\|x_k - x_0\|} \geq \varepsilon$$

for some $\varepsilon > 0$. Let $C = \{x_k : k = 1, 2, \ldots\} \cup \{x_0\}$. Then C is a compact set that we may assume to be in V. Let $\{f_i\}$ be a sequence of continuously differentiable functions stated in the definition of unbounded derivative containers. For each $k = 1, 2, \ldots$ with $\|x_k - x_0\| < \delta_c$, let $i_k > i_C$ be an index sufficiently large so that

$$\|f_{i_k}(x) - f(x)\| \leq \|x_k - x_0\|^2 \quad \text{for every } x \in C.$$

Applying the classical mean value theorem, we find for each k, a matrix $M_k \in \mathrm{co}(\nabla f_{i_k}[x_0, x_k])$ such that

$$f_{i_k}(x_k) - f_{i_k}(x_0) = M_k(x_k - x_0) .$$

For k with $\|x_k - x_0\| < \delta_c$, one has $\nabla f_{i_k}[x_0, x_k] \subseteq \Lambda^\varepsilon f(x_0)$. Hence we derive $M_k \in \mathrm{co}(\Lambda^\varepsilon f(x_0))$. For such k, we have

$$f(x_k) - f(x_0) = f(x_k) - f_{i_k}(x_k) + f_{i_k}(x_k) - f_{i_k}(x_0) + f_{i_k}(x_0) - f(x_0)$$
$$= f(x_k) - f_{i_k}(x_k) + f_{i_k}(x_0) - f(x_0) + M_k(x_k - x_0).$$

Hence

$$\frac{d\big(f(x_k) - f(x_0),\ \overline{\mathrm{co}}(\Lambda^\varepsilon f(x_0))(x_k - x_0)\big)}{\|x_k - x_0\|} \le 2\|x_k - x_0\| .$$

This is impossible when $\|x_k - x_0\| < \varepsilon/2$.   □

A more restrictive pseudo-Jacobian can be required as follows. We say that a nonempty subset $\partial f(x) \subseteq L(\mathbb{R}^n, \mathbb{R}^m)$ is a *strict pseudo-Jacobian* of f at x_0 if for every x and y there is some matrix $M_{x,y} \in \partial f(x)$ such that

$$f(x) - f(y) = M_{x,y}(x - y) + o(\|x - y\|),$$

where $o(\|x - y\|)/\|x - y\| \to 0$, as $x \to x_0, y \to x_0$, and $x \ne y$.

It is evident that any strict pseudo-Jacobian is a Fréchet pseudo-Jacobian. The converse is not true. For instance, the function $f : \mathbb{R} \to \mathbb{R}$ given by

$$f(x) = \begin{cases} x^2 \sin(1/x) & \text{if } x \ne 0; \\ 0 & \text{else} \end{cases}$$

admits $\{0\}$ as a Fréchet pseudo-Jacobian at $x = 0$, but this set is not a strict pseudo-Jacobian.

Proposition 1.7.12 *Let $f : \mathbb{R}^n \to \mathbb{R}^m$ be strictly differentiable at x_0. Then the set $\{\nabla f(x_0)\}$ is a strict pseudo-Jacobian of f at x_0. Conversely, if f admits a singleton strict pseudo-Jacobian $\{A\}$ at x_0, then it is strictly differentiable at x_0 and $\nabla f(x_0) = A$.*

Proof. This follows directly from the definitions of strict pseudo-Jacobians and strict differentiability.   □

Proposition 1.7.13 *Assume that $f : \mathbb{R}^n \to \mathbb{R}^m$ is locally Lipschitz at x_0. Then the Clarke generalized Jacobian is a strict pseudo-Jacobian of f at x_0.*

Proof. Let $\varepsilon > 0$. By the upper semicontinuity of the Clarke generalized Jacobian map, there is some $\delta > 0$ such that

$$\partial^C f(x) \subseteq \partial^C f(x_0) + \varepsilon B_{m \times n}$$

for every $x \in x_0 + \delta B_n$. In view of Lebourg's mean value theorem, for every $x, y \in x_0 + \delta B_n$ there exist some matrices $M_{x,y} \in \partial f(x_0)$ and $P_{x,y} \in B_{m \times n}$ such that

$$f(x) - f(y) = M_{x,y} + \varepsilon P_{x,y}(x - y).$$

This implies that $\partial f(x_0)$ is a strict pseudo-Jacobian of f at x_0.   □

Corollary 1.7.14 *A continuous function* $f : \mathbb{R}^n \to \mathbb{R}^m$ *is locally Lipschitz at* x_0 *if and only if it admits a bounded strict pseudo-Jacobian at* x_0.

Proof. According to Proposition 1.7.13 it suffices to show the "if" part. Let $\partial f(x_0)$ be a bounded strict pseudo-Jacobian of f at x_0. There is a convex neighborhood U of x_0 such that $-1 \leq o(\|x - y\|) \leq 1$. Let $\alpha = \sup_{M \in \partial f(x_0)} \|M\|$. Then for every $x, y \in U$, one has

$$\|f(x) - f(y)\| \leq (\alpha + 1)\|x - y\|$$

as requested. $\qquad\square$

Using a strict pseudo-Jacobian at a point, we obtain pseudo-Jacobians in a neighborhood of the point.

Proposition 1.7.15 *Suppose that* $f : \mathbb{R}^n \to \mathbb{R}^m$ *is continuous and that* $\partial f(x_0)$ *is a bounded strict pseudo-Jacobian of* f *at* x_0. *Then, for every* $\varepsilon > 0$, *there exists* $\delta > 0$ *such that the set* $\partial f(x_0) + \varepsilon B_{m \times n}$ *is a pseudo-Jacobian of* f *at every* $x \in x_0 + \delta B_n$.

Proof. Suppose to the contrary that for some fixed $\varepsilon > 0$, there are points x_k converging to x_0 such that $\partial f(x_0) + \varepsilon B_{m \times n}$ is not a pseudo-Jacobian of f at x_k. We can find vectors $v_k \in R^m$ and $u_k \in R^m$ with $\|v_k\| = 1$ and $\|u_k\| = 1$ such that

$$((v_k \circ f)^+(x_k), u_k) > \sup_{M \in \partial f(x_0), N \in B_{m \times n}} \langle v_n, M(u_n) + \varepsilon N(u_n) \rangle.$$

We may assume that $\{v_k\}_{k=1}^{\infty}$ and $\{u_k\}_{k=1}^{\infty}$ converge respectively to $v \neq 0$ and $u \neq 0$. It follows from the definition of the upper directional derivative that there are positive numbers t_k converging to 0 such that

$$\left\langle v_k, \frac{f(x_k + t_k u_k) - f(x_k)}{t_k} \right\rangle \geq \sup_{M \in \partial f(x_0)} \langle v_k, M(u_k) \rangle + \frac{\delta}{2} \qquad (1.7)$$

for $k \geq 1$. Because $\partial f(x_0)$ is a strict pseudo-Jacobian of f at x_0, there are matrices $M_k \in \partial f(x_0)$, which may be assumed to converge to some $M \in \partial f(x_0)$, such that

$$f(x_k + t_k u_k) - f(x_k) = M_k(t_k u_k) + o(\|t_k u_k\|).$$

Substituting this expression into (1.7) and passing to the limit as $k \to +\infty$, we derive

$$(v, f)^+(x_0; u) \geq \sup_{M \in \partial f(x_0)} \langle v, M(u) \rangle + \frac{\delta}{2}$$

which is a contradiction. $\qquad\square$

2

Calculus Rules for Pseudo-Jacobians

In this chapter we develop a number of generalized calculus rules for pseudo-Jacobians, including various forms of chain rules. In particular, the diversity of chain rules together with the fact that most of the rules are available without regularity conditions permits us to employ a variety of generalized derivatives to study a variational problem. This feature facilitates a wide range of applications of the rules to different classes of problems.

2.1 Elementary Rules

We first proceed to provide elementary calculus rules for pseudo-Jacobians, that allow us to treat the simplest combinations of continuous functions.

Scalar Multiples and Sums

Theorem 2.1.1 *Let f and $g\colon \mathbb{R}^n \to \mathbb{R}^m$ be continuous functions. If $\partial f(x)$ and $\partial g(x)$ are pseudo-Jacobians of f and g, respectively, at x, then*

(i) $\alpha \partial f(x)$ *is a pseudo-Jacobian of* αf *at* x *for every* $\alpha \in R$.
(ii) $\mathrm{cl}(\partial f(x) + \partial g(x))$ *is a pseudo-Jacobian of* $f + g$ *at* x.

Proof. Let $\alpha \in \mathbb{R}$. If $\alpha \geq 0$, then for every $u \in \mathbb{R}^n$ and $v \in \mathbb{R}^m$ we have

$$(v(\alpha f))^+(x; u) = \alpha(vf)^+(x; u) \leq \alpha \sup_{M \in \partial f(x)} \langle v, M(u) \rangle$$

$$\leq \sup_{M \in \partial f(x)} \langle v, \alpha M(u) \rangle \leq \sup_{N \in \alpha \partial f(x)} \langle v, N(u) \rangle.$$

This and the fact that the set $\alpha \partial f(x)$ is closed show that $\alpha \partial f(x)$ is a pseudo-Jacobian of αf at x. When $\alpha < 0$, we similarly have

$$(v(\alpha f))^+(x; u) = -\alpha(-vf)^+(x; u) \leq -\alpha \sup_{M \in \partial f(x)} \langle -v, M(u) \rangle$$

$$\leq \sup_{M \in \partial f(x)} \langle v, \alpha M(u) \rangle \leq \sup_{N \in \alpha \partial f(x)} \langle v, N(u) \rangle,$$

and arrive at the same conclusion.

For the second part, let $u \in \mathbb{R}^n$ and $v \in \mathbb{R}^m$. We have

$$(v(f+g))^+(x; u) \leq (vf)^+(x; u) + (vg)^+(x; u)$$

$$\leq \sup_{M \in \partial f(x)} \langle v, M(u) \rangle + \sup_{N \in \partial g(x)} \langle v, N(u) \rangle$$

$$\leq \sup_{P \in \partial f(x) + \partial g(x)} \langle v, P(u) \rangle,$$

which shows that the closure of the set $\partial f(x) + \partial g(x)$ is a pseudo-Jacobian of $f + g$ at x. $\qquad \square$

When f and g are locally Lipschitz, the second assertion of Theorem 2.1.1 gives a known sum rule of the Clarke generalized Jacobian.

Corollary 2.1.2 *Assume that f and g are locally Lipschitz functions from $\mathbb{R}^n$ to $\mathbb{R}$. Then*

$$\partial^C(f + g)(x) \subseteq \partial^C f(x) + \partial^C g(x).$$

Proof. According to Theorem 2.1.1, the set $\partial^C f(x) + \partial^C g(x)$ is a pseudo-Jacobian of $f + g$ at x. Moreover, the set-valued map $x \mapsto \partial^C f(x) + \partial^C g(x)$ is compact, convex-valued, and upper semicontinuous. By Corollary 1.6.8, $\partial^C(f + g)(x) \subseteq \partial^C f(x) + \partial^C g(x)$. $\qquad \square$

Cartesian Products

We agree that by writing $M \times N$ for $M \in L(\mathbb{R}^n, \mathbb{R}^m)$ and $N \in L(\mathbb{R}^n, \mathbb{R}^\ell)$ we mean the $(m + \ell) \times n$-matrix $\binom{M}{N} \in L(\mathbb{R}^n, \mathbb{R}^{m+\ell})$.

Theorem 2.1.3 *Let $f \colon \mathbb{R}^n \to \mathbb{R}^m$ and $g \colon \mathbb{R}^n \to \mathbb{R}^\ell$ be continuous functions. If $\partial f(x) \subseteq L(\mathbb{R}^n, \mathbb{R}^m)$ and $\partial g(x) \subseteq L(\mathbb{R}^n, \mathbb{R}^\ell)$ are pseudo-Jacobians of f and g at x, respectively, then $\partial f(x) \times \partial g(x)$ is a pseudo-Jacobian of (f, g) at x. If $f = (f_1, \ldots, f_m)$ and $\partial f_1(x), \ldots, \partial f_m(x)$ are pseudo-differentials of the scalar component functions $f_1, \ldots, f_m$ at x, respectively, then $\partial f_1(x) \times \cdots \times \partial f_m(x)$ is a pseudo-Jacobian of f at that point.*

Proof. Let $u \in \mathbb{R}^n$ and $(v, w) \in \mathbb{R}^{m+\ell}$. Then

$$((v, w)f \times g)^+(x; u) = (vf + wg)^+(x; u)$$
$$\leq (vf)^+(x; u) + (wg)^+(x; u)$$
$$\leq \sup_{M \in \partial f(x)} \langle v, M(u) \rangle + \sup_{N \in \partial g(x)} \langle w, N(u) \rangle$$
$$\leq \sup_{M \times N \in \partial f(x) \times \partial g(x)} \langle (v, w), (M \times N)(u) \rangle.$$

This shows that $\partial f(x) \times \partial g(x)$ is a pseudo-Jacobian of $f \times g$ at x. The second part is immediate from the first one. □

Note that, in general, $\partial f(x) \times \partial g(x)$ is not the smallest among all possible pseudo-Jacobians of $f \times g$ at x even if $\partial f(x)$ and $\partial g(x)$ are.

Example 2.1.4 Let $f(x) = |x|$ for $x \in \mathbb{R}$ and let $h \colon \mathbb{R} \to \mathbb{R}^2$ be the product $f \times f$. The set $\partial f(0) = \{1, -1\}$ is a pseudo-differential of f at 0. It is not hard to see that this is the smallest one; that is, any pseudo-differential of f at 0 contains $\partial f(0)$ in its convex hull. It follows from Theorem 2.1.3 that the set

$$\partial f(0) \times \partial f(0) = \left\{ \begin{pmatrix} 1 \\ 1 \end{pmatrix}, \begin{pmatrix} 1 \\ -1 \end{pmatrix}, \begin{pmatrix} -1 \\ 1 \end{pmatrix}, \begin{pmatrix} -1 \\ -1 \end{pmatrix} \right\}$$

is a pseudo-Jacobian of $h = f \times f$ at 0. It is clear that this pseudo-Jacobian is not the smallest because the smaller set

$$\partial h(0) = \left\{ \begin{pmatrix} 1 \\ 1 \end{pmatrix}, \begin{pmatrix} -1 \\ -1 \end{pmatrix} \right\}$$

is also a pseudo-Jacobian of h at 0.

Products and Quotients

Theorem 2.1.5 *Let $f, g : \mathbb{R}^n \to \mathbb{R}$ be continuous functions. Let $\partial f(x)$ and $\partial g(x)$ be pseudo-differentials of f and g, respectively, at x. If at least one of the values $f(x)$ and $g(x)$ is nonzero whenever both $\partial f(x)$ and $\partial g(x)$ are unbounded, then the closure of the set*

$$f(x)\partial g(x) + g(x)\partial f(x)$$

is a pseudo-differential of the product fg at x.

Proof. Let $\alpha \in \mathbb{R}$ and $u \in \mathbb{R}^n$. Let $\{t_k\}_{k=1}^{\infty}$ be a sequence of positive numbers converging to 0 such that

$$(\alpha fg)^+(x; u) = \lim_{k \to \infty} \frac{(\alpha fg)(x + t_k u) - (\alpha fg)(x)}{t_k}.$$

Let $f(x) \neq 0$, say $f(x) > 0$. In view of the continuity of f, we may assume that $f(x + t_k u) > 0$ for all $k \geq 1$. Expressing

$$(\alpha f g)(x + t_k u) - (\alpha f g)(x) = f(x + t_k u)[(\alpha g)(x + t_k u) - (\alpha g)(x)]$$
$$+ g(x)[(\alpha f)(x + t_k u) - (\alpha f)(x)],$$

we obtain

$$(\alpha f g)^+(x; u) = \lim_{k \to \infty} \left(f(x + t_k u) \frac{(\alpha g)(x + t_k u) - (\alpha g)(x)}{t_k} \right.$$
$$\left. + \frac{(\alpha g(x)) f(x + t_k x) - (\alpha g(x)) f(x)}{t_k} \right). \tag{2.1}$$

By the definition of $\partial f(x)$,

$$\limsup_{k \to \infty} \frac{(\alpha g(x)) f(x + t_k u) - (\alpha g(x)) f(x)}{t_k} \leq \sup_{M \in \partial f(x)} \langle \alpha g(x), M(u) \rangle. \tag{2.2}$$

Consider the sequence

$$\left\{ \frac{((\alpha g)(x + t_k u) - (\alpha g)(x))}{t_k} \right\}_{k \geq 1}.$$

If it is bounded, then

$$\limsup_{k \to \infty} f(x + t_k u) \frac{(\alpha g)(x + t_k u) - (\alpha g)(x)}{t_k}$$
$$= \limsup_{k \to \infty} f(x) \frac{(\alpha g)(x + t_k u) - (\alpha g)(x)}{t_k}$$
$$\leq \sup_{N \in \partial g(x)} \langle \alpha f(x), N(u) \rangle.$$

This combined with (2.1) and (2.2) yields

$$(\alpha f g)^+(x; u) \leq \sup_{N \in \partial g(x)} \langle \alpha f(x), N(u) \rangle + \sup_{M \in \partial f(x)} \langle \alpha g(x), M(u) \rangle$$
$$\leq \sup_{N \in \partial g(x), M \in \partial f(x)} \alpha \langle f(x) N^{tr} + g(x) M^{tr}, u \rangle, \tag{2.3}$$

which shows that the closure of the set $f(x) \partial g(x) + g(x) \partial f(x)$ is a pseudo-Jacobian of fg at x.

If the sequence

$$\left\{ \frac{((\alpha g)(x + t_k u) - (\alpha g)(x))}{t_k} \right\}_{k \geq 1}$$

is unbounded, then the upper limit

$$q := \limsup_{k\to\infty} \frac{(\alpha g)(x + t_k u) - (\alpha g)(x)}{t_k}$$

may take either the value $+\infty$ or $-\infty$. Because $f(x) > 0$, it follows that the limit

$$\limsup_{k\to\infty} f(x + t_k u) \frac{(\alpha g)(x + t_k u) - (\alpha g)(x)}{t_k}$$

takes the same value $+\infty$ or $-\infty$. If $q = +\infty$, then

$$\sup_{N\in\partial g(x)} \alpha f(x)\langle N, u\rangle = \sup_{N\in\partial g(x)} \alpha\langle N, u\rangle = +\infty$$

and

$$\sup_{N\in\partial g(x), M\in\partial f(x)} \langle \alpha f(x)N^{tr} + \alpha g(x)M^{tr}, u\rangle = +\infty$$

which implies (2.3) as well.

If $q = -\infty$, then

$$f(x + t_k u)\frac{(\alpha g)(x + t_k u) - (\alpha g)(x)}{t_k} \leq \sup_{N\in\partial g(x)} \alpha f(x)\langle N, u\rangle$$

for k sufficiently large. This proves (2.3). In this way, the closure of the set $f(x)\partial g(x) + g(x)\partial f(x)$ is a pseudo-Jacobian of fg at x. □

Theorem 2.1.6 *Let $f, g: \mathbb{R}^n \to \mathbb{R}$ be continuous functions with $g(x) \neq 0$. Let $\partial f(x)$ and $\partial g(x)$ be pseudo-differentials of f and g at x respectively. Then the closure of the set*

$$\frac{g(x)\partial f(x) - f(x)\partial g(x)}{g^2(x)}$$

is a pseudo-differential of the quotient function f/g at x.

Proof. Apply the same method of proof as in Theorem 2.1.5. □

A product and quotient formula for the Clarke generalized subdifferential can also be obtained when f and g are locally Lipschitz.

Corollary 2.1.7 *Let $f, g : \mathbb{R}^n \to \mathbb{R}$ be locally Lipschitz. Then we have*

$$\partial^C(fg)(x) \subseteq f(x)\partial^C g(x) + g(x)\partial^C f(x),$$

$$\partial^C(f/g)(x) \subseteq \frac{g(x)\partial^C f(x) - f(x)\partial^C g(x)}{g^2(x)} \quad \text{when } g(x) \neq 0.$$

Proof. Use the same argument as in the proof of Corollary 2.1.2.   □

The next example shows that Theorem 2.1.5 may fail without the condition that at least one of the values of $f(x)$ and $g(x)$ is nonzero.

Example 2.1.8 Let f and $g : \mathbb{R} \to \mathbb{R}$ be defined by

$$f(x) = x^{1/3} \quad \text{and} \quad g(x) = x^{2/3}.$$

Let

$$\partial f(x) = \begin{cases} \{(1/3)x^{-2/3}\} & \text{if } x \neq 0; \\ \{\alpha \in \mathbb{R} : \alpha \geq 1\} & \text{if } x = 0, \end{cases}$$

$$\partial g(x) = \begin{cases} \{(2/3)x^{-1/3}\} & \text{if } x \neq 0; \\ \{\alpha \in \mathbb{R} : |\alpha| \geq 1\} & \text{if } x = 0. \end{cases}$$

A simple calculation confirms that $\partial f(x)$ and $\partial g(x)$ are pseudo-differentials of f and g, respectively, and they are upper semicontinuous at $x = 0$. The set $g(0)\partial f(0) + f(0)\partial g(0)$ consists of zero only, which evidently is not a pseudo-differential of fg at 0.

Max-Functions and Min-Functions

Let $f_i, i = 1, \ldots, k$ be scalar continuous functions on $\mathbb{R}^n$. Let us define, respectively, the *max-function* and the *min-function* f and $g: \mathbb{R}^n \to \mathbb{R}$ by

$$f(x) := \max\{f_i(x) : i = 1, \ldots, k\},$$

$$g(x) := \min\{f_i(x) : i = 1, \ldots, k\}.$$

Denote by $I(x)$ the set of all indices $i \in \{1, ldots, k\}$ such that $f_i(x) = f(x)$ and by $J(x)$ the set of all indices $j \in \{1, ldots, k\}$ such that $f_j(x) = g(x)$.

Theorem 2.1.9 *Assume that $\partial f_1(x), \ldots, \partial f_k(x)$ are pseudo-differentials of $f_1, ldots, f_k$ respectively at x. Then the union $\bigcup_{i \in I(x)} \partial f_i(x)$ (respectively, $\bigcup_{j \in J(x)} \partial f_j(x)$) is a pseudo-differential of f (respectively, g) at x.*

Proof. We first observe that being the max-function of a finite family of continuous functions, f is continuous. Now let $u \in \mathbb{R}^n$. Let $t_k > 0$ converging to 0 be such that

$$f^+(x; u) = \lim_{k \to \infty} \frac{f(x + t_k u) - f(x)}{t_k}.$$

It follows from the continuity of f_i that there is $k_0 > 0$ such that

$$I(x + t_k u) \subseteq I(x) \quad \text{for all } k \geq k_0.$$

Because $I(x)$ is finite, there is at least one index $i_0 \in I(x)$ and a subsequence $\{t_{i_0(k)}\}$ such that

$$f(x + t_{i_0(k)}u) = f_{i_0}(x + t_{i_0(k)}u) \quad \text{for all } i_0(k).$$

Then we can write $f^+(x; u)$ as

$$f^+(x; u) = \lim_{k \to \infty} \frac{f_{i_0}(x + t_{i_0(k)}u) - f_{i_0}(x)}{t_{i_0(k)}}$$

$$\leq f_{i_0}^+(x; u) \leq \sup_{\xi \in \partial f_{i_0}(x)} \langle \xi, u \rangle$$

$$\leq \sup_{\xi \in \bigcup_{i \in I(x)} \partial f_i(x)} \langle \xi, u \rangle.$$

In a similar way we obtain

$$f^-(x; u) \geq \inf_{\xi \in \bigcup_{i \in I(x)} \partial f_i(x)} \langle \xi, u \rangle.$$

By this $\bigcup_{i \in I(x)} \partial f_i(x)$ is a pseudo-differential of f at x. The proof for the min-function is similar. $\qquad \square$

Here is a formula to calculate the Clarke subdifferential of the max-function when f_i are locally Lipschitz.

Corollary 2.1.10 *Assume that $f_1, \ldots, f_k$ are locally Lipschitz. Then*

$$\partial^C f(x) \subseteq \text{co}\left(\bigcup_{i \in I(x)} \partial^C f_i(x) \right).$$

Proof. Apply Theorem 2.1.9 and Corollary 1.6.8. $\qquad \square$

The Gâteaux differentiability of the max-function can also be obtained in certain circumstances.

Corollary 2.1.11 *Assume that $f_1, \ldots, f_k : \mathbb{R}^n \to \mathbb{R}$ are Gâteaux differentiable at x. If x is a maximum or a minimum point of $f_i, i \in I(x)$, then f is Gâteaux differentiable at x and $\nabla f(x) = 0$.*

Proof. It follows that $\nabla f_i(x) = 0$ for $i \in I(x)$. Hence the singleton $\{0\}$ is a pseudo-differential of f at x. According to Proposition 1.2.2, f is Gâteaux differentiable at this point and its derivative is $\{0\}$. $\qquad \square$

Note that the conclusion of the preceding theorem is no longer true when f is a max-function of an infinite number of continuous functions.

Example 2.1.12 Suppose that $f_k : \mathbb{R} \to \mathbb{R}$ is given by

$$f_k(x) = \begin{cases} x & \text{if } x \geq 2^{-k}, \\ 2x - 2^{-k} & \text{if } 2^{-k} > x \geq 2^{-(k+1)}, \\ 0 & \text{otherwise.} \end{cases}$$

The max-function of the family $\{f_1, f_2, \ldots\}$ is given by

$$f(x) = \begin{cases} x & \text{if } x \geq 0, \\ 0 & \text{otherwise.} \end{cases}$$

By taking $\partial f_i(0) = \{0\}$, we see that it is a pseudo-differential of f_i at 0 for every $i = 1, 2, \ldots$. Moreover, $I(0) = \{1, 2, \ldots\}$ and $\bigcup_{i \in I(0)} f_i(0) = \{0\}$. It is evident that $\{0\}$ cannot be a pseudo-differential of f at 0.

Optimality Conditions

Let $f : \mathbb{R}^n \to \mathbb{R}$ be a continuous function. A point $x_0 \in \mathbb{R}^n$ is said to be a *local minimizer* of f if there is a neighborhood U of x_0 in $\mathbb{R}^n$ such that $f(x) \geq f(x_0)$ for all $x \in U$. Next we give a necessary condition for a point to be a local minimizer.

Theorem 2.1.13 *If x_0 is a local minimizer of f and $\partial f(x_0)$ is a pseudo-differential of f at x_0, then*

$$0 \in \overline{\mathrm{co}}(\partial f(x_0)).$$

Proof. Because x_0 is a local minimizer of f, one has

$$f^+(x_0; u) \geq 0 \quad \text{for every } u \in \mathbb{R}^n.$$

It follows from the definition of pseudo-differential that

$$0 \leq f^+(x_0; u) \leq \sup_{\xi \in \partial f(x_0)} \langle \xi, u \rangle, \quad \text{for } u \in \mathbb{R}^n.$$

Consequently $0 \in \overline{\mathrm{co}}(\partial f(x_0))$.   □

We deduce from the above theorem some familiar results when the function is differentiable or locally Lipschitz.

Corollary 2.1.14 *If x_0 is a local minimizer of f, then*

i) $\nabla f(x_0) = 0$ provided f is Gâteaux differentiable at x_0.
ii) $0 \in \partial^{MP} f(x_0)$ provided f is locally Lipschitz.

Proof. The first assertion is clear because $\{\nabla f(x_0)\}$ is a pseudo-differential of f at x_0. The second assertion is obtained from Theorem 2.1.13 and the fact that when f is locally Lipschitz the Michel–Penot subdifferential is a convex compact pseudo-differential. $\square$

The optimality condition given in Theorem 2.1.13 is quite sharp in comparison with the one expressed in terms of Michel–Penot's subdifferential and Mordukhovich's basic differential.

Example 2.1.15 For $x > 0$, define

$$f(x) = \begin{cases} 2^{-\frac{1}{2}} & \text{if } 2^{-2} \leq x, \\ 2^{-\frac{2k+1}{2}} & \text{if } 2^{-2(k+1)} \leq x < 2^{-(2k+1)}, k = 1, 2, \ldots, \\ (2^{\frac{3k+2}{2}} - 2^{\frac{2k+1}{2}})x + a & \text{if } 2^{-(2k+1)} \leq x < 2^{-2k}, k = 1, 2, \ldots, \end{cases}$$

where $a = 2^{-((2k-1)/2)} - 2^{-(k/2)}$; and $f(x) = -f(-x)$ for $x < 0$, and $f(0) = 0$. This function is neither locally Lipschitz nor directionally differentiable at $x = 0$. Direct calculation shows that the Michel–Penot subdifferential of f at 0 is the set $[0, \infty)$, the Mordukhovich basic subdifferential of f at 0 is the singleton $\{0\}$, and the singular subdifferential is the set $[0, \infty)$. All these subdifferentials contain 0, which means that the necessary optimality condition expressed by them is satisfied at $x = 0$. However, it is not difficult to see that the set $[1, \infty)$ provides a pseudo-differential of f at $x = 0$, for which the optimality condition is not fulfilled.

Given a nonempty subset C of $\mathbb{R}^n$ and $x \in \text{cl}(C)$, the *cone of feasible directions* of C at x is the set

$$T_0(C, x) := \{u \in \mathbb{R}^n : \text{there is } t > 0 \text{ such that } x + su \in C \text{ for } s \in (0, t)\}.$$

When C is convex, the closure of the cone $T_0(C, x)$ coincides with the *tangent cone* of C at x which is defined by

$$T(C, x) := \text{cl}\{t(c - x) : x \in C, t \geq 0\}.$$

For functions defined on the subset C, the optimality condition above can be generalized as follows.

Theorem 2.1.16 *Let C be a nonempty set in $\mathbb{R}^n$ and let $f \colon \mathbb{R}^n \to \mathbb{R}$ be a continuous function. If $x \in C$ is a local minimum point of f on C and if $\partial f(x)$ is a pseudo-differential of f at x, then*

$$\sup_{\xi \in \partial f(x)} \langle \xi, u \rangle \geq 0 \quad \text{for all } u \in \text{cl}(T_0(C, x)).$$

Proof. It suffices to show the inequality for those $u \in T_0(C, x)$ of the form $u = c - x$, where $c \in C$. Suppose to the contrary that the inequality does not hold for some $u = c - x$, $c \in C$; that is,

$$\sup_{\xi \in \partial f(x)} \langle \xi, c - x \rangle < 0.$$

It follows that

$$f^+(x; c - x) = \limsup_{t \downarrow 0} \frac{f(x + t(c - x)) - f(x)}{t} < 0.$$

Hence for t sufficiently small, we derive

$$f(x + t(c - x)) - f(x) < 0,$$

which contradicts the hypothesis. $\qquad \square$

2.2 The Mean Value Theorem and Taylor's Expansions

We establish in this section some mean value theorems for continuous vector functions in terms of pseudo-Jacobians and derive related results. To this end, let us prove a result on separation of convex sets that we have already mentioned in Section 1.1.

Lemma 2.2.1 *Suppose that $C \subseteq \mathbb{R}^n$ is a convex set, and that the point y does not belong to C. Then there exists a nonzero vector ξ of $\mathbb{R}^n$ such that*

$$\langle \xi, y \rangle \leq \inf_{x \in C} \langle \xi, x \rangle.$$

If, in addition, C is closed, then the vector ξ can be chosen so that the above inequality is strict.

Proof. We may suppose that $y = 0$. Consider the convex cone generated by C,

$$\text{cone}(C) = \{tx : x \in C, t \geq 0\}.$$

By passing to a space of less dimension if necessary, we may assume that this cone has a nonempty interior; say e is one of its elements. Then the vector $-e$ does not belong to the closed convex cone $\text{cl}(\text{cone}(C))$ because C does not contain 0. Consider the function

$$h(x) := \|x + e\| \quad \text{for } x \in C.$$

This function is strictly convex in the sense that for every $x, y \in \mathbb{R}^n$ with $x \neq y$ and $\lambda \in (0, 1)$ one has $h(\lambda x + (1 - \lambda)y) < \lambda h(x) + (1 - \lambda)h(y)$. Therefore, it attains its unique minimum on the closed convex set $\text{cl}(\text{cone}(C))$ at some point $\overline{x}$. In view of Theorem 2.1.16, one has

$$\langle \nabla h(\overline{x}), x - \overline{x} \rangle \geq 0 \quad \text{for every } x \in \text{cl}(\text{cone}(C)).$$

Because $\overline{x} \in \text{cl}(\text{cone}(C))$ and $\nabla h(\overline{x}) = 2(\overline{x} + e) \neq 0$, we deduce from the above inequality that

$$\langle \nabla h(\overline{x}), x \rangle \geq \langle \nabla h(\overline{x}), \overline{x} \rangle = 0$$

for every $x \in C$. The vector $\xi = \nabla h(\overline{x})$ is the one for which we are looking.

If C is closed, there is a positive ε such that $0 \notin C + \varepsilon B_n$. By applying the first part of the proof, one finds some nonzero vector ξ of $\mathbb{R}^n$ such that

$$\langle \xi, x + \varepsilon b \rangle \geq 0$$

for every $x \in C$ and $b \in B_n$. This gives

$$\langle \xi, x \rangle \geq \varepsilon \|\xi\| > 0$$

for every $x \in C$ and the proof is complete. $\qquad\square$

The Mean Value Theorem

Theorem 2.2.2 *Let $a, b \in \mathbb{R}^n$ and let $f : \mathbb{R}^n \to \mathbb{R}^m$ be a continuous function. Assume that for each $x \in [a, b]$, $\partial f(x)$ is a pseudo-Jacobian of f at x. Then*

$$f(b) - f(a) \in \overline{\text{co}}\{\partial f([a, b])(b - a)\}.$$

Proof. Let us first note that the right-hand side above is the closed convex hull of all points of the form $M(b - a)$, where $M \in \partial f(c)$ for some $c \in [a, b]$. Let $v \in \mathbb{R}^m$ be arbitrary and fixed. Consider the real-valued function $g : [0, 1] \to \mathbb{R}$,

$$g(t) = \langle v, f(a + t(b - a)) - f(a) + t(f(a) - f(b)) \rangle.$$

Then g is continuous on $[0, 1]$ with $g(0) = g(1)$. So, g attains a minimum or a maximum at some $t_0 \in (0, 1)$. Suppose that t_0 is a minimum point. Then, for each $\alpha \in \mathbb{R}$, $g^+(t_0; \alpha) \geq 0$. It now follows from direct calculations that

$$g^+(t_0; \alpha) = (vf)^+(a + t_0(b - a); \alpha(b - a)) + \alpha \langle v, f(a) - f(b) \rangle.$$

Hence for each $\alpha \in \mathbb{R}$,

$$(vf)^+(a + t_0(b - a); \alpha(b - a)) \geq \alpha \langle v, f(b) - f(a) \rangle.$$

Now, by taking $\alpha = 1$ and $\alpha = -1$, we obtain that

$$-(vf)^+(a + t_0(b - a); a - b) \leq \langle v, f(b) - f(a) \rangle \leq (vf)^+(a + t_0(b - a); b - a).$$

By the definition of pseudo-Jacobian, we get

$$\inf_{M \in \partial f(a + t_0(b-a))} \langle v, M(b-a) \rangle \leq \langle v, f(b) - f(a) \rangle \leq \sup_{M \in \partial f(a + t_0(b-a))} \langle v, M(b-a) \rangle$$

Consequently,

$$\langle v, f(b) - f(a) \rangle \in \overline{\text{co}}(\langle v, \partial f(a + t_0(b - a))(b - a) \rangle)$$

and so,

$$\langle v, f(b) - f(a) \rangle \in \overline{\text{co}}(\langle v, \partial f([a, b])(b - a) \rangle). \tag{2.4}$$

If t_0 is a maximum point, then it provides a minimum point of the function $-g$ on $(0, 1)$. Using the same line of arguments as above, we arrive at the conclusion

$$\langle -v, f(b) - f(a) \rangle \in \overline{\text{co}}(\langle -v, \partial f([a, b])(b - a) \rangle),$$

which is equivalent to (2.4). Because v is arbitrary, we deduce that

$$f(b) - f(a) \in \overline{\text{co}}\{\partial f([a, b])(b - a)\}.$$

In fact, if this is not so, then it follows from the separation theorem that

$$\langle p, f(b) - f(a) \rangle - \varepsilon > \sup_{u \in \overline{\text{co}}\{\partial f([a,b])(b-a)\}} \langle p, u \rangle,$$

for some $p \in \mathbb{R}^m$ because $\overline{\text{co}}\{\partial f([a, b])(b - a)\}$ is a closed convex subset of $\mathbb{R}^m$. This implies

$$\langle p, f(b) - f(a) \rangle > \sup\{\alpha : \alpha \in \langle p, \overline{\text{co}}\{\partial f([a, b])(b - a)\} \rangle\}$$
$$\geq \sup\{\alpha : \alpha \in \overline{\text{co}}(\langle p, \partial f([a, b])(b - a) \rangle)\},$$

which contradicts (2.4). $\qquad \square$

Corollary 2.2.3 *Let $a, b \in \mathbb{R}^n$ and $f : \mathbb{R}^n \to \mathbb{R}^m$ be a continuous function. Assume that ∂f is a bounded pseudo-Jacobian of f which as a set-valued map on $[a, b]$ is upper semicontinuous on this segment. Then*

$$f(b) - f(a) \in \{\text{co}(\partial f([a, b]))\}(b - a).$$

Proof. Because for each $x \in [a, b]$, $\partial f(x)$ is compact, and the set-valued map ∂f is upper semicontinuous, we obtain that the set $\partial f([a, b]) \subset L(\mathbb{R}^n, \mathbb{R}^m)$ is compact, hence the set $\partial f([a, b])(b - a) \subset \mathbb{R}^m$ is compact too. Consequently,

$$\overline{co}\{\partial f([a, b])(b - a)\} = co\{\partial f([a, b])(b - a)\} = \{co(\partial f([a, b]))\}(b - a),$$

and so the conclusion follows from Theorem 2.2.2. □

In the following corollary we deduce the mean value theorem for locally Lipschitz functions as a special case of Theorem 2.2.2.

Corollary 2.2.4 *Let $a, b \in \mathbb{R}^n$ and let $f: \mathbb{R}^n \to \mathbb{R}^m$ be locally Lipschitz. Then*

$$f(b) - f(a) \in \{co(\partial^C f([a, b]))\}(b - a).$$

Proof. We know that the Clarke generalized Jacobian map $\partial^C f$ is a compact valued, upper semicontinuous pseudo-Jacobian map of f. Hence the conclusion follows from Corollary 2.2.3. □

Note that even for the case where f is locally Lipschitz, Corollary 2.2.3 provides a stronger mean value condition than the one of Corollary 2.2.4.

Example 2.2.5 Let $f : \mathbb{R}^2 \to \mathbb{R}$ be defined by

$$f(x, y) = |x| - |y|,$$

and let $a = (-1, -1)$ and $b = (1, 1)$. Then the conclusion of Corollary 2.2.1 is verified by

$$\partial f(x, y) = \{(1, -1), (-1, 1)\}$$

for every $(x, y) \in [a, b]$. However, the condition of Corollary 2.2.4 holds for $\partial^C f(0, 0)$, where

$$\partial^C f(0, 0) = co(\{(1, 1), (-1, -1), (1, -1), (-1, 1)\}) \supset \partial f([a, b]).$$

As a special case of the above theorem we see that if f is real valued, then an asymptotic mean value equality is obtained.

Corollary 2.2.6 *Let $a, b \in X$ and $f: \mathbb{R}^n \to \mathbb{R}$ be a continuous function. Assume that, for each $x \in [a, b]$, $\partial f(x)$ is a pseudo-differential of f. Then there exist $c \in (a, b)$ and a sequence $\{\xi_k\} \subset co(\partial f(c))$ such that*

$$f(b) - f(a) = \lim_{k \to \infty} \langle \xi_k, \ b - a \rangle.$$

In particular, when f is locally Lipschitz, we obtain Lebourg's mean value theorem: there is some $\xi \in \partial^C f(c)$ such that

$$f(b) - f(a) = \langle \xi, b - a \rangle.$$

Proof. The conclusion follows from the proof of Theorem 2.2.2. The particular case is derived from Corollary 2.2.4.   □

We notice that for a continuous function which is not necessarily locally Lipschitz, the exact mean value equality (Lebourg's mean value theorem) does not hold as shown in the next example.

Example 2.2.7 Let $f : \mathbb{R}^2 \to \mathbb{R}$ be defined by

$$f(x) = \sqrt{|x|} + \sqrt[3]{y}.$$

Define

$$\partial f(x,y) = \begin{cases} \{ (\frac{\text{sign}(x)}{2\sqrt{|x|}}, \frac{1}{\sqrt[3]{y^2}}) \} & \text{if } x \neq 0 \text{ or } y \neq 0, \\ \{ (\frac{\text{sign}(x)}{2\sqrt{|x|}}, \alpha) : \alpha \geq 1 \} & \text{if } x \neq 0 \text{ and } y = 0, \\ \{ (\alpha, \frac{1}{\sqrt[3]{y^2}}) : |\alpha| \geq 1 \} & \text{if } x = 0 \text{ and } y \neq 0, \\ \{ (\frac{1}{\alpha}, |\alpha|) : |\alpha| \geq 1 \} & \text{if } x = 0 \text{ and } y = 0. \end{cases}$$

It is not hard to see that $\partial f(x,y)$ is a pseudo-differential of f at (x,y). For the points $a = (-1,0)$ and $b = (1,0)$, there is no $c \in [a,b]$ such that

$$0 = f(b) - f(a) \in \text{co}(\partial f(c))(b - a).$$

By choosing $\xi_k = (1/k, k) \in \text{co}(\partial f(0,0))$, we do have

$$0 = f(b) - f(a) = \lim_{k \to \infty} \langle \xi_k, \; b - a \rangle = \lim_{k \to \infty} \frac{2}{k}$$

as expected by Corollary 2.2.6.

Characterizing Locally Lipschitz Continuity in Terms of Pseudo-Jacobians

In this section we describe how locally Lipschitz functions can be characterized in terms of pseudo-Jacobians using the mean value theorem. We recall that a set-valued map $G : \mathbb{R}^n \rightrightarrows L(\mathbb{R}^n, \mathbb{R}^m)$ is *locally bounded* at x if there exist a neighborhood U of x and a positive α such that $\|A\| \leq \alpha$, for each $A \in G(U)$. Clearly, if G is upper semicontinuous at x and if $G(x)$ is bounded, then G is locally bounded at x.

Proposition 2.2.8 *Let $f: \mathbb{R}^n \to \mathbb{R}^m$ be a continuous function. Then, the following conditions are equivalent.*

(i)   f is locally Lipschitz at x.
(ii)   f admits a locally bounded pseudo-Jacobian map at x.
(iii) f admits a pseudo-Jacobian map whose recession upper limit at x is trivial.

Proof. Assume that $\partial f(y)$ is a pseudo-Jacobian of f for each y in a neighborhood U of x and that ∂f is locally bounded on U. Without loss of generality, we may assume that U is convex. Then there exists $\alpha > 0$ such that $\|A\| \leq \alpha$ for each $A \in \partial f(U)$. Let $x, y \in U$. Then $[x, y] \subset U$ and by the mean value theorem

$$f(x) - f(y) \in \overline{\text{co}}(\partial f([x, y])(x - y)) \subset \overline{\text{co}}(\partial f(U)(x - y)).$$

Hence

$$\|f(x) - f(y)\| \leq \|x - y\| \max\{\|A\| : A \in \partial f(U)\}.$$

This gives us that

$$\|f(x) - f(y)\| \leq \alpha \|x - y\|$$

and so, f is locally Lipschitz at x.

Conversely, if f is locally Lipschitz at x, then the Clarke generalized Jacobian can be chosen as a locally bounded pseudo-Jacobian map of f at the point x. This proves the equivalence between (i) and (ii). The equivalence of (ii) and (iii) is clear.   □

As we have seen in Example 1.7.7, a non-Lipschitz function may have a bounded pseudo-Jacobian. In view of the above proposition, a pseudo-Jacobian map of such a function cannot be locally bounded.

For a continuous function f one defines the Lipschitz modulus at a point a by

$$\text{lipf}(a) := \limsup_{\text{x,y} \to a, \text{x} \neq \text{y}} \frac{\|\text{f(x)} - \text{f(y)}\|}{\|\text{x} - \text{y}\|}.$$

It is clear that f is locally Lipschitz at a if and only if it has the finite Lipschitz modulus at that point. The latter can be evaluated by pseudo-Jacobians around a. Let us denote by $\mathcal{G}(x)$ the collection of all pseudo-Jacobians of f at x and set

$$|\mathcal{G}(x)| := \inf_{G \in \mathcal{G}(x)} \sup_{M \in G} \|M\|.$$

Corollary 2.2.9 *Let $f: \mathbb{R}^n \to \mathbb{R}^m$ be a continuous function. Then it is locally Lipschitz at a if and only if $\limsup_{x \to a} |\mathcal{G}(x)|$ is finite in which case*

$$lipf(a) = \limsup_{x \to a} |\mathcal{G}(x)|.$$

Proof. Assume that f is locally Lipschitz at a. Then for every x and y close to a and for every pseudo-Jacobian map ∂f of f, by the mean-value theorem, one has

$$\frac{\|f(x) - f(y)\|}{\|x - y\|} \leq \sup_{M \in \partial f([x,y])} \|M\|,$$

which implies

$$\frac{\|f(x) - f(y)\|}{\|x - y\|} \leq \sup_{z \in [x,y]} |\mathcal{G}(z)|.$$

When x and y tend to a we derive

$$\mathrm{lip} f(a) \leq \limsup_{x \to a} |\mathcal{G}(x)|$$

and deduce that f is locally Lipschitz at a. The converse implication is immediate.

The equality follows from the fact that the Clarke generalized Jacobian belongs to the collection $\mathcal{G}(x)$.   □

Partial Pseudo-Jacobians

In order to show that the partial pseudo-Jacobians of a function form a pseudo-Jacobian we need the following continuity property of a sup-function.

Lemma 2.2.10 *Let $F : \mathbb{R}^n \rightrightarrows L(\mathbb{R}^n, \mathbb{R}^m)$ be a set-valued map, that has nonempty closed values and is upper semicontinuous at x. Then for each $u \in \mathbb{R}^n$ and $v \in \mathbb{R}^m$, the sup-function*

$$f(x') := \sup_{M \in F(x')} \langle v, M(u) \rangle$$

is upper semicontinuous at x.

Proof. First observe that because

$$|\langle v, M(u) \rangle| \leq \|v\| \|M(u)\| \leq \|v\| \|u\| \|M\|$$

for $u \in \mathbb{R}^n$ and $v \in \mathbb{R}^m$ fixed, one has

$$\sup_{\|M\| \leq 1} \langle v, M(u) \rangle \leq \|v\| \|u\|.$$

For every $\varepsilon > 0$, by the upper semicontinuity of F, there is $\delta > 0$ such that

$$F(x') \subseteq F(x) + \varepsilon B_{m \times n} \quad \text{for } x' \text{ with } \|x - x'\| < \delta.$$

It follows that

$$\limsup_{x' \to x} f(x') = \limsup_{x' \to x} \sup_{M \in F(x')} \langle v, M(u) \rangle$$

$$\leq \limsup_{x' \to x} \sup_{M \in F(x) + \varepsilon B_{m \times n}} \langle v, M(u) \rangle$$

$$\leq \sup_{M \in F(x)} \langle v, M(u) \rangle + \varepsilon \|v\| \|u\|$$

$$\leq f(x) + \varepsilon \|v\| \|u\|.$$

Because $\varepsilon > 0$ is arbitrary, we conclude the upper semicontinuity of f. $\quad \square$

Proposition 2.2.11 *Let* $f \colon \mathbb{R}^n \times \mathbb{R}^k \to \mathbb{R}^m$ *be a continuous function. Let* $\partial_x f(x,y) \subseteq L(\mathbb{R}^n, \mathbb{R}^m)$ *and* $\partial_y f(x,y) \subseteq L(\mathbb{R}^k, \mathbb{R}^m)$ *be partial pseudo-Jacobians of* f *at* (x,y). *If the set-valued map* $x' \mapsto \partial_y f(x', y)$ *is upper semicontinuous at* x, *then the set* $(\partial_x f(x,y), \partial_y f(x,y))$ *is a pseudo-Jacobian of* f *at* (x,y).

Proof. Let $(u,v) \in \mathbb{R}^n \times \mathbb{R}^k$ and $w \in \mathbb{R}^m$. Then

$$(wf)^+((x,y); (u,v)) = \limsup_{t \downarrow 0} \frac{(wf)(x+tu, y+tv) - (wf)(x,y)}{t}$$

$$\leq \limsup_{t \downarrow 0} \frac{(wf)(x+tu, y+tv) - (wf)(x+tu, y)}{t}$$

$$+ \limsup_{t \downarrow 0} \frac{(wf)(x+tu, y) - (wf)(x,y)}{t}$$

$$\leq \limsup_{t \downarrow 0} \frac{(wf)(x+tu, y+tv) - (wf)(x+tu, y)}{t} + \sup_{M \in \partial_x f(x,y)} \langle w, M(u) \rangle.$$

Applying the mean value theorem to $f(x+tu, \cdot)$ on the interval $[y, y+tv]$, we obtain

$$(wf)(x+tu, y+tv) - (wf)(x+tu, y) \in t\overline{co}(\partial_y f(x+tu, y)(v)).$$

Under the hypothesis of the theorem, the set-valued map $t \mapsto \partial_y f(x+tu, y)$ is upper semicontinuous at $t = 0$. By Lemma 2.2.10 this implies the following inequality concerning the first term of the latter inequality

$$\limsup_{t \downarrow 0} \frac{(wf)(x+tu, y+tv) - (wf)(x+tu, y)}{t}$$

$$\leq \limsup_{t \downarrow 0} \sup_{N \in \overline{co}(\partial_y f(x+tu, y))} \langle w, N(v) \rangle$$

$$\leq \limsup_{t \downarrow 0} \sup_{N \in \partial_y f(x+tu, y)} \langle w, N(v) \rangle$$

$$\leq \sup_{N \in \partial_y f(x,y)} \langle w, N(v) \rangle.$$

We deduce that

$$(wf)^+((x,y);(u,v)) \leq \sup_{N \in \partial_y f(x,y)} \langle w, Nv \rangle + \sup_{M \in \partial_x f(x,y)} \langle w, M(u) \rangle$$

$$\leq \sup_{(MN) \in (\partial_x f(x,y), \partial_y f(x,y))} \langle w, (MN)(u,v) \rangle.$$

This shows that $(\partial_x f(x,y), \partial_y f(x,y))$ is a pseudo-Jacobian of f at (x,y). □

It is known from mathematical analysis that a function may have partial derivatives at a point without being Gâteaux differentiable at that point. Next we derive a sufficient condition for a function of two variables to be Gâteaux differentiable provided that it is Gâteaux differentiable with respect to each of its variables separately.

Corollary 2.2.12 *Assume that f is Gâteaux differentiable with respect to x at (x,y) and Gâteaux differentiable with respect to y at every (x',y), where x' is in a neighborhood of x, and that the partial derivative $\nabla_y f(x',y)$ is continuous in the first variable at x. Then f is Gâteaux differentiable at (x,y) and $\nabla f(x,y) = (\nabla_x f(x,y), \nabla_y f(x,y))$.*

Proof. By Proposition 2.2.11, the singleton set $\{(\nabla_x f(x,y), \nabla_y f(x,y))\}$ is a pseudo-Jacobian of f at (x,y). The conclusion follows then from Proposition 1.2.2. □

That the conclusion of Proposition 2.2.11 may fail without the upper semicontinuity of at least one of the partial pseudo-Jacobians is illustrated by the following example.

Example 2.2.13 Let $f: \mathbb{R}^2 \to \mathbb{R}^2$ be given by

$$f(x,y) = \begin{cases} (|x|, \frac{x^2 y}{x^2+y^2}) & \text{if } (x,y) \neq (0,0), \\ (0,0) & \text{else.} \end{cases}$$

It is easily seen that the sets

$$\partial_x f(0,0) = \left\{ \begin{pmatrix} 1 \\ 0 \end{pmatrix}, \begin{pmatrix} -1 \\ 0 \end{pmatrix} \right\}, \quad \partial_y f(0,0) = \left\{ \begin{pmatrix} 0 \\ 0 \end{pmatrix} \right\}$$

are partial pseudo-Jacobians of f at $(0,0)$. By taking $u = (1,1)$ and $v = (0,1)$, we obtain

$$(vf)^+((0,0);u) = \limsup_{t \downarrow 0} \frac{(vf)(tu)}{t} = \frac{1}{2}.$$

On the other hand, a simple calculation confirms

$$\sup_{M \in (\partial_x f(0,0), \partial_y f(0,0))} \langle v, M(u) \rangle = 0,$$

which shows that $\{(\partial_x f(0,0), \partial_y f(0,0))\}$ is not a pseudo-Jacobian of f at $(0,0)$.

Let $\partial f(x,y)$ be a pseudo-Jacobian of f at (x,y). The function f is differentiable at $(x,y) \neq (0,0)$, thus in view of Proposition 1.2.3, one has

$$[\nabla f(x,y)]^{tr}(v) \in \text{co}\{M^{tr}(v) : M \in \partial f(x,y)\}$$

for every $v \in \mathbb{R}^2$, where the derivative $\nabla f(x,y)$ is given by

$$\nabla f(x,y) = \begin{pmatrix} \text{sign}(x) & 0 \\ \frac{2xy^3}{(x^2+y^2)^2} & \frac{x^2(x^2-y^2)}{(x^2+y^2)^2} \end{pmatrix}.$$

By choosing $v = (0,1)$ we obtain

$$\text{co}\{M^{tr}(v) : M \in (\partial_0 f(0,0), \partial_y f(0,0))\} = \{(0,0)\},$$

$$[\nabla f(x,y)]^{tr}(v) = (0, \frac{x^2(x^2-y^2)}{(x^2+y^2)^2}).$$

These equalities show that the pseudo-Jacobian $\partial f(x,y)$ cannot be upper semicontinuous once taking the value $(\partial_x f(0,0), \partial_y f(0,0))$ at $(0,0)$.

Gâteaux and Fréchet Pseudo-Jacobians

As we have seen in the first chapter, every Fréchet pseudo-Jacobian is a Gâteaux pseudo-Jacobian, and in its turn every Gâteaux pseudo-Jacobian is a pseudo-Jacobian, and in general the converse is not true. Here we provide a method of constructing a Fréchet pseudo-Jacobian from a given pseudo-Jacobian.

Proposition 2.2.14 *Let $f : \mathbb{R}^n \to \mathbb{R}^m$ be a continuous function. If ∂f is a pseudo-Jacobian map of f that is upper semicontinuous at x_0, then $\overline{\text{co}}(\partial f(x_0))$ is a Fréchet pseudo-Jacobian (hence a Gâteaux pseudo-Jacobian) of f at x_0.*

Proof. For every $\varepsilon > 0$, by the upper semicontinuity of ∂f, there is some $\delta > 0$ such that

$$\overline{\text{co}}\{\partial f([x, x_0])(x - x_0)\} \subseteq \{\overline{\text{co}}(\partial f([x, x_0]))\}(x - x_0) + \varepsilon B_{m \times n}(x - x_0)$$

whenever $\|x - x_0\| < \delta$. This and the mean value theorem imply that there exist a matrix $M_x \in \overline{\text{co}}(\partial f(x_0))$ and $P_x \in B_{m \times n}$ such that

$$f(x) - f(x_0) = M_x(x - x_0) + \varepsilon P_x(x - x_0).$$

Consequently,

$$\frac{\|f(x) - f(x_0) - M_x(x - x_0)\|}{\|x - x_0\|} < \varepsilon$$

whenever $\|x - x_0\| < \delta$ and the conclusion follows. $\square$

The Clarke generalized Jacobian is convex, compact-valued, and upper semicontinuous, therefore the first conclusion of Proposition 1.7.4 is an immediate corollary of Proposition 2.2.14.

Next we give a method to find a Fréchet pseudo-Jacobian of inverse functions.

Proposition 2.2.15 *Let* $f: \mathbb{R}^n \to \mathbb{R}^n$ *be a continuous function. Assume that* f^{-1} *is the inverse of* f *in a neighborhood of* $f(x_0)$ *which is Lipschitz at* $f(x_0)$. *If* $\partial f(x_0)$ *is a Fréchet pseudo-Jacobian of* f *at* x_0 *and consists of invertible matrices only, then the set*

$$\Gamma := \{M^{-1} : M \in \partial f(x_0)\}$$

is a Fréchet pseudo-Jacobian of f^{-1} *at* $f(x_0)$.

Proof. Set $y_0 = f(x_0)$ and let y be a point in a small neighborhood of y_0 in which the inverse function f^{-1} is defined. Set $x = f^{-1}(y)$. There exists an element $M_y \in \partial f(x_0)$ such that

$$f(x) - f(x_0) = M_y(x - x_0) + o(\|x - x_0\|),$$

where $o(\|x - x_0\|)/\|x - x_0\| \to 0$ as x tends to x_0. We derive

$$f^{-1}(y) - f^{-1}(y_0) = x - x_0 = M_y^{-1}(y - y_0) + M_y^{-1}(o(\|x - x_0\|)).$$

Because $\partial f(x_0)$ is closed and its elements are invertible, there is a positive number δ such that $\|M^{-1}\| \leq \delta$ for every $M \in \partial f(x_0)$. This and the Lipschitz continuity of f^{-1} imply

$$\lim_{y \to y_0} \frac{\|M_y^{-1}(o(\|x - x_0\|))\|}{\|y - y_0\|} \leq \lim_{y \to y_0} \delta \frac{\|o(\|f^{-1}(y) - f^{-1}(y_0)\|)\|}{\|y - y_0\|} = 0.$$

This shows that the set Γ is a Fréchet pseudo-Jacobian of f^{-1} at $f(x_0)$. $\square$

Sup-Functions and Inf-Functions

We consider the case in which the max-function and the min-function are defined by an infinite family of continuous functions. Let Λ be a topological space and let $f : \mathbb{R}^n \times \Lambda \to \mathbb{R}$ be given. The *sup-function* and the *inf-function* of the family $\{f(.,\lambda) : \lambda \in \Lambda\}$ are defined by

$$p(x) := \sup\{f(x,\lambda) : \lambda \in \Lambda\},$$

$$q(x) := \inf\{f(x,\lambda) : \lambda \in \Lambda\}.$$

Let x be fixed and let $\varepsilon > 0, \delta > 0$. Denote by

$$\Lambda(\varepsilon,\delta) := \{\lambda \in \Lambda : f(y,\lambda) \geq p(x) - \varepsilon, \text{for } y \in x + \delta B_n\},$$

$$\Gamma(\varepsilon,\delta) := \{\lambda \in \Lambda : f(y,\lambda) \leq q(x) + \varepsilon, \text{for } y \in x + \delta B_n\}.$$

Theorem 2.2.16 *Let $x \in \mathbb{R}^n$ be given. Assume that the sup-function p (respectively, inf-function q) is continuous and that for some positive $\varepsilon > 0$ and $\delta > 0$, the set $\partial_x f(y,\lambda)$ is a pseudo-differential of $f(.,\lambda)$ at $y \in x + \delta B_n$, where $\lambda \in \Lambda(\varepsilon,\delta)$ (respectively, $\lambda \in \Gamma(\varepsilon,\delta)$), and is such that the set-valued map $y \mapsto \bigcup_{\lambda \in \Lambda(\varepsilon,\delta)} \partial_x f(y,\lambda)$ (respectively, $y \mapsto \bigcup_{\lambda \in \Gamma(\varepsilon,\delta)} \partial_x f(y,\lambda)$) is upper semicontinuous at x. Then the closure of the set*

$$\bigcup_{\lambda \in \Lambda(\varepsilon,\delta)} \partial_x f(x,\lambda) \text{ (respectively, } \bigcup_{\lambda \in \Gamma(\varepsilon,\delta)} \partial_x f(x,\lambda))$$

is a pseudo-differential of p (respectively q) at x.

Proof. Let $u \in \mathbb{R}^n$ and let $\{t_k\}$ be a sequence of positive numbers converging to 0 such that

$$p^+(x;u) = \lim_{k\to\infty} \frac{p(x + t_k u) - p(x)}{t_k}.$$

We may assume that $\|t_k u\| \leq \delta$ for each $k = 1, 2, \ldots$. Then

$$
\begin{aligned}
p(x + t_k u) - p(x) &= \sup_{\lambda \in \Lambda} f(x + t_k u, \lambda) - \sup_{\lambda \in \Lambda} f(x, \lambda) \\
&= \sup_{\lambda \in \Lambda(\varepsilon,\delta)} f(x + t_k u, \lambda) - \sup_{\lambda \in \Lambda(\varepsilon,\delta)} f(x, \lambda) \\
&\leq \sup_{\lambda \in \Lambda(\varepsilon,\delta)} (f(x + t_k u, \lambda) - f(x, \lambda)).
\end{aligned}
$$

Let $r > 0$ be arbitrary. By the upper semicontinuity assumption, there is some positive $s > 0$ such that

$$\bigcup_{y \in x + s B_n} \bigcup_{\lambda \in \Lambda(\varepsilon,\delta)} \partial_x f(y,\lambda) \subset \bigcup_{\lambda \in \Lambda(\varepsilon,\delta)} \partial_x f(x,\lambda) + r B_n.$$

Consequently,

$$\bigcup_{y \in x + s B_n} \bigcup_{\lambda \in \Lambda(\varepsilon,\delta)} \text{co}\{\partial_x f(y,\lambda)\} \subset \text{co}\{ \bigcup_{\lambda \in \Lambda(\varepsilon,\delta)} \partial_x f(x,\lambda)\} + r B_n.$$

Denote the set on the left-hand side P and the set on the right-hand side Q. Without loss of generality we may assume that $t_k < s$ for all $k = 1, 2, \ldots$. Then applying the mean value theorem, we find $y_k \in (x, x + t_k u) \subset x + s B_n$, $\lambda_k \in \Lambda(\varepsilon, \delta)$, and $\xi_k \in \mathrm{co}(\partial_x f(y_k, \lambda_k))$ such that

$$p(x + t_k u) - p(x) \leq f(x + t_k u, \lambda_k) - f(x, \lambda_k) + t_k r \leq \langle \xi_k, t_k u \rangle + t_k r$$

for $k = 1, 2, \ldots$. It follows that

$$\frac{p(x + t_k u) - p(x)}{t_k} \leq \langle \xi_k, u \rangle + r$$

$$\leq \sup_{\lambda \in \Lambda(\varepsilon, \delta)} \sup_{\xi \in \mathrm{co}(\partial_x f(y_k, \lambda))} \langle \xi, u \rangle + r$$

$$\leq \sup_{\xi \in P} \langle \xi, u \rangle + r$$

$$\leq \sup_{\xi \in Q} \langle \xi, u \rangle + r$$

$$\leq \sup_{\xi \in \bigcup_{\lambda \in \Lambda(\varepsilon, \delta)} \partial_x f(x, \lambda)} \langle \xi, u \rangle + r(1 + \|u\|).$$

By passing to the limit in the above inequalities when k tends to ∞, we obtain

$$p^+(x; u) \leq \sup_{\xi \in \bigcup_{\lambda \in \Lambda(\varepsilon, \delta)} \partial_x f(x, \lambda)} \langle \xi, u \rangle + r(1 + \|u\|).$$

Because $r > 0$ is arbitrary, we have

$$p^+(x; u) \leq \sup_{\xi \in \bigcup_{\lambda \in \Lambda(\varepsilon, \delta)} \partial_x f(x, \lambda)} \langle \xi, u \rangle,$$

and similarly,

$$p^-(x; u) \geq \inf_{\xi \in \bigcup_{\lambda \in \Lambda(\varepsilon, \delta)} \partial_x f(x, \lambda)} \langle \xi, u \rangle$$

which shows that the closure of the set $\bigcup_{\lambda \in \Lambda(\varepsilon, \delta)} \partial_x f(x, \lambda)$ is a pseudo-differential of p at x. For the inf-function the proof is analogous.   □

Lemma 2.2.17 *Let $x \in \mathbb{R}^n$ be given. Assume that Λ is a compact space and f is a continuous function. Then for every $\varepsilon > 0$, there is some $\delta > 0$ such that*

$$p(y) = \max\{f(y, \lambda) : \lambda \in \Lambda(\varepsilon, 0)\} \quad \text{for } y \in x + \delta B_n.$$

Proof. Suppose to the contrary that there is some $\varepsilon_0 > 0$ and x_k converging to x such that

$$p(x_k) > \max\{f(x_k, \lambda) : \lambda \in \Lambda(\varepsilon, 0)\}.$$

Let $\lambda_k \in \Lambda$ be such that

$$p(x_k) = f(x_k, \lambda_k).$$

Then $\lambda_k \notin \Lambda(\varepsilon, 0)$. Without loss of generality we may assume that the sequence $\{\lambda_k\}$ converges to $\lambda_0 \in \Lambda$ as k tends to ∞. It is clear that $\lambda_0 \in \Lambda(\varepsilon, 0)$ and $p(x) = f(x, \lambda_0)$. It follows from the continuity of f that there is $\delta > 0$ and a neighborhood V of λ_0 in Λ such that

$$f(y, \lambda) \geq p(x) - \varepsilon \quad \text{for all} \quad y \in x + \delta B_n, \lambda \in V.$$

In particular, $f(x, \lambda_k) \geq p(x) - \varepsilon$ for k so large that $\lambda_k \in V$. This shows that $\lambda_k \in \Lambda(\varepsilon, 0)$, a contradiction. □

Lemma 2.2.18 Let $x \in \mathbb{R}^n$ be given. Assume that Λ is a compact space, f is a continuous function, and the set-valued map $y \mapsto \partial_x f(y, \lambda)$ is a pseudo-differential map of $f(., \lambda)$, which is upper semicontinuous in two variables y and λ at $(x, \lambda), \lambda \in \Lambda(\varepsilon, 0)$. Then the set-valued map

$$y \mapsto \bigcup_{\lambda \in \Lambda(\varepsilon, 0)} \partial_x f(y, \lambda)$$

is upper semicontinuous at x.

Proof. Let $r > 0$ be given. For each $\lambda \in \Lambda(\varepsilon, 0)$, there is $s(\lambda) > 0$ and a neighborhood $V(\lambda) \subseteq \Lambda(\varepsilon, 0)$ of λ such that

$$\partial_x f(y, \lambda') \subseteq \partial_x f(x, \lambda) + s(\lambda) B_n \quad \text{for } y \in x + r B_n \quad \text{and} \quad \lambda' \in V(\lambda).$$

It follows from the hypothesis of the lemma that $\Lambda(\varepsilon, 0)$ is compact. Hence there exist $\lambda_1, \ldots, \lambda_k \in \Lambda(\varepsilon, 0)$ such that $\Lambda(\varepsilon, 0)$ is covered by $V(\lambda_1), \ldots, V(\lambda_k)$. By choosing

$$s = \min\{s(\lambda_1), \ldots, s(\lambda_k)\}$$

we obtain

$$\partial_x f(y, \lambda) \subseteq \partial_x f(x, \lambda) + s B_n \quad \text{for } y \in x + r B_n \quad \text{and} \quad \lambda \in \Lambda(\varepsilon, 0).$$

By taking the union of the above sets over $\lambda \in \Lambda(\varepsilon, 0)$, we deduce the conclusion. □

Corollary 2.2.19 Let $x \in \mathbb{R}^n$ be given. Assume that Λ is a compact space, f is a continuous function, and that the set-valued map $\partial_x f(., \lambda)$ is a pseudo-differential map of $f(., \lambda)$ which is upper semicontinuous in the two variables at x. Then for every $\varepsilon > 0$, the closure of the set

$$\bigcup_{\lambda \in \Lambda: f(x,\lambda) \geq p(x) - \varepsilon} \partial_x f(x, \lambda) \quad (\text{respectively,} \quad \bigcup_{\lambda \in \Lambda: f(x,\lambda) \leq q(x) + \varepsilon} \partial_x f(x, \lambda))$$

is a pseudo-differential of p (respectively, q) at x.

Proof. According to Lemma 2.2.17, in a sufficiently small neighborhood of x, the sup-function p can be defined by the family of functions $f(., \lambda)$ with $\lambda \in \Lambda(\varepsilon, 0)$ only. This and Lemma 2.2.18 allow us to apply Theorem 2.2.16 to conclude the corollary. $\square$

Taylor's Expansion

In this part, we see how Taylor's expansions can be obtained for C^1- functions using pseudo-Hessians.

Theorem 2.2.20 Let $f : \mathbb{R}^n \to \mathbb{R}$ be continuously differentiable on $\mathbb{R}^n$; let $x, y \in \mathbb{R}^n$. Suppose that for each $z \in [x, y]$, $\partial^2 f(z)$ is a pseudo-Hessian of f at z. Then there exists $c \in (x, y)$ such that

$$f(y) \in f(x) + \langle \nabla f(x), y - x \rangle + \frac{1}{2}\overline{co}(\langle \partial^2 f(c)(y - x), (y - x) \rangle).$$

Proof. Let us define a real function h on $\mathbb{R}$ by

$$h(t) = f(y + t(x - y)) + t\langle \nabla f(y + t(x - y)), y - x \rangle + \frac{1}{2}at^2 - f(y),$$

where $a = -2(f(x) - f(y) + \langle \nabla f(x), y - x \rangle)$. Then h is continuous and $h(0) = h(1) = 0$. So, h attains its extremum at some $\gamma \in (0, 1)$. Suppose that γ is a minimum point of h. Now, by necessary conditions, we have for all $v \in R$,

$$h^-(\gamma; v) \geq 0.$$

By setting $u := x - y$, we derive

$$0 \leq h^-(\gamma; v)$$
$$= \liminf_{\lambda \to 0^+} \frac{h(\gamma + \lambda v) - h(\gamma)}{\lambda}$$
$$= \lim_{\lambda \to 0^+} \frac{f(y + (\gamma + \lambda v)u) - f(y + \gamma u)}{\lambda}$$
$$+ \frac{1}{2} \lim_{\lambda \to 0^+} \frac{a(\gamma + \lambda v)^2 - a\gamma^2}{\lambda}$$
$$+ \liminf_{\lambda \to 0^+} \frac{(\gamma + \lambda v)\langle \nabla f(y + (\gamma + \lambda v)u), -u \rangle - \gamma\langle \nabla f(y + \gamma u), -u \rangle}{\lambda}.$$

So,

$$0 \leq h^-(\gamma; v)$$
$$= v\langle \nabla f(y + \gamma u), u\rangle + a\gamma v + v\langle \nabla f(y + \gamma u), -u\rangle$$
$$+ \gamma \liminf_{\lambda \to 0^+} \frac{\langle \nabla f(y + (\gamma + \lambda v)u), -u\rangle - \langle \nabla f(y + \gamma u), -u\rangle}{\lambda}$$
$$= a\gamma v + \gamma \liminf_{\lambda \to 0^+} \frac{\langle \nabla f(y + (\gamma + \lambda v)u), -u\rangle - \langle \nabla f(y + \gamma u), -u\rangle}{\lambda}.$$

Let $c = y + \gamma(x - y)$. Then $c \in (x, y)$ and for $v = 1$, we get

$$0 \leq a\gamma + \gamma \liminf_{\lambda \to 0^+} \frac{\langle \nabla f(y + \gamma u + \lambda u), -u\rangle - \langle \nabla f(y + \gamma u), -u\rangle}{\lambda}$$
$$\leq a\gamma + \sup_{M \in \partial^2 f(c)} \langle M(-u), u\rangle.$$

This gives us

$$a \geq \inf_{M \in \partial^2 f(c)} \langle M(-u), -u\rangle.$$

Similarly, for $v = -1$, we obtain

$$0 \leq -a\gamma + \gamma \liminf_{\lambda \to 0^+} \frac{\langle \nabla f(y + \gamma u + \lambda(-u)), -u\rangle - \langle \nabla f(y + \gamma u), -u\rangle}{\lambda}$$
$$\leq -a\gamma + \sup_{M \in \partial^2 f(c)} \langle M(-u), -u\rangle;$$

thus

$$a \leq \sup_{M \in \partial^2 f(c)} \langle M(-u), -u\rangle.$$

Hence, it follows that

$$\inf_{M \in \partial^2 f(c)} \langle M(-u), -u\rangle \leq a \leq \sup_{M \in \partial^2 f(c)} \langle M(-u), -u\rangle,$$

and so,

$$a \in \overline{\mathrm{co}}(\langle \partial^2 f(c)(-u), -u\rangle).$$

Recalling that $u = x - y$, we obtain

$$f(y) - f(x) - \langle \nabla f(x), y - x\rangle = \frac{a}{2} \in \frac{1}{2}\overline{\mathrm{co}}(\langle \partial^2 f(c)(y - x), (y - x)\rangle).$$

The reasoning is similar in the case when γ is a maximum point of h. The details are left to the reader. $\qquad \square$

Corollary 2.2.21 *Let $f: \mathbb{R}^n \to \mathbb{R}$ be continuously differentiable on $\mathbb{R}^n$ and $x, y \in \mathbb{R}^n$. Suppose that for each $z \in [x, y]$, $\partial^2 f(z)$ is a convex and compact pseudo-Hessian of f at z. Then there exist $c \in (x, y)$ and $M \in \partial^2 f(c)$ such that*

$$f(y) = f(x) + \langle \nabla f(x), y - x\rangle + \frac{1}{2}\langle M(y - x), y - x\rangle.$$

Proof. It follows from the hypothesis that for each $z \in [x, y]$, $\partial^2 f(z)$ is convex and compact, and so the $\overline{co}$ in the conclusion of the previous theorem is superfluous. Thus the inequalities

$$\inf_{M \in \partial^2 f(c)} \langle M(y-x), x-y \rangle \leq a \leq \sup_{M \in \partial^2 f(c)} \langle M(y-x), x-y \rangle$$

give us that

$$a \in \langle \partial^2 f(c)(y-x), (y-x) \rangle.$$

$\square$

Corollary 2.2.22 Let $f \colon \mathbb{R}^n \to \mathbb{R}$ be $C^{1,1}$ and $x, y \in \mathbb{R}^n$. Then there exist $c \in (x, y)$ and $M \in \partial_H^2 f(c)$ such that

$$f(y) = f(x) + \langle \nabla f(x), y-x \rangle + \frac{1}{2} \langle M(y-x), y-x \rangle.$$

Proof. The conclusion follows from the above corollary by choosing the generalized Hessian $\partial_H^2 f(x)$ as a pseudo-Hessian of f for each x. $\square$

2.3 A General Chain Rule

Some chain rules are now developed for computing pseudo-Jacobians of composite functions. We begin with the following formula for the convex hull of compositions of matrices.

Lemma 2.3.1 Let $\Gamma_2 \subseteq L(\mathbb{R}^n, \mathbb{R}^m)$ and $\Gamma_1 \subseteq L(\mathbb{R}^m, \mathbb{R}^k)$ be nonempty. Then we have

$$(co(\Gamma_1)) \circ (co(\Gamma_2)) \subseteq co(\Gamma_1 \circ \Gamma_2).$$

Proof. Let $M \in co(\Gamma_1)$ and $N \in co(\Gamma_2)$. There are matrices $M_i \in \Gamma_1$, $N_i \in \Gamma_2$ and positive numbers λ_i and μ_i, $i = 1, \ldots, l$ such that $\sum_{i=1}^l \lambda_i = \sum_{j=1}^l \mu_j = 1$ and $M = \sum_{i=1}^l \lambda_i M_i$, $N = \sum_{i=1}^l \mu_i N_i$. Then

$$M \circ N = \sum_{i=1}^l \lambda_i M_i \circ \sum_{j=1}^l \mu_j N_j = \sum_{i=1}^l \lambda_i \left\{ \sum_{j=1}^l \mu_j M_i \circ N_j \right\},$$

which shows that $M \circ N \in co(co(\Gamma_1 \circ \Gamma_2)) = co(\Gamma_1 \circ \Gamma_2)$. $\square$

Fuzzy Chain Rules

The chain rule for a composite function proved presently involves pseudo-Jacobians around the given point. For this reason, it is called a fuzzy chain rule.

Theorem 2.3.2 *Let $f: \mathbb{R}^n \to \mathbb{R}^m$ and $g: \mathbb{R}^m \to \mathbb{R}^k$ be continuous functions. Let ∂f and ∂g be pseudo-Jacobian maps of f and g, respectively. Then for each $\varepsilon_1, \varepsilon_2 > 0$, the closure of the set*

$$\bigcup_{x \in x_0 + \varepsilon_1 B_n, y \in f(x_0) + \varepsilon_2 B_m} \partial g(y) \circ \partial f(x)$$

is a pseudo-Jacobian of the composite function $g \circ f$ at x_0.

Proof. Let $\varepsilon_1, \varepsilon_2 > 0$ be given. Denote by $D_1 := x_0 + \varepsilon_1 B_n$, $D_2 := f(x_0) + \varepsilon_2 B_m$ and

$$\Gamma_1 := \bigcup_{x \in D_1} \partial f(x) \quad \text{and} \quad \Gamma_2 := \bigcup_{y \in D_2} \partial g(y).$$

We have to show that for every $u \in \mathbb{R}^n$ and $w \in \mathbb{R}^k$,

$$(w(g \circ f))^+ (x_0; u) \leq \sup_{M \in \Gamma_1 \circ \Gamma_2} \langle w, M(u) \rangle.$$

To this purpose, let $\{t_i\}$ be a sequence of positive numbers converging to 0 such that

$$(w(g \circ f))^+ (x_0; u) = \lim_{i \to \infty} \frac{w(g \circ f)(x_0 + t_i u) - w(g \circ f)(x_0)}{t_i}.$$

Applying the mean value theorem to f and g we obtain

$$f(x_0 + t_i u) - f(x_0) \in \overline{\mathrm{co}}(\partial f[x_0, x_0 + t_i u](t_i u))$$
$$g(f(x_0 + t_i u)) - g(f(x_0)) \in \overline{\mathrm{co}}(\partial g[f(x_0), f(x_0 + t_i u)](f(x_0 + t_i u) - f(x_0))).$$

Denote the sets on the right-hand sides above P_i and Q_i, respectively, and observe that as f is continuous, there is $i_0 \geq 1$ such that

$$[x_0, x_0 + t_i u] \subseteq D_1,$$
$$[f(x_0), f(x_0 + t_i u)] \subseteq D_2 \text{ for } i \geq i_0.$$

Thus, in view of Lemma 2.3.1, we conclude

$$(w(g \circ f))^+ (x_0; u) \leq \lim_{i \to \infty} \sup_{\xi \in Q_i \circ P_i} \frac{1}{t_i} \langle w, \xi \rangle$$

$$\leq \lim_{i \to \infty} \sup_{\xi \in \overline{\mathrm{co}}(\bigcup_{x \in D_1, y \in D_2} \partial g(y) \circ \partial f(x)(t_i u))} \frac{1}{t_i} \langle w, \xi \rangle$$

$$\leq \sup\{\langle w, A(u) \rangle : A \in \Gamma_1 \circ \Gamma_2\}.$$

This shows that the closure of the set $\Gamma_1 \circ \Gamma_2$ is a pseudo-Jacobian of $g \circ f$ at x_0. $\qquad \square$

Chain Rules for Upper Semicontinuous Pseudo-Jacobians

An interesting case arises when f and g admit upper semicontinuous pseudo-Jacobians. A chain rule that involves perturbed sets of pseudo-Jacobians of f and g at a point under consideration replaces the fuzzy rule.

Theorem 2.3.3 *Let $f\colon \mathbb{R}^n \to \mathbb{R}^m$ and $g\colon \mathbb{R}^m \to \mathbb{R}^k$ be continuous functions. Let ∂f and ∂g be pseudo-Jacobian maps of f and g that are upper semicontinuous at x_0 and at $f(x_0)$ respectively. Then for each $\varepsilon_1, \varepsilon_2 > 0$, the closure of the set*

$$(\partial g(f(x_0)) + \varepsilon_2 B_{k \times m}) \circ (\partial f(x_0) + \varepsilon_1 B_{m \times n})$$

is a pseudo-Jacobian of the composite function $g \circ f$ at x_0.

Proof. By the hypothesis on the upper semicontinuity of ∂f and ∂g, we can find for every $\varepsilon_1, \varepsilon_2 > 0$ a positive δ such that

$$\partial f(x) \subseteq \partial f(x_0) + \varepsilon_1 B_{m \times n} \quad \text{for } x \text{ with } \|x - x_0\| \leq \delta,$$
$$\partial g(y) \subseteq \partial g(f(x_0)) + \varepsilon_2 B_{k \times m} \quad \text{for } y \text{ with } \|y - f(x_0)\| \leq \delta.$$

It follows that

$$\bigcup_{x \in x_0 + \delta B_n, y \in f(x_0) + \delta B_m} \partial g(y) \circ \partial f(x) \subseteq (\partial g(f(x_0)) + \varepsilon_2 B_{k \times m}) \circ (\partial f(x_0) + \varepsilon_1 B_{m \times n}).$$

We apply Theorem 2.3.2 to complete the proof. $\qquad\square$

When g admits a bounded pseudo-Jacobian, for instance, when it is differentiable or locally Lipschitz, Theorem 2.3.3 takes a simpler form.

Corollary 2.3.4 *Assume that ∂f is a pseudo-Jacobian map of f which is upper semicontinuous at x and ∂g is a pseudo-Jacobian of g which is bounded and upper semicontinuous at $f(x)$. Then for every $\varepsilon > 0$, the closure of the set*

$$(\partial g(f(x)) + \varepsilon B_{k \times m}) \circ \partial f(x)$$

is a pseudo-Jacobian of the composite function $g \circ f$ at x.

Proof. According to the preceding theorem, for every $\varepsilon_1, \varepsilon_2 > 0$ one has

$$
\begin{aligned}
(w(g \circ f))^+(x; u) &\leq \sup_{M \in \Gamma_1, N \in \Gamma_2} \langle w, (M \circ N)(u) \rangle \\
&\leq \sup_{M \in \Gamma_1, N \in \partial f(x_0)} \langle w, (M \circ N)(u) \rangle \\
&\quad + \varepsilon_2 \sup_{M \in \Gamma_1, N \in B_{m \times n}} \langle w, (M \circ N)(u) \rangle,
\end{aligned}
$$

where $\Gamma_1 = \partial g(f(x)) + \varepsilon_2 B_{k \times m}$ and $\Gamma_2 = \partial f(x) + \varepsilon_1 B_{m \times n}$. Because the set on which the supremum of the second term in the latter inequality is taken is bounded and ε_2 is arbitrary, we derive

$$(w(g \circ f))^+(x; u) \leq \sup_{M \in \Gamma_1, N \in \partial f(x_0)} \langle w, (M \circ N)(u) \rangle,$$

and obtain the desired pseudo-Jacobian. $\square$

As a special case of Theorem 2.3.3, when both functions f and g admit bounded pseudo-Jacobians, we obtain the following exact chain rule.

Corollary 2.3.5 *Assume that ∂f and ∂g are pseudo-Jacobian maps of f and g which are bounded and upper semicontinuous at x and $f(x)$, respectively. Then the set*

$$\partial g(f(x)) \circ \partial f(x)$$

is a pseudo-Jacobian of the composite function $g \circ f$ at x.

Proof. Use the method of the proof of Corollary 2.3.1 and the hypothesis that $\partial f(x)$ is bounded. $\square$

We notice that under the hypothesis of this corollary, the pseudo-Jacobian maps of f and g are locally bounded at x and $f(x)$, respectively. Hence, in view of Proposition 2.2.9 the functions f and g are locally Lipschitz near these points.

2.4 Chain Rules Using Recession Pseudo-Jacobian Matrices

It should be noted that Theorem 2.3.3 provides us with a construction of a pseudo-Jacobian of the composite function $g \circ f$ by using perturbed sets of pseudo-Jacobians of f and g. As we show, when $\partial f(x)$ and $\partial g(f(x))$ are not bounded, the exact chain rule as that of Corollary 2.3.5 is no longer true. The concept of recession directions (Section 1.5) is of great help in obtaining a chain rule in which only the recession Jacobian is perturbed. First we give some auxiliary results.

Lemma 2.4.1 *Let F be a set-valued map from $\mathbb{R}^n$ to $\mathbb{R}^k$ that is upper semicontinuous at $x_0 \in \mathbb{R}^n$. Let $\{t_i\}$ be a sequence of positive numbers converging to 0, $q_i \in \overline{co}(F(x_0 + t_i B_n))$ with $\lim_{i \to \infty} \|q_i\| = \infty$ and $\lim_{i \to \infty} q_i / \|q_i\| = q_0$ for some $q_0 \in \mathbb{R}^k$. Then $q_0 \in [co(F(x_0))]_\infty$. Moreover, if the cone $co(F(x_0)_\infty)$ is pointed, then $q_0 \in co(F(x_0)_\infty) = [co(F(x_0))]_\infty$.*

Proof. By the upper semicontinuity of F, for every $\varepsilon > 0$, there is i_0 sufficiently large such that

$$F(x_0 + t_i B_n) \subseteq F(x_0) + \varepsilon B_k \quad i \geq i_0.$$

Hence we have

$$q_i \in \overline{\text{co}}(F(x_0) + \varepsilon B_k) \subseteq \text{co}(F(x_0) + \varepsilon B_k) + \varepsilon B_k \quad \text{for } i \geq i_0.$$

Consequently,

$$q_0 \in [\text{co}(F(x_0) + \varepsilon B_k) + \varepsilon B_k]_\infty$$
$$\subseteq [\text{co}(F(x_0) + \varepsilon B_k)]_\infty \subseteq [\text{co}(F(x_0))]_\infty$$

(see Lemma 1.5.1). For the second part of the lemma the inclusion $\text{co}(F(x_0)_\infty) \subseteq [\text{co}(F(x_0))]_\infty$ always holds because $F(x_0) \subseteq \text{co}(F(x_0))$ and $[\text{co}(F(x_0))]_\infty$ is a closed convex cone. For the inverse inclusion, let $p \in [\text{co}(F(x_0))]_\infty$, $p \neq 0$. By Caratheodory's theorem, one can find convex combinations $p_i = \sum_{j=1}^{k+1} \lambda_{ij} p_{ij}$ with $\lambda_{ij} \geq 0$, $p_{ij} \in F(x_0)$ and $\sum_{j=1}^{k+1} \lambda_{ij} = 1$ such that

$$p/\|p\| = \lim_{i \to \infty} p_i/\|p_i\| \quad \text{and} \quad \lim_{i \to \infty} \|p_i\| = \infty.$$

Without loss of generality we may assume that $\lim_{i \to \infty} \lambda_{ij} = \lambda_j \geq 0$ for $j = 1, \ldots, k+1$ and $\sum_{j=1}^{k+1} \lambda_j = 1$. For every j, consider the sequence $\{\lambda_{ij} p_{ij}/\|p_i\|\}_{i \geq 1}$. We claim that this sequence is bounded, hence we may assume that it converges to some $p_{oj} \in (F(x_0))_\infty$. Then $p = \sum_{j=1}^{k+1} p_{oj} \in \text{co}(F(x_0)_\infty)$ as wanted. To achieve the proof we suppose to the contrary that $\{\lambda_{ij} p_{ij}/\|p_i\|\}_{i \geq 1}$ is unbounded. Denote $a_{ij} = \lambda_{ij} p_{ij}/\|p_i\|$. One may assume by taking a subsequence if necessary, that $\|a_{ij_0}\| = \max\{\|a_{ij}\|, j = 1, \ldots, k+1\}$, for every i. Hence $\lim_{i \to \infty} \|a_{ij_0}\| = \infty$. Because $p_i/\|p_i\| = \sum_{j=1}^{k+1} a_{ij}$, we have

$$0 = \lim_{i \to \infty} p_i/(\|p_i\| \|a_{ij_0}\|) = \lim_{i \to \infty} \sum_{j=1}^{k+1} a_{ij}/\|a_{ij_0}\|.$$

Again we may assume that $\{a_{ij}/\|a_{ij_0}\|\}_{i \geq 0}$ converges to some $a_{oj} \in F(x_0)_\infty$ for $j = 1, \ldots, k+1$ because these sequences are bounded. As $a_{oj_0} \neq 0$, the equality $0 = \sum_{j=1}^{k+1} a_{oj}$ shows that $\text{co}(F(x_0)_\infty)$ is not pointed, a contradiction. $\qquad \square$

Lemma 2.4.2 *Let K be a straight line cone in $\mathbb{R}^k$. Then for every $\varepsilon > 0$, the convex hull of the conic ε-neighborhood K^ε of K is the entire space $\mathbb{R}^k$.*

Proof. It is obvious that the interior of K^ε is nonempty. It contains for instance $K \setminus \{0\}$. Hence for every $x \in \mathbb{R}^k$, one has $(x+K) \cap \text{int}(K^\varepsilon) \neq \emptyset$. Let $y = x + k \in \text{int}(K^\varepsilon)$ for some $k \in K$. Then $x = y - k \in \text{int}(K^\varepsilon) + (-K) \subseteq \text{int}(K^\varepsilon) + K \subseteq \text{co}(K^\varepsilon)$. $\qquad\square$

It is well known in linear algebra that a linear transformation can be represented by a matrix, and every matrix determines a linear transformation. For this reason, we say that a *matrix is surjective* (respectively, *injective*) if the associated linear transformation is surjective (respectively, injective).

Theorem 2.4.3 *Let $f \colon \mathbb{R}^n \to \mathbb{R}^m$ and $g \colon \mathbb{R}^m \to \mathbb{R}^k$ be continuous functions. Let ∂f and ∂g be pseudo-Jacobian maps of f and g that are upper semicontinuous at x and at $f(x)$, respectively. Assume further that*

(i)   Elements of $\partial g(f(x))$ are surjective whenever $(\partial f(x))_\infty$ is nontrivial.
(ii)   Elements of $\partial f(x)$ are injective whenever $(\partial g(f(x)))_\infty$ is nontrivial.

Then for every $\varepsilon > 0$, the closure of the set

$$[\partial g(f(x)) + (\partial g(f(x)))_\infty^\varepsilon] \circ [\partial f(x) + (\partial f(x))_\infty^\varepsilon]$$

is a pseudo-Jacobian of the composite function $g \circ f$ at x.

Proof. This theorem can be derived from Theorem 2.3.3. However, we provide here a direct proof. We wish to show that for every $u \in \mathbb{R}^n$, $w \in \mathbb{R}^k$,

$$\langle w, g \circ f \rangle^+(x; u) \leq \sup_{M \in P, N \in Q} \langle w, MN(u) \rangle, \tag{2.5}$$

where $P := \partial g(f(x)) + (\partial g(f(x)))_\infty^\varepsilon$ and $Q := \partial f(x) + (\partial f(x))_\infty^\varepsilon$. The case $u = 0$ or $w = 0$ being obvious, we assume $u \neq 0$ and $w \neq 0$. Let $\{t_i\}$ be a sequence of positive numbers converging to 0 such that

$$\langle w, g \circ f \rangle^+(x; u) = \lim_{i \to \infty} \frac{\langle w, g(f(x + t_i u)) - g(f(x)) \rangle}{t_i}. \tag{2.6}$$

It follows from the mean value theorem that for each t_i there exist some $M_i \in \overline{\text{co}}(\partial g[f(x), f(x + t_i u)])$ and $N_i \in \overline{\text{co}}(\partial f[x, x + t_i u])$ such that

$$f(x + t_i u) - f(x) = t_i N_i(u) \tag{2.7}$$
$$g(f(x + t_i u)) - g(f(x)) = M_i(f(x + t_i u) - f(x)).$$

By taking a subsequence, if necessary, we need to deal with four cases.

(a) $\{N_i\}$ converges to some N_0 and $\{M_i\}$ converges to some M_0.
(b) $\{N_i\}$ converges to some N_0 and $\lim_{i \to \infty} \|M_i\| = \infty$ with $\{M_i/\|M_i\|\}$ converging to some M_*.

(c) $\lim_{i \to \infty} \|N_i\| = \infty$ with $\{N_i/\|N_i\|\}$ converging to some N_* and $\{M_i\}$ converges to some M_0.

(d) $\lim_{i \to \infty} \|N_i\| = \infty$ with $\{N_i/\|N_i\|\}$ converging to some N_* and $\lim_{i \to \infty} \|M_i\| = \infty$ with $\{M_i/\|M_i\|\}$ converging to some M_*.

It follows from (2.6) and (2.7) that

$$\langle w, g \circ f \rangle^+(x; u) = \lim_{i \to \infty} \langle w, M_i N_i(u) \rangle.$$

In (a) one has $N_0 \in \overline{co}(\partial f(x))$, $M_0 \in \overline{co}(\partial g(f(x)))$ by the upper semicontinuity of ∂f and ∂g. Therefore,

$$\langle w, g \circ f \rangle^+(x; u) = \langle x, M_0 N_0(u) \rangle \leq \sup_{M \in P,\, N \in Q} \langle w, MN(u) \rangle.$$

Case (b). By Lemma 2.4.1, $M_* \in [co(\partial g(f(x)))]_\infty$. If $co\{[\partial g(f(x))]_\infty\}$ is not pointed, then by Lemma 2.4.2, $co\{[\partial g(f(x))]_\infty^\varepsilon\}$ coincides with the whole space $L(\mathbb{R}^m, \mathbb{R}^k)$. This and the injectivity of $N \in \partial f(x)$ imply

$$\sup_{M \in P,\, N \in Q} \langle w, MN(u) \rangle \geq \sup_{M \in L(\mathbb{R}^m, \mathbb{R}^\ell),\, N \in Q} \langle w, MN(u) \rangle = \infty$$

(because $u \neq 0$), and (2.5) holds obviously. If the cone $co\{[\partial g(f(x))]_\infty\}$ is pointed, then by Lemma 2.4.1 it contains M_*. Let

$$\alpha := \langle w, M_* N_0(u) \rangle.$$

If $\alpha > 0$, then from the fact that $\lambda M_* \in co\{[\partial g(f(x))]_\infty\}$ for all $\lambda \geq 0$, we derive the following relation which subsumes (2.5),

$$\sup_{M \in P,\, N \in Q} \langle w, MN(u) \rangle \geq \sup_{M \in M_r + co\{[\partial g(f(x))]_\infty^\varepsilon\}} \langle w, MN_0(u) \rangle$$
$$\geq \limsup_{\lambda \to \infty} \langle w, (\lambda M_* + M_r) N_0(u) \rangle \geq \infty,$$

where M_r is an arbitrary element of $\partial g(f(x))$.

If $\alpha < 0$, then for i sufficiently large, one has

$$\left\langle w, \frac{M_i}{\|M_i\|} N_i(u) \right\rangle < \frac{\alpha}{2} < 0.$$

Hence

$$\langle w, g \circ f \rangle^+(x; u) = \lim_{i \to \infty} \langle w, M_i N_i(u) \rangle \leq \lim_{i \to \infty} \|M_i\| \frac{\alpha}{2} = -\infty.$$

This shows that (2.5) is true.

If $\alpha = 0$, then observe that $M_* \in \text{int}\{co[(\partial g(f(x)))_\infty^\varepsilon]\}$. Let

$$K := \{\mathrm{co}[(\partial g(f(x)))^\varepsilon_\infty]\}^{tr} \circ w.$$

Then K consists of all elements $M^{tr}w \in \mathbb{R}^m$, where $M \in \mathrm{co}[(\partial g(f(x)))^\varepsilon_\infty]$. We claim that $M^{tr}_* w \in \mathrm{int}(K)$. Indeed, if this is not the case, then one can find a nonzero vector $v \in \mathbb{R}^m$ such that

$$\langle v, (M^{tr} - M^{tr}_*)(w) \rangle \geq 0 \quad \text{for every } M \in \mathrm{co}[(\partial g(f(x)))^\varepsilon_\infty].$$

Because M_* is an interior point, the above inequality must hold for every $M \in L(\mathbb{R}^m, \mathbb{R}^\ell)$. Moreover, as $v \neq 0$, this is possible only when $w = 0$, a contradiction. Recalling that N_0 is injective, hence $N_0 u \neq 0$ and because $M^{tr}_*(w) \in \mathrm{int}(K)$, we can find a matrix $M_1 \in \mathrm{int}\{\mathrm{co}[(\partial g(f(x)))^\varepsilon_\infty]\}$ sufficiently close to M_* such that $\langle M^{tr}_1(w), N_0(u) \rangle > 0$. We deduce that

$$\sup_{M \in P, N \in Q} \langle w, MN(u) \rangle \geq \sup_{M \in M_r + \mathrm{co}[(\partial g(f(x)))^\varepsilon_\infty]} \langle w, MN_0(u) \rangle$$
$$\geq \lim_{\lambda \to \infty} \langle (\lambda M_1 + M_r)^{tr}(w), N_0(u) \rangle \geq \infty,$$

where M_r is an arbitrary element of $\partial g(f(x))$. Hence (2.5) holds.

The case (c) is proven in a similar manner with noting that $M \in \partial g(f(x))$ is surjective if and only if M^{tr} is injective.

Finally, let us proceed to the case (d). In virtue of Lemma 2.4.1, we have $M_* \in [\mathrm{co}(\partial g(f(x)))]_\infty$ and $N_* \in [\mathrm{co}(\partial f(x))]_\infty$. We distinguish four possible subcases according to the pointedness of the recession cones of the pseudo-Jacobians.

Subcase (d1): $\mathrm{co}\{(\partial g(f(x)))_\infty\}$ and $\mathrm{co}\{(\partial f(x))_\infty\}$ are pointed. By Lemma 2.4.1, $M_* \in \mathrm{co}\{(\partial g(f(x)))_\infty\}$ and $N_* \in \mathrm{co}\{(\partial f(x))_\infty\}$. Let us consider

$$\beta := \langle w, M_* N_*(u) \rangle.$$

If $\beta > 0$, then for $\lambda \geq 0$, one has $\lambda M_* \in \mathrm{co}\{(\partial g(f(x)))^\varepsilon_\infty\}$ and $\lambda N_* \in \mathrm{co}\{(\partial f(x))^\varepsilon_\infty\}$. Hence

$$\sup_{M \in P, N \in Q} \langle w, MN(u) \rangle \geq \sup_{M \in M_r + \mathrm{co}\{(\partial f(x))^\varepsilon_\infty\}, N \in N_r + \mathrm{co}\{(\partial f(x))^\varepsilon_\infty\}} \langle w, MN(u) \rangle$$
$$\geq \lim_{\lambda \to \infty} \langle w, (\lambda M_* + M_r)(\lambda N_* + N_r)(u) \rangle = \infty,$$

where M_r and N_r are arbitrary elements of $\partial g(f(x))$ and $\partial f(x)$ respectively. This shows that (2.5) is true.

If $\beta < 0$, then for i sufficiently large,

$$\left\langle w, \frac{M_i}{\|M_i\|} \frac{N_i}{\|N_i\|}(u) \right\rangle < \frac{\beta}{2} < 0.$$

Consequently

$$\langle w, g \circ f \rangle^+(x; u) = \lim_{i \to \infty} \langle w, M_i N_i(u) \rangle \le \lim_{i \to \infty} \frac{\beta}{2} \|M_i\| \|N_i\| = -\infty$$

which also implies (2.5).

If $\beta = 0$, then, as in the subcase (b3), one has $M_* \in \text{int}\{[\text{co}(\partial g(f(x)))]_\infty^\varepsilon\}$ and $N_* \in \text{int}\{[\text{co}(\partial f(x))]_\infty^\varepsilon\}$ for $\lambda > 0$. The relation

$$\beta = \langle M_*^{tr}(w), N_*(u) \rangle = 0$$

implies the existence of two elements $M_1 \in \text{int}\{[\text{co}(\partial g(f(x)))]_\infty^\varepsilon\}$ and $N_1 \in \text{int}\{[\text{co}(\partial f(x))]_\infty^\varepsilon\}$ sufficiently close to M_* and N_* such that

$$\langle M_1^{tr} w, N_1(u) \rangle > 0.$$

Then

$$\sup_{M \in P, N \in Q} \langle w, M N(u) \rangle \ge \sup_{M \in M_r + \text{co}\{(\partial g f(x))_\infty^\varepsilon\}, N \in N_r + \text{co}\{(\partial f(x))_\infty^\varepsilon\}} \langle p^{tr}(w), N(u) \rangle$$

$$\ge \lim_{\lambda \to \infty} \langle (\lambda M_1 + M_r)^{tr}(w), (\lambda N_1 + N_r)(u) \rangle \ge \infty,$$

where M_r and N_r are arbitrary elements of $\partial g(f(x))$ and $\partial f(x)$, respectively. This again implies (2.5) as well.

Subcase (d2): $\text{co}\{(\partial g(f(x)))_\infty\}$ is pointed and $\text{co}\{(\partial f(x))_\infty\}$ is not pointed. By Lemma 2.4.1, $M_* \in \text{co}\{(\partial g(f(x)))_\infty\}$, and by Lemma 2.4.2, Q may be replaced by $L(\mathbb{R}^n, \mathbb{R}^m)$. As shown before, $M_*^{tr} w \in \text{int}\{[\text{co}(\partial g(f(x)))_\infty^\varepsilon]^{tr} w\}$. Hence there is $M_1 \in \text{int}\{\text{co}[(\partial g(f(x)))_\infty^\varepsilon]\}$ sufficiently close to M_* such that $M_1^{tr} w \ne 0$. Then we obtain

$$\sup_{M \in P, N \in Q} \langle w, M N(u) \rangle \ge \sup_{N \in L(\mathbb{R}^n, \mathbb{R}^m)} \langle w, M_1 N(u) \rangle = \infty,$$

which shows that (2.5) holds.

Subcase (d3): $(\partial g(f(x)))_\infty$ is not pointed and $\text{co}\{(\partial f(x))_\infty\}$ is pointed. This case is proven similarly to the subcase (d2).

Subcase (d4): Both of $\text{co}\{(\partial g(f(x)))_\infty\}$ and $\text{co}\{(\partial f(x))_\infty\}$ are not pointed. By Lemma 2.4.2, P may be replaced by $L(\mathbb{R}^m, \mathbb{R}^k)$ and Q may be replaced by $L(\mathbb{R}^n, \mathbb{R}^m)$. Therefore, we have

$$\sup_{M \in P, N \in Q} \langle w, M N(u) \rangle \ge \sup_{M \in L(\mathbb{R}^m, \mathbb{R}^k), N \in L(\mathbb{R}^n, \mathbb{R}^m)} \langle w, M N(u) \rangle = \infty,$$

which implies (2.5).   □

Proposition 2.4.4 *Under the hypothesis of Theorem 2.4.3, for every $\varepsilon > 0$, the closure of the set*

$$[\partial g(f(x)) \cup \{(\partial g(f(x)))_\infty^\varepsilon\} \setminus \text{int}(B_{k \times n})\}] \circ [\partial f(x) \cup \{(\partial f(x))_\infty^\varepsilon \setminus \text{int}(B_{m \times n})\}]$$

is a pseudo-Jacobian of the composite function $g \circ f$ at x.

Proof. The proof is similar to the proof of the preceding theorem and so it is omitted here. $\square$

The particular case of Theorem 2.4.3, presented below, is useful in the applications later.

Corollary 2.4.5 *Assume that ∂f is a pseudo-Jacobian of f which is upper semicontinuous at x and g is differentiable with ∇g continuous at $f(x)$ and $\nabla g(f(x)) \neq 0$. Then for every $\varepsilon > 0$, the set*

$$\nabla g(f(x)) \circ [\partial f(x) + (\partial f(x))_\infty^\varepsilon]$$

is a pseudo-Jacobian of the composite function $g \circ f$ at x.

Proof. We know that ∇g is a pseudo-Jacobian of g. Moreover, if $\nabla g(f(x)) \neq 0$, then it is a surjective map from $\mathbb{R}^m$ to $\mathbb{R}$. The hypotheses of Theorem 2.4.3 are satisfied and so the conclusion holds. $\square$

The following modified version of Theorem 2.4.3 is useful in practice, especially when each component of f has its own generalized derivative that is easy to compute. Let ∂g be a pseudo-Jacobian map of $g : \mathbb{R}^{m_1} \times \mathbb{R}^{m_2} \to \mathbb{R}^k$. Then $\partial_1 g$ and $\partial_2 g$ denote the projections of ∂g on $L(\mathbb{R}^{m_1}, \mathbb{R}^k)$ and on $L(\mathbb{R}^{m_2}, \mathbb{R}^k)$, respectively.

Proposition 2.4.6 *Let $f_1 \colon \mathbb{R}^n \to \mathbb{R}^{m_1}$, $f_2 \colon \mathbb{R}^n \to \mathbb{R}^{m_2}$, and $g \colon \mathbb{R}^{m_1+m_2} \to \mathbb{R}^k$ be continuous functions. Let $\partial f_1, \partial f_2$, and ∂g be pseudo-Jacobians of f_1, f_2 and g that are upper semicontinuous at x and at $y := (f_1(x), f_2(x))$, respectively. Further assume that for $j = 1, 2$,*

(i)   Elements of $\partial_j g(y)$ are surjective whenever $(\partial f_j(x))_\infty$ is nontrivial.
(ii)   Elements of $\partial f_j(x)$ are injective whenever $(\partial_j g(y))_\infty$ is nontrivial.

Then for every $\varepsilon_1, \varepsilon_2 > 0$, the closure of the set

$$[\partial_1 g(y) + (\partial_1 g(y))_\infty^{\varepsilon_1}] \circ [\partial f_1(x) + (\partial f_1(x))_\infty^{\varepsilon_2}]$$

$$[+\partial_2 g(y) + (\partial_2 g(y))_\infty^{\varepsilon_1}] \circ [\partial f_2(x) + (\partial f_2(x))_\infty^{\varepsilon_2}]$$

is a pseudo-Jacobian of the composite function $g \circ f$ at x.

Proof. We wish to apply Theorem 2.3.2 to the functions $f = (f_1, f_2)$ and g. First observe that by Theorem 2.1.3, $\partial f_1 \times \partial f_2$ is a pseudo-Jacobian map of f which is upper semicontinuous at x. For every $i \geq 1$, the closure of the set

$$(\partial g(f(x)) + (1/i)B_{k \times m}) \circ ((\partial f_1 \times \partial f_2)(x) + (1/i)B_{m \times n}),$$

where $m = m_1 + m_2$, is a pseudo-Jacobian of $g \circ f$ at x. Therefore, for each $u \in \mathbb{R}^n$ and $w \in \mathbb{R}^k$, there exist matrices $N_{ji} \in \partial_j g(y) + (1/i)B_{k \times m_j}$, $M_{ji} \in \partial f_j(x) + (1/i)B_{m_j \times n}$ such that

$$\langle w, g \circ f \rangle^+(x; u) \leq \lim_{i \to \infty} \langle w, (N_{1i}M_{1i} + N_{2i}M_{2i})(u) \rangle$$
$$\leq \lim_{i \to \infty} \langle w, (N_{1i}M_{1i})(u) \rangle + \lim_{i \to \infty} \langle w, (N_{2i}M_{2i})(u) \rangle.$$

Further observe that the pseudo-Jacobian maps $\partial_1 g$ and $\partial_2 g$ are upper semicontinuous as is the map ∂g. Hence the argument of the proof of Theorem 2.3.3 applied to each of the terms on the right-hand side of the latter inequality produces the following relations,

$$\limsup_{i \to \infty} \langle w, (N_{ji}M_{ji})(u) \rangle \leq \sup_{N \in Q_j, M \in P_j} \langle w, (MN)(u) \rangle,$$

where $j = 1, 2$; and

$$P_j := \partial f_j(x) + (\partial f_j(x))_\infty^{\varepsilon_1} \quad \text{and} \quad Q_j := \partial_j g(y) + (\partial_j g(y))_\infty^{\varepsilon_2}.$$

Consequently,

$$\langle w, g \circ f \rangle^+(x; u) \leq \sup_{N \in Q_1, M \in P_1} \langle w, (MN)(u) \rangle + \sup_{N \in Q_2, M \in P_2} \langle w, (MN)(u) \rangle$$
$$\leq \sup_{N \in Q_1 Q_2, M \in P_1 \times P_2} \langle w, (MN)(u) \rangle,$$

which shows that the closure of the set $Q_1 \circ P_1 + Q_2 \circ P_2$ is a pseudo-Jacobian of $g \circ f$ at x. $\qquad \square$

A close inspection of the above chain rule raises some interesting questions:
1. Does the result in Corollary 2.4.5 remain valid without $\nabla g(f(x)) \neq 0$?
2. Is it possible to eliminate $\varepsilon > 0$ in Corollary 2.4.5?
The next two examples show that in general the answers to the above questions are in the negative.

Example 2.4.7 Let $n = m = l = 1$. Let $f(x) = \sqrt[3]{x}$ and $g(y) = y^3$. An upper semicontinuous pseudo-Jacobian of f is given by

$$\partial f(x) = \begin{cases} (1/3)x^{-2/3} & \text{if } x \neq 0, \\ [\alpha, \infty) & \text{if } x = 0, \end{cases}$$

where $\alpha \in \mathbb{R}$. Then $g \circ f(x) = x$ and $\nabla g(f(0)) \circ (\partial f(0) + (\partial f(0))_\infty^\varepsilon) = \{0\}$ and hence it cannot be a pseudo-Jacobian of $g \circ f$ at $x = 0$. Note that $\nabla g(f(0)) = 0$.

Example 2.4.8 Let $n = 2$, $m = 2$, and $\ell = 1$. Let f and g be defined by

$$f(x, y) = (x^{1/3}, y)$$
$$g(u, v) = u^3 + v.$$

Then $g \circ f(x, y) = x + y$. A pseudo-Jacobian of f is given by

$$\partial f(x, y) = \begin{pmatrix} (1/3)x^{-2/3} & 0 \\ 0 & 1 \end{pmatrix} \quad \text{if} \quad x \neq 0,$$

$$\partial f(0, y) = \left\{ \begin{pmatrix} \alpha & 0 \\ 0 & 1 \end{pmatrix} : \alpha \geq 0 \right\} \quad \text{if} \quad x = 0.$$

The function g is continuously differentiable with

$$\nabla g(u, v) = (3u^2, 1).$$

The map $(u, v) \mapsto \nabla g(0, 0)(u, v)$ is a surjective map from $\mathbb{R}^2$ onto $\mathbb{R}$. The recession cone of $\partial f(0, 0)$ is

$$(\partial f(0, 0))_\infty = \left\{ \begin{pmatrix} \alpha & 0 \\ 0 & 0 \end{pmatrix} : \alpha \geq 0 \right\}.$$

Then

$$\nabla g(0, 0) \circ (\partial f(0, 0) + (\partial f(0, 0))_\infty) = \{(0, 1)\}.$$

It is obvious that this set cannot be a pseudo-Jacobian of the composite function $g \circ f$ at $(0, 0)$.

2.5 Chain Rules for Gâteaux and Fréchet Pseudo-Jacobians

Theorem 2.5.1 Let $f \colon \mathbb{R}^n \to \mathbb{R}^m$ and let $g \colon \mathbb{R}^m \to \mathbb{R}^k$ be continuous functions. Assume that

(i) $\partial f(x_0)$ is a Gâteaux pseudo-Jacobian of f at x_0;
(ii) ∂g is a pseudo-Jacobian map of g that is locally bounded at $y_0 = f(x_0)$.

Then for every $\varepsilon > 0$, the closure of the set

$$\partial g(y_0 + \varepsilon B_m) \circ \partial f(x_0)$$

is a pseudo-Jacobian of $g \circ f$ at x. In particular, when $\partial f(x_0)$ is bounded, the set $\partial g(y_0) \circ \partial f(x_0)$ is a pseudo-Jacobian of $g \circ f$ at x_0.

Proof. Let $\varepsilon > 0$ and let $u \in \mathbb{R}^n, u \neq 0$, and $w \in \mathbb{R}^k, w \neq 0$. We have to show that

$$\langle w, g \circ f \rangle^+(x_0; u) \leq \sup_{N \in \partial g(y_0 + \varepsilon B_m), M \in \partial f(x_0)} \langle w, N \circ M(u) \rangle. \qquad (2.8)$$

Let $\{t_i\}$ be a sequence of positive numbers converging to 0 and such that

$$\langle w, g \circ f \rangle^+(x_0; u) = \lim_{i \to \infty} \frac{\langle w, g(f(x_0 + t_i u)) - g(f(x_0)) \rangle}{t_i}.$$

Without loss of generality we may assume, by the continuity of f, that $f(x_0 + t_i u) \in y_0 + \varepsilon B_m$ for all i. Applying the mean value theorem to the function g on $[f(x_0), f(x_0 + t_i u)]$, we have

$$g(f(x_0 + t_i u)) - g(f(x_0)) \in \overline{\text{co}}\{\partial g[f(x_0), f(x_0 + t_i u)](f(x_0 + t_i u) - f(x_0))\}$$
$$\subseteq \overline{\text{co}}\{\partial g(y_0 + \varepsilon B_m)(f(x_0 + t_i u) - f(x_0))\}.$$

Moreover, it follows from the definition of the Gâteaux pseudo-Jacobian that there exists $M_i \in \partial f(x_0)$ such that

$$f(x_0 + t_i u) - f(x_0) = M_i(t_i u) + o(t_i),$$

where $(o(t_i)/t_i) \to 0$ as $t_i \to 0$. So, we deduce that

$$g(f(x_0 + t_i u)) - \alpha g(f(x_0)) \in \overline{\text{co}}\partial\{g(y_0 + \varepsilon B_m)(f(x_0 + t_i u) - f(x_0))\},$$

which implies that

$$\frac{1}{t_i}\langle w, g(f(x_0 + t_i u)) - \alpha g(f(x_0))\rangle \leq \sup_{N \in \partial g(y_0 + \varepsilon B_m)} \langle w, N \circ (M_i(u) + \tfrac{o(t_i)}{t_i})\rangle$$

$$\leq \sup_{N \in \partial g(y_0 + \varepsilon B_m), M \in \partial f(x_0)} \langle w, N \circ M(u) + N \circ \tfrac{o(t_i)}{t_i}\rangle.$$

Because ∂g is bounded, we may assume that $\partial g(y_0 + \varepsilon B_m)$ is bounded. By letting $t_i \to 0$ in the above inequality, we obtain (2.8). Now if $\partial f(x_0)$ is bounded, then the sequence $\{M_i\}_{i \geq 1}$ is bounded, which may be assumed to converge to some $M_0 \in \partial f(x_0)$. According to Proposition 2.2.9, g is locally Lipschitz. Hence there is $\alpha > 0$ such that

$$\|g(f(x_0 + t_i u)) - g(f(x_0))\| \leq \alpha \|t_i(M_i - M_0)(u) + o(t_i)\|.$$

We deduce that

$$\langle w, g \circ f \rangle^+(x_0; u) = \lim_{i \to \infty} \frac{1}{t_i}\langle w, g(f(x_0 + t_i u)) - g(f(x_0))\rangle$$

$$\leq \sup_{N \in \partial g(y_0)} \langle w, N \circ M_0(u)\rangle + \lim_{i \to \infty} \alpha \|w\| . \|(M_i - M_0)(u) + o(t_i)/t_i\|$$

$$\leq \sup_{N \in \partial g(y_0), M \in \partial f(x_0)} \langle w, N \circ M(u)\rangle.$$

This shows that $\partial g(y_o) \circ f(x_0)$ is a pseudo-Jacobian of $g \circ f$ at x_0. $\square$

Next we present a chain rule for Gâteaux differentiable functions.

Corollary 2.5.2 *Assume that $f \colon \mathbb{R}^n \to \mathbb{R}^m$ is a continuous and Gâteaux differentiable function at x_0. If $g \colon \mathbb{R}^m \to \mathbb{R}$ is locally Lipschitz and Gâteaux differentiable at $y_0 = f(x_0)$, then the composite function $g \circ f$ is Gâteaux differentiable at x_0 and $\nabla(g \circ f)(x_0) = \nabla g(y_0) \circ \nabla f(x_0)$.*

Proof. Because a Gâteaux derivative is a pseudo-Jacobian, in view of Theorem 2.5.1, the singleton set $\{\nabla g(y_0) \circ \nabla f(x_0)\}$ is a pseudo-Jacobian of $g \circ f$ at x_0. By Proposition 1.2.2, $g \circ f$ is Gâteaux differentiable at x_0 and its derivative is $\nabla g(y_0) \circ \nabla f(x_0)$. $\square$

When both g and f are locally Lipschitz, we derive a chain rule for the Clarke generalized Jacobian.

Corollary 2.5.3 *Assume that $f \colon \mathbb{R}^n \to \mathbb{R}^m$ and $g \colon \mathbb{R}^m \to \mathbb{R}$ are locally Lipschitz functions. Then*

$$\partial^C g \circ f(x) \subseteq \mathrm{co}(\partial^C g(y_0) \circ \partial^C f(x_0)).$$

Proof. When g and f are locally Lipschitz, the composite function $g \circ f$ is locally Lipschitz too. Moreover, as $\partial^C g$ and $\partial^C f$ are upper semicontinuous pseudo-Jacobian maps, the set-valued map $x \mapsto \mathrm{co}(\partial^C g(f(x)) \circ \partial^C f(x))$ is upper semicontinuous and convex-valued, and it is also a pseudo-Jacobian map of $g \circ f$. In view of Corollary 1.6.8, the conclusion follows. $\square$

We say that $f \colon \mathbb{R}^n \to \mathbb{R}^m$ is *radially Lipschitz* at x_0 if for each $u \in \mathbb{R}^n, u \neq 0$, there are $\alpha > 0$ and $t_0 > 0$ such that

$$\|f(x_0 + tu) - f(x_0)\| \leq \alpha \|tu\| \quad \text{for } 0 \leq t \leq t_0.$$

Theorem 2.5.4 *Let $f \colon \mathbb{R}^n \to \mathbb{R}^m$ be continuous and radially Lipschitz at x_0 and let $g \colon \mathbb{R}^m \to \mathbb{R}^k$ be continuous. Assume that $\partial g(y_0)$ is a Fréchet pseudo-Jacobian of g at $y_0 = f(x_0)$ and ∂f is a pseudo-Jacobian map of f. Then for every $\varepsilon > 0$, the closure of the set*

$$\partial g(y_0) \circ \partial f(x_0 + \varepsilon B_n)$$

is a pseudo-Jacobian of $g \circ f$ at x_0. In particular, when $\partial g(y_0)$ is bounded, the set $\partial g(y_0) \circ \partial f(x_0)$ is a pseudo-Jacobian of $g \circ f$ at x_0.

Proof. Let $\varepsilon > 0$ be given. Let $u \in \mathbb{R}^n, u \neq 0$ and $w \in \mathbb{R}^k, w \neq 0$. As in the proof of the preceding theorem, $\{t_i\}$ is a sequence of positive numbers converging to 0 such that

$$\langle w, g \circ f \rangle^+(x_0; u) = \lim_{i \to \infty} \frac{\langle w, g(f(x_0 + t_i u)) - g(f(x_0)) \rangle}{t_i}.$$

By the radial Lipschitzianity of f, there is $\alpha > 0$ such that

$$\|f(x_0 + t_i u) - f(x_0)\| \leq \alpha t_i \|u\| \quad \text{for every } i \geq 1.$$

We may assume $x_0 + t_i u \in x_0 + \varepsilon B_n$ for all $i \geq 1$. It follows from the definition of Fréchet pseudo-Jacobian and the mean value theorem that

$$g(f(x_0 + t_i u)) - g(f(x_0)) = N_i(f(x_0 + t_i u) - f(x_0)) + o(f(x_0 + t_i u) - f(x_0))$$
$$f(x_0 + t_i u) - f(x_0) \in \overline{co}\{\partial f(x_0 + \varepsilon B_n)(t_i u)\},$$

where $N_i \in \partial g(y_0)$ and $o(f(x_0 + t_i u) - f(x_0))/\|f(x_0 + t_i u) - f(x_0)\| \to 0$ as $f(x_0 + t_i u) \to f(x_0)$. The radial Lipschitzianity of f implies also that $o(f(x_0 + t_i u) - f(x_0))/t_i \to 0$ as $i \to \infty$. By the above, we obtain

$$\langle w, g(f(x_0 + t_i u) - f(x_0)) \rangle \leq \sup_{M \in \partial f(x_0 + \varepsilon B_n)} \langle w, N_i \circ M(t_i u)$$
$$+ o(f(x_0 + t_i u) - f(x_0)) \rangle$$

which yields

$$\langle w, g \circ f \rangle^+(x_0; u) \leq \sup_{N \in \partial g(y_0), M \in \partial f(x_0 + \varepsilon B_n)} \langle w, N \circ M(u) \rangle$$

as requested. If $\partial g(y_0)$ is bounded, then so is the sequence $\{N_i\}_{i \geq 1}$ which may be assumed to converge to some $N_0 \in \partial g(y_0)$. It follows that

$$\|(N_i - N_0)(f(x_0 + t_i u) - f(x_0))\| \leq \alpha t_i \|u\| \|N_i - N_0\|$$

and consequently

$$\langle w, g(f(x_0 + t_i u)) - g(f(x_0)) \rangle = \langle N_0^*(w), f(x_0 + t_i u) - f(x_0) \rangle$$
$$+ \langle w, (N_i - N_0)(f(x_0 + t_i u) - f(x_0)) \rangle + o(f(z_0 + t_i u) - f(x_0)).$$

This yields

$$\langle w, g \circ f \rangle^+(x_0; u) \leq \sup_{M \in \partial f(x_0)} \langle w, N_0 \circ M(u) \rangle \leq \sup_{N \in \partial g(y_0), M \in \partial f(x_0)} \langle w, N \circ M(u) \rangle.$$

$\square$

Observe that when a function is Gâteaux differentiable at a point, then it is radially Lipschitz at that point. We now derive another chain rule for the Gâteaux derivative of composite functions.

Corollary 2.5.5 *Suppose that* $f: \mathbb{R}^n \to \mathbb{R}^m$ *is continuous, Gâteaux differentiable at* x_0*, and* $g: \mathbb{R}^m \to \mathbb{R}^k$ *is Fréchet differentiable at* $y_0 = f(x_0)$*. Then the composite function* $g \circ f$ *is Gâteaux differentiable at* x_0 *and*

$$\nabla(g \circ f)(x_0) = \nabla g(f(x_0)) \circ \nabla f(x_0).$$

Proof. As we have noticed, f is radially Lipschitz at x_0. Moreover, $\{\nabla g(f(x_0))\}$ is a Fréchet pseudo-Jacobian of g at $f(x_0)$. Hence Theorem 2.5.4 applies and we infer that the singleton set $\{\nabla g(f(x_0)) \circ \nabla f(x_0)\}$ is a pseudo-Jacobian of $g \circ f$ at x_0. Therefore, by Proposition 1.2.2, $g \circ f$ is Gâteaux differentiable at x_0, and its derivative coincides with $\nabla g(f(x_0)) \circ \nabla f(x_0)$.   □

For Fréchet pseudo-Jacobians we also have the following simple chain rule.

Proposition 2.5.6 *Let* $f: \mathbb{R}^n \to \mathbb{R}^m$ *and* $g: \mathbb{R}^m \to \mathbb{R}^k$ *be continuous functions. If* $\partial f(x_0)$ *is a bounded Fréchet pseudo-Jacobian of* f *at* x_0 *and* $\partial g(f(x_0))$ *is a Fréchet pseudo-Jacobian of* g *at* $f(x_0)$*, then the closure of the set* $\partial g(f(x_0)) \circ \partial f(x_0)$ *is a Fréchet pseudo-Jacobian of the composite function* $g \circ f$ *at* x_0*.*

Proof. Let x be a point in a neighborhood of x_0. Then $f(x) \to f(x_0)$ as x tends to x_0. There exist $M_x \in \partial f(x_0)$ and $N_y \in \partial g(f(x_0))$ such that

$$f(x) - f(x_0) = M_x(x - x_0) + o_1(\|x - x_0\|),$$

$$g(f(x)) - g(f(x_0)) = N_x(f(x) - f(x_0)) + o_2(\|f(x) - f(x_0)\|),$$

where $o_1(\|x - x_0\|)/\|x - x_0\|$ and $o_2(\|f(x) - f(x_0)\|)/\|f(x) - f(x_0)\|$ converge to 0 as x tends to x_0. We deduce that

$$g(f(x)) - g(f(x_0)) = N_x \circ M_x(x - x_0) + o_2(\|M_x(x - x_0) + o_1(\|x - x_0\|)\|). \tag{2.9}$$

Because $\partial f(x_0)$ is bounded, the value $M_x(x - x_0) + o_1(\|x - x_0\|)$ converges to 0 as x tends to x_0 and $\|M_x(x - x_0) + o_1(\|x - x_0\|)\|/\|x - x_0\|$ is bounded. Consequently, $\lim_{x \to x_0} o_2(\|M_x(x - x_0) + o_1(\|x - x_0\|)\|)/\|x - x_0\| = 0$. This and (2.9) achieve the proof.   □

3

Openness of Continuous Vector Functions

In this chapter we develop sufficient conditions for openness of continuous vector functions by using pseudo-Jacobians. Related topics such as inverse functions, implicit functions, convex interior mappings, metric regularity, and pseudo-Lipschitzianity are also examined. The pseudo-Jacobian-based approach provides an elementary and classical scheme for studying these topics, allows combined use of different generalized derivatives, and hence offers a useful complement to the existing methods of modern variational analysis [94, 107].

3.1 Equi-Invertibility and Equi-Surjectivity of Matrices

Let M be an invertible $n \times n$-matrix. Then there is a positive α such that

$$\|M(u)\| \geq \alpha \|u\| \text{ for every } u \in \mathbb{R}^n. \tag{3.1}$$

Clearly, the converse is also true; that is, if the above inequality holds, then M is invertible. Furthermore, let $\Gamma \subset L(\mathbb{R}^n, \mathbb{R}^n)$ be a nonempty set. We say that Γ is *equi-invertible* if (3.1) is satisfied for every $M \in \Gamma$. It is clear that if a matrix is invertible, then it has a neighborhood which is equi-invertible. As a consequence, a compact set of invertible matrices is equi-invertible. A noncompact set of invertible matrices is not necessarily equi-invertible. For instance, the closed set $\Gamma \subseteq L(\mathbb{R}^2, \mathbb{R}^2)$ consists of matrices

$$M_k = \begin{pmatrix} 1 & 0 \\ k & 1/k \end{pmatrix}, k = 1, 2, \ldots$$

that are invertible. However, it is not equi-invertible, for $\|M_k(u)\|$ with $u = (0, 1)$ tends to 0 as $k \to \infty$.

The next lemma gives a sufficient condition for the equi-invertibility of an unbounded set of invertible matrices. We recall that the recession cone of a set A is denoted A_∞.

Lemma 3.1.1 *Let Γ be a closed set of $n \times n$-matrices. If every element of $\Gamma \cup (\Gamma_\infty \setminus \{0\})$ is invertible, then Γ is equi-invertible.*

Proof. Suppose to the contrary that for each k, there is $M_k \in \Gamma$ and $u_k \neq 0$ such that

$$\|M_k(u_k)\| \leq \frac{1}{k}\|u_k\|. \tag{3.2}$$

Without loss of generality we may assume that $\|u_k\| = 1$ and $\lim_{k\to\infty} u_k = u \neq 0$. Let us consider the sequence $\{M_k\}$. If it is bounded, we may assume that it converges to some $M \in \Gamma$. Then (3.2) implies $\|M(u)\| = 0$, which contradicts the hypothesis. If the sequence $\{M_k\}$ is unbounded, we may assume $\lim_{k\to\infty}\|M_k\| = \infty$ and $\lim_{k\to\infty} M_k/\|M_k\| = M_* \in \Gamma_\infty \cap B_{n\times n}$. Again (3.2) implies $\|M_*(u)\| = 0$, and a contradiction is obtained as well. $\square$

We now give a modified version of this lemma that is more suitable when dealing with those families of matrices in which certain components are bounded. Given a set $\Gamma \subseteq L(\mathbb{R}^n, \mathbb{R}^n)$ and $1 \leq m < n$, denote by $\Gamma_1 \subseteq L(\mathbb{R}^n, \mathbb{R}^m)$ and $\Gamma_2 \subseteq L(\mathbb{R}^n, \mathbb{R}^{n-m})$ the collections of matrices such that for every $M_1 \in \Gamma_1$ there is some $M_2 \in \Gamma_2$ such that the matrix $[M_1 M_2]$ belongs to Γ and vice versa. Here $[M_1 M_2]$ stands for the matrix whose first m rows are those of M_1, followed by rows of M_2. In other words, Γ_1 and Γ_2 are the projections of Γ on $L(\mathbb{R}^n, \mathbb{R}^m)$ and $L(\mathbb{R}^n, \mathbb{R}^{n-m})$, respectively.

Lemma 3.1.2 *Let Γ be a closed set of invertible $n \times n$-matrices. If the matrices of the form $[M_1 M_2]$, where $M_1 \in \Gamma_1 \cup ((\Gamma_1)_\infty \setminus \{0\})$, $M_2 \in \Gamma_2 \cup ((\Gamma_2)_\infty \setminus \{0\})$, and at least one of them is a recession matrix, are invertible, then Γ is equi-invertible.*

Proof. As in the proof of Lemma 3.1.1, by supposing the contrary one can find a sequence of matrices $M_k = [M_{1k}\ M_{2k}]$ and vectors $u_k \in \mathbb{R}^n$ with $\|u_k\| = 1$ such that $u_k \to u_0$ and $\|M_k(u_k)\|^2 = \|M_{1k}(u_k)\|^2 + \|M_{2k}(u_k)\|^2 \to 0$ as $k \to \infty$. If $\{M_k\}$ is bounded, then we may assume that it converges to some $M_0 \in \Gamma$ because Γ is closed, and arrive at a contradiction $M_0(u_0) = 0$. If $\{M_k\}$ is not bounded, then at least one of the components $\{M_{1k}\}$ and $\{M_{2k}\}$ is unbounded. Let $\{M_{1k}\}$ be unbounded with $\|M_{1k}\| \to \infty$ as $k \to \infty$. We may assume $\{M_{1k}/\|M_{1k}\|\}$ converges to some $M_1 \in (\Gamma_1)_\infty \setminus \{0\}$. For $\{M_{2k}\}$, we may assume either it is bounded and converges to some $M_2 \in \Gamma_2$ or $\|M_{2k}\| \to \infty$ as $k \to \infty$ and $\{M_{2k}/\|M_{2k}\|\}$ converges to some $M_2 \in (\Gamma_2)_\infty \setminus \{0\}$. In all cases we obtain $M_1(u) = 0$ and $M_2(u) = 0$ with $M_1 \in (\Gamma_1)_\infty \setminus \{0\}$ and $M_2 \in \Gamma_2 \cup ((\Gamma_2)_\infty \setminus \{0\})$. This shows that $[M_1 \ M_2]$ is not invertible which contradicts the hypothesis.   □

Example 3.1.3 Consider the set Γ consisting of matrices M_k, $k = 1, 2, \ldots$ given by

$$M_k = \begin{pmatrix} k & 1/k \\ 0 & k + 1/k \end{pmatrix}.$$

The recession cone Γ_∞ consists of matrices

$$M = \begin{pmatrix} s & 0 \\ 0 & s \end{pmatrix} \quad \text{with } s \geq 0.$$

Then each element of $\Gamma \cup (\Gamma_\infty \backslash \{0\})$ is invertible. In view of Lemma 3.1.1, Γ is equi-invertible.

Example 3.1.4 Consider the set Γ consisting of matrices M_k, given by

$$M_k = \begin{pmatrix} k & 1 \\ 0 & k^2 \end{pmatrix}, \quad k = 1, 2, \ldots.$$

The recession cone Γ_∞ consists of matrices

$$M = \begin{pmatrix} 0 & 0 \\ 0 & \alpha \end{pmatrix} \quad \text{with } \alpha \geq 0.$$

In this case, Lemma 3.1.1 does not apply. Now consider

$$\Gamma_1 = \{(k, 1) : k = 1, 2, \ldots\} \subseteq L(\mathbb{R}^2, R),$$
$$\Gamma_2 = \{(0, k^2) : k = 1, 2, \ldots\} \subseteq L(\mathbb{R}^2, R).$$

We have

$$(\Gamma_1)_\infty = \{(s, 0) : s \geq 0\},$$
$$(\Gamma_2)_\infty = \{(0, \alpha) : \alpha \geq 0\}.$$

Hence the condition of Lemma 3.1.2 is verified, by which Γ is equi-invertible.

Proposition 3.1.5 *Let $F: \mathbb{R}^n \rightrightarrows L(\mathbb{R}^n, \mathbb{R}^n)$ be a set-valued map. Let $x_0 \in \mathbb{R}^n$ be given. If there is $\beta > 0$ such that every element of the set $\overline{\text{co}}(F(x_0 + \beta B_n)) \cup \{[\text{co}(F(x_0 + \beta B_n))]_\infty \backslash \{0\}\}$ is invertible, then the set $\overline{\text{co}}(F(x_0 + \beta B_n))$ is equi-invertible.*

Proof. This follows immediately from Lemma 3.1.1. □

When F is an upper semicontinuous map, the equi-invertibility of $F(x_0 + \beta B_n)$ can be guaranteed by the invertibility of $F(x_0)$ and of its recession matrices.

Proposition 3.1.6 *Suppose that* $F: \mathbb{R}^n \rightrightarrows L(\mathbb{R}^n, \mathbb{R}^n)$ *is upper semicontinuous at* x_0. *If each element of the set* $\overline{\mathrm{co}}(F(x_0)) \cup \mathrm{co}(F(x_0)_\infty \backslash \{0\})$ *is invertible, then there exists* $\beta > 0$ *such that the set* $\mathrm{co}(F(x_0 + \beta B_n))$ *is equi-invertible.*

Proof. Suppose to the contrary that there is no $\beta > 0$ such that the set $\mathrm{co}(F(x_0 + \beta B_n))$ is equi-invertible. For each $i \geq 1$, there is a matrix $M_i \in \mathrm{co}(F(x_0 + (1/i)B_n))$ and a vector u_i with $\|u_i\| = 1$ such that

$$\|M_i(u_i)\| \leq \frac{1}{i}.$$

We may assume $\lim_{i \to \infty} u_i = u \neq 0$. By the Caratheodory theorem, there exist positive numbers λ_{il} with $\sum_{l=1}^{n^2+1} \lambda_{il} = 1$ and matrices $N_{il} \in F(x_0 + (1/i)B_n), l = 1, \ldots, n^2 + 1$, satisfying

$$M_i = \sum_{l=1}^{n^2+1} \lambda_{il} N_{il}.$$

Because ∂f is upper semicontinuous at x_0, we may also assume that

$$N_{il} = M_{il} + \frac{1}{i} P_{il} \quad \text{for some } M_{il} \in \partial f(x_0), P_{il} \in B_{n \times n}.$$

It follows that

$$\lim_{i \to \infty} \sum_{l=1}^{n^2+1} \lambda_{il} M_{il}(u_i) = \lim_{i \to \infty} \left(M_i(u_i) - \frac{1}{i} \sum_{l=1}^{n^2+1} \lambda_{il} P_{il} \right) = 0. \qquad (3.3)$$

Consider the convex combination $\sum_{l=i}^{n^2+1} \lambda_{il} M_{il}$. By taking a subsequence if necessary, we may decompose the index set $\{1, \ldots, n^2 + 1\}$ into three subsets I_1, I_2, I_3 with the following properties

(i)   For $l \in I_1, \lim_{i \to \infty} M_{il} = M_{0l} \in F(x_0)$ and $\lim_{i \to \infty} \lambda_{il} = \lambda_{0l}$.
(ii) For $l \in I_2, \lim_{i \to \infty} \|M_{il}\| = \infty$ and $\lim_{i \to \infty} \lambda_{il} M_{il} = M_{*l} \in (F(x_0))_\infty$.
(iii) For $l \in I_3, \lim_{i \to \infty} \|\lambda_{il} M_{il}\| = \infty$, and $\lim_{i \to \infty} \lambda_{il} M_{il} / \|\lambda_{il_0} M_{il_0}\| = M_{*l} \in (F(x_0))_\infty$, where $l_0 \in I_3$, with $\|\lambda_{il_0} M_{il_0}\| \geq \|\lambda_{il} M_{il}\|$ for $i \geq 1$ and $l \in I_3$.

Let us first consider the case where $I_3 \neq \emptyset$. By dividing the above-mentioned convex combination by $\|\lambda_{il_0} M_{il_0}\|$ and passing to the limit when $i \to \infty$, and by observing that $M_{*l_o} \neq 0$, we deduce

$$\sum_{l \in I_3} M_{*l} \in \mathrm{co}((F(x_0))_\infty \backslash \{0\}).$$

This and (3.3) yield a contradiction

$$\sum_{l \in I_3} M_{*l}(u) = 0.$$

It remains to consider the case $I_3 = \emptyset$. It follows from (ii) that $\lim_{i \to \infty} \lambda_{il} = 0$ for $l \in I_2$ and $\sum_{l \in I_1} \lambda_{ol} = 1$. Consequently,

$$\lim_{i \to \infty} \sum_{l=1}^{n^2+1} \lambda_{il} M_{il} = \sum_{l \in I_1} \lambda_{ol} M_{ol} + \sum_{l \in I_2} M_{*l} \in \overline{\mathrm{co}}(F(x_0)),$$

which together with (3.3) yields a contradiction

$$\left(\sum_{l \in I_1} \lambda_{ol} M_{ol} + \sum_{l \in I_2} M_{*l} \right)(u) = 0.$$

The proof is complete. $\square$

The following modified version of the preceding proposition is more practical when some of the components of F are bounded.

Proposition 3.1.7 *Let* $F = (F_1, F_2)$ *where* $F_i \colon \mathbb{R}^n \rightrightarrows L(\mathbb{R}^n, \mathbb{R}^{n_i})$, $i = 1, 2$, *are set-valued maps, and* $n_1 + n_2 = n$. *Assume that* F_1 *and* F_2 *are upper semicontinuous at* x_0. *If each matrix of the form* $[M_1 \ M_2]$ *where* $M_i \in \overline{\mathrm{co}}(F_i(x_0)) \cup \mathrm{co}((F_i(x_0))_\infty \backslash \{0\})$, $i = 1, 2$, *is invertible, then there exists* $\beta > 0$ *such that the set* $\mathrm{co}(F(x_0 + \beta B_n))$ *is equi-invertible, where* $F(x_0 + \beta B_n)$ *consists of matrices* $[M \ N]$ *with* $M \in F_1(x_0 + \beta B_n)$ *and* $N \in F_2(x_0 + \beta B_n)$.

Proof. Use the same technique as in the proof of Proposition 3.1.6 and Lemma 3.1.2. $\square$

Equi-Surjectivity

Let $C \subset \mathbb{R}^n$ be a nonempty set and let M be an $m \times n$-matrix. We say that M is *surjective on* C *at* $x \in \mathrm{cl}(C)$ if

$$M(x) \in \mathrm{int}(M(C)),$$

or equivalently, there is some $\alpha > 0$ such that

$$\alpha B_m \subseteq M(C - x).$$

Now let $\Gamma \subset L(\mathbb{R}^n, \mathbb{R}^m)$ be a nonempty set. We say that Γ is *equi-surjective on* C *around* $x \in \mathrm{cl}(C)$ if there are positive numbers α and δ such that

$$\alpha B_m \subseteq M(C - x')$$

for every $x' \in \overline{C} \cap (x + \delta B_n)$ and for every $M \in \Gamma$.

We have the following remarks on the above definitions

(i) A particular case of the surjectivity on C is when $C = \mathbb{R}^n$ and $x = 0$. A matrix M is surjective on $\mathbb{R}^n$ at $x = 0$ if $0 \in \mathrm{int}(M(\mathbb{R}^n))$, or equivalently $M(\mathbb{R}^n) = \mathbb{R}^m$. As a consequence, $m \leq n$ and the matrix M has a maximal rank. The converse is also true; that is, if $m \leq n$ and the rank of M equals m, then M is surjective on $\mathbb{R}^n$ at $x = 0$, hence at any $x \in \mathbb{R}^n$ as well. When $C \neq \mathbb{R}^n$ this conclusion is no longer true. For instance, consider $M = (1, 0) \in L(\mathbb{R}^2, \mathbb{R})$. This 1×2-matrix has rank 1, which is maximal. Let $C = \{(0, y) \in \mathbb{R}^2 : y \geq 0\}$. Then $M(C) = \{0\}$ and M is not surjective on C at $x = 0$.

(ii) Another particular case is when $n = m$. If there exists a set $C \in \mathbb{R}^n$ and a point $x \in \mathrm{cl}(C)$ such that M is surjective on C at x, then M is necessarily an invertible matrix. In this situation x must be an interior point of C.

(iii) When C is convex, the dimension of C is the dimension of the smallest affine subspace that contains C. It follows immediately from the definition that if M is surjective on a convex set C at $x \in \mathrm{cl}(C)$, then M has a maximal rank that is equal to $m \leq n$ and the dimension of C is at least m.

It is clear that every element of an equi-surjective set on C around x is surjective on C at x. A set of matrices that are surjective on C at x is not always equi-surjective on C around x except for some particular cases when the set is compact, or more generally, when the set has surjective recession matrices.

Proposition 3.1.8 *Let $C \subseteq \mathbb{R}^n$ be a nonempty convex set with $0 \in \mathrm{cl}(C)$. Let $F: \mathbb{R}^n \rightrightarrows L(\mathbb{R}^n, \mathbb{R}^m)$ be a set-valued map with closed values, that is upper semicontinuous at 0. If every element of the set $\overline{\mathrm{co}}(F(0)) \cup \mathrm{co}((F(0))_\infty \backslash \{0\})$ is surjective on C at 0, then there exists some $\delta > 0$ such that the set*

$$\bigcup_{y \in \delta B_n} \overline{\mathrm{co}} \left[F(y) + (F(y))_\infty^\delta \right]$$

is equi-surjective on C around 0.

Proof. Suppose to the contrary that the conclusion is not true. Thus, for each $k \geq 1$ and $\delta = 1/k$, there exist $x_k \in ((1/k)B_n) \cap \mathrm{cl}(C)$, $v_k \in B_m$, and

$$M_k \in \bigcup_{y \in (1/k)B_n} \overline{\mathrm{co}} \left[F(y) + (F(y))_\infty^\delta \right] \text{ such that}$$

$$v_k \notin k M_k [B_n \cap (C - x_k)]. \tag{3.4}$$

Without loss of generality we may assume that

$$\lim_{k \to \infty} v_k = v_0 \in B_m.$$

We claim that by taking a subsequence if necessary, it can be assumed that either

$$\lim_{k \to \infty} M_k = M_0 \in \overline{\text{co}}F(0) \tag{3.5}$$

or

$$\lim_{k \to \infty} t_k M_k = M_* \in \text{co}\left[(F(0))_\infty \backslash \{0\}\right], \tag{3.6}$$

where $\{t_k\}$ is some sequence of positive numbers converging to 0.

Let us first see that (3.5) or (3.6) leads to a contradiction. If (3.5) holds, then by the surjectivity of M_0 there is some $\varepsilon > 0$ and $k_0 \geq 1$ such that

$$v_0 + \varepsilon B_m \subseteq k_0 M_0[B_n \cap C]. \tag{3.7}$$

Moreover, there is $k_1 \geq k_0$ such that

$$\|M_k - M_0\| < \varepsilon/4 \quad \text{for } k \geq k_1. \tag{3.8}$$

We want to show that there is $k_2 \geq k_1$ such that

$$v_0 + \frac{\varepsilon}{2} B_m \subseteq k_0 M_0[B_n \cap (C - x_k)] \quad \text{for } k \geq k_2. \tag{3.9}$$

Indeed, if this is not the case, then one may assume that for each x_k there is some $b_k \in (\varepsilon/2)B_m$ satisfying

$$v_0 + b_k \notin k_0 M_0[B_n \cap (C - x_k)].$$

The set $B_n \cap (C - x_k)$ is convex, therefore there exists some $\xi_k \in \mathbb{R}^m$ with $\|\xi_k\| = 1$ such that

$$\langle \xi_k, \, v_0 + b_k \rangle \leq \langle \xi_k, k_0 M_0(x) \rangle \quad \text{for all } x \in B_n \cap (C - x_k).$$

Using subsequences if needed, one may again assume that

$$\lim_{k \to \infty} b_k = b_0 \in \tfrac{\varepsilon}{2} B_m,$$
$$\lim_{k \to \infty} \xi_k = \xi_0 \quad \text{with } \|\xi_0\| = 1.$$

It follows then

$$\langle \xi_0, v_0 + b_0 \rangle \leq \langle \xi_0, k_0 M_0(x) \rangle \quad \text{for all } x \in B_n \cap C.$$

This inequality contradicts (3.7) because the point $v_0 + b_0$ is an interior point of the set $v_0 + \varepsilon B_m$. Thus (3.9) holds for some $k_2 \geq k_1$. Now using (3.8) and (3.9) we derive the following inclusions for $k \geq k_2$.

$$v_0 + \frac{\varepsilon}{2} B_m \subseteq k_0 M_0 [B_n \cap (C - x_k)]$$
$$\subseteq k_0 \{ M_k [B_n \cap (C - x_k)] + (M_0 - M_k)[B_n \cap (C - x_k)] \}$$
$$\subseteq k_0 M_k [B_n \cap (C - x_k)] + (\varepsilon/4) B_m. \tag{3.10}$$

This gives us

$$v_0 + \frac{\varepsilon}{4} B_m \subseteq k_0 M_k [B_n \cap (C - x_k)] \quad \text{for } k \geq k_2. \tag{3.11}$$

Now we choose $k \geq k_2$ so large that $v_k \in v_0 + (\varepsilon/4) B_m$. Then (3.11) yields

$$v_k \in k M_k [B_n \cap (C - x_k)], \tag{3.12}$$

which contradicts (3.4).

Now we assume (3.6). Again, because M_* is surjective, relations (3.7) through (3.10) remain true when we replace M_0 by M_* and M_k by $t_k M_k$. Then relation (3.11) becomes

$$v_0 + \frac{\varepsilon}{4} B_m \subseteq k_0 t_k M_k [B_n \cap (C - x_k)] \quad \text{for } k \geq k_2.$$

By choosing $k \geq k_2$ sufficiently large so that $v_k \in v_0 + (\varepsilon/4) B_m$ and $0 < t_k \leq 1$, we arrive at the same contradiction as (3.12).

The proof will be then completed if we show that either (3.5) or (3.6) holds.

Let
$$M_k \in \overline{co} \left[F(y_k) + (F(y_k))_\infty^{1/k} \right] \quad \text{for some } y_k \in \tfrac{1}{k} B_n.$$

Because F is upper semicontinuous at 0, there is $k_0 \geq 1$ such that

$$(F(y_k))_\infty \subseteq (F(0))_\infty \qquad k \geq k_0.$$

We may assume without loss of generality that this inclusion is true for all $k = 1, 2, \ldots$. Thus, for each $k \geq 1$, there exist $M_{kj} \in F(y_k), N_{kj} \in (F(0))_\infty$, P_{kj}, and P_k with

$$\|P_{kj}\| \leq 1, \|P_k\| \leq 1, \quad \text{and} \quad \lambda_{kj} \in [0,1], \ j = 1, \ldots, nm + 1$$

such that $\sum_{j=1}^{mn+1} \lambda_{kj} = 1$ and

$$M_k = \sum_{j=1}^{mn+1} \lambda_{kj} \left(M_{kj} + N_{kj} + \frac{1}{k} \|N_{kj}\| P_{kj} \right) + \frac{1}{k} P_k.$$

If all the sequences $\{\lambda_{kj} M_{kj}\}_{k \geq 1}$, $\{\lambda_{kj} N_{kj}\}_{k \geq 1}$, and $j = 1, \ldots, mn + 1$ are bounded, then so is the sequence $\{M_k\}$. By passing to subsequences if necessary, we may assume

$$\lim_{k \to \infty} M_k = M_0, \quad \lim_{k \to \infty} \lambda_{kj} = \lambda_{0j},$$

$$\lim_{j \to \infty} \lambda_{kj} N_{kj} = N_{0j}, \quad \lim_{k \to \infty} \lambda_{kj} M_{kj} = M_{0j}$$

for $j = 1, \ldots, mn + 1$. Because $(F(0))_\infty$ is a closed cone, we have

$$N_{0j} \in (F(0))_\infty, \quad \sum_{j=1}^{nm+1} N_{0j} \in \mathrm{co}(F(0))_\infty.$$

Moreover, we also have $\sum_{j=1}^{nm+1} \lambda_{0j} = 1$. Decompose the sum $\sum_{j=1}^{nm+1} \lambda_{kj} M_{kj}$ into two sums: the first sum $\sum_1$ consists of those terms with $\{M_{kj}\}_{k \geq 1}$ bounded, and the second sum $\sum_2$ consists of those terms with $\{M_{kj}\}_{k \geq 1}$ unbounded. Then the limits λ_{0j} with j in the second sum are all zero and the corresponding limits M_{0j} are recession directions of $F(0)$. Hence $\sum_1 \lambda_{0j} = 1$ and

$$\lim_{k \to \infty} \sum_1 \lambda_{kj} M_{kj} = \sum_1 M_{0j} \in \mathrm{co}(F(0))$$

by the upper semicontinuity of F at 0, and

$$\lim_{k \to \infty} \sum_2 \lambda_{kj} M_{kj} = \sum_2 M_{0j} \in \mathrm{co}(F(0)_\infty).$$

Thus, $M_0 \in \mathrm{co}(F(0)) + \mathrm{co}(F(0)_\infty) \subseteq \overline{\mathrm{co}}(F(0))$ and (3.5) is fulfilled.

If among the sequences $\{\lambda_{kj} M_{kj}\}_{k \geq 1}$, $\{\lambda_{kj} N_k\}_{k \geq 1}$, $j = 1, \ldots, mn + 1$ there are unbounded ones, then again by taking subsequences instead, we may choose one of them, say $\{\lambda_{kj_0} M_{kj_0}\}_{k \geq 1}$ for some $j_0 \in \{1, \ldots, mn+1\}$, such that $\|\lambda_{kj_0} M_{kj_0}\| = \max_{j=1,\ldots,mn+1} \{\|\lambda_{kj} M_{kj}\|, \|\lambda_{kj} N_{kj}\|\}$. The same argument works when the maximum is attained for some $\{\lambda_{kj_0} N_{kj_0}\}$. Consider the sequence $\{M_k / \|\lambda_{kj_0} M_{kj_0}\|\}_{k \geq 1}$. It is clear that this sequence is bounded, and we may assume it converges to some matrix M_*. We have then $M_* \in \mathrm{co}(F(0))_\infty$. Note that the cone $\mathrm{co}(F(0)_\infty)$ is pointed, otherwise $\mathrm{co}[(F(0))_\infty \backslash \{0\}]$ should contain the zero matrix, which is certainly not surjective and this should contradict the hypothesis. As before, we may assume that each term in the sum of $\{M_k / \|\lambda_{kj_0} M_{kj_0}\|\}$ is convergent. Then M_* is a finite sum of elements from $\mathrm{co}(F(0))_\infty$. At least one of the terms of this sum is nonzero (the term corresponding to the index j_0 has a unit norm), and the cone $\mathrm{co}(F(0))_\infty$ is pointed, thus we deduce that M_* is nonzero, and so (3.6) holds. Hence the proof is complete. $\qquad \square$

Proposition 3.1.9 *Let $C \subseteq \mathbb{R}^{n_1+n_2}$ be a nonempty convex set with $0 \in \overline{C}$. Let $F_i : \mathbb{R}^{n_1+n_2} \rightrightarrows L(\mathbb{R}^{n_i}, \mathbb{R}^m)$, $i = 1, 2$ be closed set-valued maps that are upper semicontinuous at 0. If for each pair of matrices $M \in \overline{\mathrm{co}}(F_1(0)) \cup$*

$\operatorname{co}[(F_1(0))_\infty \backslash \{0\}]$ *and* $N \in \overline{\operatorname{co}}(F_2(0)) \cup \operatorname{co}[(F_2(0))_\infty \backslash \{0\}]$, *the matrix* (MN) *is surjective on* C *at* 0, *then the set*

$$\bigcup_{y \in B_n(0,\delta)} (\overline{\operatorname{co}}[F_1(y) + (F_1(y))_\infty^\delta], \ \overline{\operatorname{co}}[F_2(y) + (F_2(y))_\infty^\delta]),$$

is equi-surjective on C *around* 0.

Proof. We proceed as in the proof of Lemma 3.1.1. Arguing by contradiction, we find

$$x_k \in \left(\frac{1}{k} B_n\right) \cap \overline{C}, \ v_k \in B_m, \ y_k \in \frac{1}{k} B_n$$
$$M_k \in \overline{\operatorname{co}}[F_1(y_k) + (F_1(y_k))_\infty^{\delta/k}], \ N_k \in \overline{\operatorname{co}}[F_2(y_k) + (F_2(y_k))_\infty^{\delta/k}]$$

such that

$$\lim_{k \to \infty} v_k = v_0 \in B_m,$$
$$v_k \notin k(M_k N_k)[B_n \cap (C - x_k)]. \tag{3.13}$$

For $\{M_k\}$ and $\{N_k\}$, we have two possible cases (by using a subsequence if necessary)

$$\lim_{k \to \infty} M_k = M_0 \in \overline{\operatorname{co}}(F_1(0))$$
$$\lim_{k \to \infty} t_k M_\varepsilon = M_* \in \operatorname{co}[(F_1(0))_\infty \backslash \{0\}],$$

where $\{t_k\}$ is some sequence of positive numbers converging to 0, and similar relations for $\{N_k\}$.

Then we have

$$v_0 + \varepsilon B_m \subseteq P[B_n \cap C],$$

where P is one of the four matrices $(M_0 N_0), (M_0 N_*), (M_* N_0)$, and $(M_* N_*)$.

Because P is surjective by hypothesis, for k sufficiently large, one has

$$v_0 + \frac{\varepsilon}{2} B_m \subseteq k_0 P[B_n \cap (C - x_k)]$$

and this implies

$$v_0 + \frac{\varepsilon}{2} B_m \subseteq k_0 P_k[B_n \cap (C - x_k)], \tag{3.14}$$

where P_k is among $(M_k N_k), (M_k(s_k N_k)), ((t_k M_k) N_k)$, and $((t_k M_k)(s_k N_k))$ with $\lim t_k M_k = M_*$ and $\lim s_k N_k = N_*$. Because $0 < t_k \le 1$, (3.14) yields

$$v_0 + \frac{\varepsilon}{2} B_m \subseteq k_0 (M_k N_k)[B_n \cap (C - x_k)],$$

which contradicts (3.13). $\square$

Lemma 3.1.10 *Let C be a convex set with $0 \in \text{cl}(C)$. There exists an increasing sequence of closed convex sets $\{D_k\}$ such that*

$$0 \in D_k \subseteq C \cup \{0\} \quad \text{and} \quad C \subseteq \text{cl}[\cup_{k=1}^{\infty} D_k].$$

Proof. Working in a space of lower dimension if needed, we may assume that C has an interior and contains a ball of radius $\alpha > 0$. Denote

$$C_k = \{x \in C : d(x, \mathbb{R}^n \setminus \text{int}(C)) \geq \alpha/k\} \cap (kB_n).$$

Because $\text{int}(C)$ is convex, the distance function $d(., \mathbb{R}^n \setminus \text{int}(C))$ is a continuous and concave function. Hence C_k is a convex and compact subset of $\text{int}(C)$. Let D_k be the convex hull of C_k and 0. Then D_k is closed and convex with $0 \in D_k \subset C \cup \{0\}$ and $D_k \subseteq D_{k+1}$ for $k = 1, 2, \ldots$ It is clear that if $x \in \text{int}(C)$, then there is some k such that $x \in C_k \subset D_k$. Hence $C \subseteq \text{cl}(\cup_{k=1}^{\infty} D_k)$ as desired. $\qquad\square$

Proposition 3.1.11 *Assume that the hypotheses of Proposition 3.1.8 hold. Then there is a closed convex set D containing 0 with $D \setminus \{0\} \subseteq C$ such that the set*

$$\bigcup_{y \in \delta B_n} \overline{\text{co}}[F(y) + (F(y))_\infty^\delta]$$

is equi-surjective on D around 0.

Proof. Let $\{D_k\}$ be a sequence of closed convex sets that exists by Lemma 3.1.10; that is, $0 \in D_k \subseteq C \cup \{0\}$ and $C \subseteq \text{cl}[\cup_{k=1}^{\infty} D_k]$. We show that for k sufficiently large, every matrix of the set $\overline{\text{co}}(F(0)) \cup \text{co} [(F(0))_\infty \setminus \{0\}]$ is surjective on D_k at 0. Indeed, if this is not the case, then for each $k = 1, 2, \ldots$ there is $M_k \in \overline{\text{co}}(F(0)) \cup \text{co} [(F(0))_\infty \setminus \{0\}]$ such that

$$0 \notin \text{int}(M_k(D_k \cap B_n)).$$

Because $D_k \cap B_n$ is convex, using the separation theorem, we find $\xi_k \in \mathbb{R}^m$ with $\|\xi_k\| = 1$ such that

$$0 \leq \langle \xi_k, M_k(x) \rangle \quad \text{for } x \in D_k \cap B_n. \tag{3.15}$$

Without loss of generality we may assume that

$$\lim_{k \to \infty} \xi_k = \xi_0 \quad \text{with } \|\xi_0\| = 1$$

and either

$$\lim_{k \to \infty} M_k = M_0 \in \overline{\text{co}}(F(0)) \cup \text{co}[(F(0))_\infty \setminus \{0\}]$$

or there is a sequence of positive numbers $\{t_k\}$ such that

$$\lim_{k\to\infty} t_k M_k = M_0 \in \text{co}[(F(0))_\infty \backslash \{0\}].$$

In all cases (3.15) yields

$$0 \leq \langle \xi_0, M_0(x) \rangle \quad \text{for } x \in C \cap B_n.$$

This contradicts the surjectivity of M_0 on C at 0. Thus, for k sufficiently large, Proposition 3.1.8 is applicable to the set $D = D_k$ and produces the desired result. $\square$

When f is a real-valued function, a slightly less restrictive surjectivity condition still produces the equi-surjectivity.

Proposition 3.1.12 *Let f be a continuous map from $\mathbb{R}^n$ to $\mathbb{R}$. Suppose that it admits a pseudo-Jacobian ∂f that is upper semicontinuous at a. If every matrix of the set $\overline{\text{co}}(\partial f(a)) \bigcup ([\text{co}(\partial f(a))]_\infty \backslash \{0\})$ is surjective, then the set*

$$\bigcup \{\overline{\text{co}}(\partial f(x)) : x \in a + \delta B_n\}$$

is equi-surjective on C around 0.

Proof. The proof follows along the same line of arguments as the proof of Proposition 3.1.8. In this case, we may assume $q_k = \sum_{j=1}^{j=nl+1} \lambda_{kj} q_{kj}$, where $q_{kj} \in \partial f(x_k)$ and $\lim_{k\to\infty} x_k = a$. Decompose q_k into two sums: (S1) consists of those terms with $\{q_{kj}\}$ bounded, and (S2) consists of the remaining terms. Without loss of generality we may assume that the bounded sequences $\{q_{kj}\}$ converge to q_{0j} and that for the unbounded sequences, the sequences of norms $\{\|q_{kj}\|\}$ converge to ∞. Because ∂f is upper semicontinuous at a, these limits belong to $\partial f(a)$, and so do the elements q_{kj} of the unbounded sequences whenever k is sufficiently large. Let $p_k = \sum_1 \lambda_{kj} q_{0j} + \sum_2 \lambda_{kj} q_{kj}$. Then $p_k \in \text{co}\partial f(a)$ for large k and $\lim_{k\to\infty}(q_k - p_k) = 0$. Now if $\{p_k\}$ is bounded, then one may assume it converges to q_0. Hence $\{q_k\}$ also converges to q_0 and $q_0 \in \overline{\text{co}}(\partial f(a))$. If $\{p_k\}$ is unbounded, then one may assume $\{p_k/\|p_k\|\}$ converges to p_0, implying that $\{q_k/\|p_k\|\}$ also converges to p_0 and so $p_0 \in (\text{co}(\partial f(a)))_\infty$, $p_0 \neq 0$. The contradiction is then obtained in the same way as in Proposition 3.1.8. $\square$

3.2 Open Mapping Theorems

Throughout this section, if ∂f is a pseudo-Jacobian map of f, then for $\beta \geq 0$ the set $\partial f(x + \beta B_n)$ is denoted $D^\beta f(x)$. Here we state an open mapping theorem for continuous functions.

Theorem 3.2.1 *Let $f : \mathbb{R}^n \to \mathbb{R}^n$ be a continuous function and let ∂f be a pseudo-Jacobian map of f. Let $x_0 \in \mathbb{R}^n$ be given. If there is $\beta > 0$ such that the set $\mathrm{co}(D^\beta f(x_0))$ is equi-invertible, then there is $\delta > 0$ such that*

$$\|f(x_0 + h) - f(x_0)\| \geq \delta \|h\| \quad \text{for all } h \neq 0, \|h\| < \beta, \tag{3.16}$$

and

$$f(x_0) + \frac{\beta\delta}{4}\mathrm{int}(B_n) \subseteq f(x_0 + \frac{\beta}{2}\mathrm{int}(B_n)). \tag{3.17}$$

Proof. Let $\alpha > 0$ be the positive number obtained by the equi-invertibility of the set $\mathrm{co}(D^\beta f(x_0))$. Let $h \neq 0$ with $\|h\| < \beta$. By the mean value theorem, we have

$$\begin{aligned}
f(x_0 + h) - f(x_0) &\in \overline{\mathrm{co}}(\partial f[x_0, x_0 + h](h)) \\
&\subseteq \overline{\mathrm{co}}(D^\beta f(x_0)(h)) \\
&\subseteq (\mathrm{co}(D^\beta f(x_0)))(h) + \frac{\alpha}{2}B_{n\times n}(h).
\end{aligned}$$

There is $M \in \mathrm{co}(D^\beta f(x_0)), N \in B_{n\times n}$ such that

$$f(x_0 + h) - f(x_0) = M(h) + \frac{\alpha}{2}N(h).$$

Hence

$$\begin{aligned}
\|f(x_0 + h) - f(x_0)\| &\geq \|M(h)\| - \frac{\alpha}{2}\|N(h)\| \\
&\geq \alpha\|h\| - \frac{\alpha}{2}\|h\| = \frac{\alpha}{2}\|h\|.
\end{aligned}$$

By taking $\delta = \alpha/2$, we obtain (3.16). To show (3.17), let $y \in f(x_0) + (\beta\delta/4)\mathrm{int}(B_n)$. We have to find $x \in x_0 + (\beta/2)\mathrm{int}(B_n)$ such that $y = f(x)$. To this end, consider the function

$$F(x) := \|f(x) - y\|^2.$$

It is obvious that F is continuous. Hence it attains a minimum on the compact set $x_0 + (\beta/2)B_n$ at some point x. We observe that $x \in x_0 + (\beta/2)\mathrm{int}(B_n)$, because otherwise

$$\begin{aligned}
\frac{\beta\delta}{4} > \|y - f(x_0)\| &\geq \|f(x_0) - f(x)\| - \|y - f(x)\| \\
&\geq \delta\|x_0 - x\| - \|y - f(x_0)\| \\
&\geq \delta\frac{\beta}{2} - \frac{\delta\beta}{4} = \frac{\beta\delta}{4},
\end{aligned}$$

which is impossible. If $f(x) = y$, then we are done. Hence we may assume $f(x) \neq y$. By the optimality condition, Theorem 2.1.13,

$$0 \in \overline{\mathrm{co}}(\partial F(x)),$$

if $\partial F(x)$ is a pseudo-Jacobian of F at x. To find a suitable pseudo-Jacobian of F, we notice that the function $z \mapsto \|y - z\|^2$ is continuously differentiable at $z = f(x)$. By the fuzzy chain rule, Theorem 2.3.2, the closure of the set

$$2\big(f(x) - y + \frac{1}{2}\|f(x) - y\|B_n\big) \circ D^\alpha f(x_0)$$

is a pseudo-Jacobian of F at x. We deduce

$$0 \in \overline{\mathrm{co}}\big((f(x) - y + \frac{1}{2}\|f(x) - y\|B_n) \circ D^\alpha f(x_0)\big).$$

This implies the existence of a vector $v \in f(x) - y + \frac{1}{2}\|f(x) - y\|B_n$ and a matrix $M \in D^\alpha f(x_0)$ such that

$$\|M^{tr}(v)\| \leq \frac{\alpha}{4}.$$

Observe that $\|v\| \geq \frac{1}{2}$, hence the latter inequality yields

$$\|M^{tr}(v)\| \leq \frac{\alpha}{2}\|v\|. \tag{3.18}$$

Let $u \in \mathbb{R}^n$ with $\|u\| = 1$ be such that

$$\langle v, M(u)\rangle = \|v\|\|M(u)\|.$$

Such a vector exists because M is invertible. Then, by the hypothesis one has

$$\langle v, M(u)\rangle = \|v\|\|M(u)\| \geq \alpha\|v\|.$$

On the other hand, (3.18) implies

$$\langle v, M(u)\rangle = \langle M^{tr}(v), u\rangle \leq \|M^{tr}(v)\| \leq \frac{\alpha}{2}\|v\|.$$

The contradiction shows that $f(x) = y$. The proof is complete.   □

Corollary 3.2.2 *Let $f : \mathbb{R}^n \to \mathbb{R}^n$ be a continuous function and let ∂f be a pseudo-Jacobian map of f. Let $x_0 \in \mathbb{R}^n$ be given. If there is $\beta > 0$ such that every element of the set $\overline{\mathrm{co}}(D^\beta f(x_0)) \cup ((\mathrm{co}(D^\beta f(x_0)))_\infty \setminus \{0\})$ is invertible, then the conclusion of Theorem 3.2.1 holds.*

Proof. Apply Theorem 3.2.1 and Lemma 3.1.1. □

Next we present an open mapping theorem in the case of the function admitting an upper semicontinuous pseudo-Jacobian.

Corollary 3.2.3 *Let $f : \mathbb{R}^n \to \mathbb{R}^n$ be a continuous function and let ∂f be a pseudo-Jacobian map of f that is upper semicontinuous at x_0. If the elements of the set $\overline{co}(\partial f(x_0)) \cup co((\partial f(x_0))_\infty \backslash \{0\})$ are invertible, then there exist $\beta > 0$ and $\delta > 0$ such that the relations (3.16) and (3.17) hold.*

Proof. Apply Theorem 3.2.1 and Proposition 3.1.6. □

When the function f admits a bounded pseudo-Jacobian at x_0, the recession part in Corollary 3.2.3 disappears. This is the case where f is locally Lipschitz and the Clarke generalized Jacobian is used as a pseudo-Jacobian.

Corollary 3.2.4 *Let $f : \mathbb{R}^n \to \mathbb{R}^m$ be a locally Lipschitz function. If all elements of the Clarke generalized Jacobian $\partial^C f(x)$ are invertible, then the conclusion of Corollary 3.2.3 holds true.*

Proof. This is obtained from Corollary 3.2.3 and from the fact that $\partial^C f$ is an upper semicontinuous pseudo-Jacobian map of f. □

In the case of unbounded pseudo-Jacobians, recession matrices play an important role and cannot be removed from the conclusion as shown by the next example.

Example 3.2.5 Let $f : \mathbb{R}^2 \to \mathbb{R}^2$ be defined by

$$f(x,y) = (-x + y^{1/3}, \ -x^3 + y).$$

Let us define

$$\partial f(x,y) = \left\{ \begin{pmatrix} -1 & (1/3)y^{-2/3} \\ -3x^2 & 1 \end{pmatrix} \right\} \quad \text{if } (x,y) \neq (0,0),$$

and

$$\partial f(0,0) = \left\{ \begin{pmatrix} -1 & \alpha \\ 0 & 1 \end{pmatrix} : \alpha \geq 1 \right\}.$$

A simple calculation confirms that ∂f is a pseudo-Jacobian map of f which is upper semicontinuous at $(0,0)$. Moreover, every matrix of the set $\overline{co}(\partial f(0,0))$ is invertible. Despite this, the conclusion of the open mapping theorem is not true. For instance, there is no (x, y) near $(0, 0)$ satisfying $f(x, y) = (t, 0)$ with $t > 0$, which means that $f(0, 0) \notin \text{int}(f(B_2))$. We observe that the recession cone of the set $\text{co}(\partial f(0, 0))$ is given by

$$(\partial f(0, 0))_\infty = \left\{ \begin{pmatrix} 0 & \alpha \\ 0 & 0 \end{pmatrix} : \alpha \geq 0 \right\},$$

and the condition on the invertibility of the matrices of the recession cone is violated.

The following result, which is a modification of the previous theorem, provides a useful case where some of the components of f have bounded pseudo-Jacobians.

Corollary 3.2.6 *Let $n = n_1 + n_2$ and let $f = (f_1, f_2) : \mathbb{R}^n \to \mathbb{R}^{n_2} \times \mathbb{R}^{n_2}$ be a continuous function. Assume that f_1 and f_2, respectively, admit pseudo-Jacobians ∂f_1 and ∂f_2 which are upper semicontinuous at x_0, and every matrix (p, q) where $p \in \overline{co}(\partial f_1(x_0)) \cup \text{co}((\partial f_1(x_0))_\infty \setminus \{0\})$ and $q \in \overline{co}(\partial f_2(x_0)) \cup \text{co}((\partial f_2(x_0))_\infty \setminus \{0\})$ is invertible. Then there is $\delta > 0$ and $\varepsilon > 0$ such that*

$$\|f(x_0 + h) - f(x_0)\| \geq \varepsilon \|h\| \quad \text{for all } h \neq 0, \|h\| < \delta$$

and

$$f(x_0) + \frac{\varepsilon \delta}{2} \text{int}(B_m) \subseteq f(x_0 + \delta \text{int}(B_n)).$$

Proof. We follow the same method of proof as in the previous theorem. In the proof of the first part of the conclusion, instead of the matrices q_k we have two submatrices: (p_k, q_k) with $p_k \in \overline{co}(\partial f_1[x_0, x_0 + h_k])$ and $q_k \in \overline{co}(\partial f_2[x_0, x_0 + h_k])$. Then a similar argument leads to the existence of some matrices $p \in \overline{co}(\partial f_1(x_0)) \cup \text{co}((\partial f_1(x_0))_\infty \setminus \{0\})$ and $q \in \overline{co}(\partial f_2(x_0)) \cup \text{co}((\partial f_2(x_0))_\infty \setminus \{0\})$ such that $p(h) = 0$ and $q(h) = 0$, which show that (p, q) is not invertible, a contradiction.

In the reasoning of the second part we have $x \in x_0 + \delta \text{int}(B_n)$ a local minimum of the function f. If $f(x) = y$ with $y = (y_1, y_2)$, then the conclusion follows. If $f(x) \neq y$, then we have several possible cases

Case (1): $f_1(x) \neq y_1$ and $f_2(x) \neq y_2$. By Corollary 2.4.5 and the product rule Theorem 2.1.3, and by the continuous differentiability of the function $\|f(.) - y\|$, the set (denoted by A) of matrices $2((f_1(x) - y_1)p, (f_2(x) - y_2)q)$ with $p \in \overline{co}(\partial f_1(x_0) + (\partial f_1(x_0))_\infty^\alpha)$ and $q \in \overline{co}(\partial f_2(x_0) + (\partial f_2(x_0))_\infty^\alpha)$ is a pseudo-Jacobian of f at x. Hence, in view of the optimality condition Theorem 2.1.13, it must contain zero. This contradicts the assumption by Proposition 3.1.6.

Case (2): $f_1(x) \neq y_1$ and $f_2(x) = y_2$. Then x is a local minimum of the function $\|f_1(\cdot) - y_1\|^2$ and the set $2(f_1(x) - y_1)\overline{\mathrm{co}}(\partial f_1(x) + (\partial f_1(x))_\infty^\alpha)$ is a pseudo-Jacobian of the function $\|f_1(\cdot) - y_1\|^2$. Hence the set A contains zero as well and we arrive at the same contradiction.

Case (3): $f_1(x) = y_1$ and $f_2(x) \neq y_2$. This case is treated in a similar way as Case (2). The proof is complete. $\square$

This corollary as well as Theorem 3.3.1 and other results in which the components of a function are split into subgroups of similar nature opens a remarkable perspective on the way of combining different generalized derivatives in solving practical problems.

3.3 Inverse and Implicit Function Theorems

In this section we apply the open mapping theorems to derive an inverse function theorem and an implicit function theorem for functions with possibly unbounded pseudo-Jacobians.

Let $f : \mathbb{R}^n \to \mathbb{R}^n$ be continuous and let $x_0 \in \mathbb{R}^n$ be given. We say that f admits locally an inverse at x_0 if there exist neighborhoods U of x_0 and V of $f(x_0)$, and a continuous function $g : V \to \mathbb{R}^n$ such that $g(f(x)) = x$ and $f(g(y)) = y$ for every $x \in U$ and $y \in V$.

Theorem 3.3.1 *Let $n = n_1 + n_2$ and let $f = (f_1, f_2) : \mathbb{R}^n \to \mathbb{R}^{n_2} \times \mathbb{R}^{n_2}$ be a continuous map. Assume that f_1 and f_2, respectively, admit pseudo-Jacobian maps ∂f_1 and ∂f_2 which are upper semicontinuous at x_0 and that every matrix (p, q), where $p \in \overline{\mathrm{co}}(\partial f_1(x_0)) \cup \mathrm{co}((\partial f_1(x_0))_\infty \setminus \{0\})$ and $q \in \overline{\mathrm{co}}(\partial f_2(x_0)) \cup \mathrm{co}((\partial f_2(x_0))_\infty \setminus \{0\})$ is invertible. Then f admits locally an inverse that is Lipschitz continuous at $f(x_o)$.*

Proof. Using Corollary 3.2.6, for every $y \in f(x_0) + (\varepsilon\delta/2)\mathrm{int}(B_n)$, we can find $x \in x_0 + \delta\mathrm{int}(B_n)$ such that $y = f(x)$.

Observe that f is locally one-to-one. To see this, suppose to the contrary that f is not one-to-one locally. Then there exist two sequences $\{x_k\}$ and $\{y_k\}$, both converging to x_0 such that $f(x_k) = f(y_k)$. By the mean value theorem (Theorem 2.2.2), one can find $q_k \in \overline{\mathrm{co}}(\partial f[x_k, y_k])$ such that $0 = q_k(x_k - y_k)$. We may now assume that $(x_k - y_k)/\|x_k - y_k\|$ converges to $u \neq 0$. If q_k admits a convergent subsequence with limit q, then $q \in \partial f(x_0)$, and $qu = 0$. This is a contradiction as q is invertible. If not, we may assume $q_k/\|q_k\|$ converges to some $p \in \mathrm{co}((\partial f)_\infty \setminus 0)$ with $pu = 0$ (see Lemma 2.4.1). This again is a contradiction.

Putting $f^{-1}(y) = x$, we observe that

$$\|y - y_0\| \geq \varepsilon \|x - x_0\|,$$

where $y_0 = f(x_0)$. Hence

$$\|f^{-1}(y) - f^{-1}(y_0)\| \leq \frac{1}{\varepsilon}\|y - y_0\|,$$

which means that f is Lipschitz continuous at y_0.   □

Notice that when $n_2 = 0$, by using the Clarke generalized Jacobian in the role of pseudo-Jacobian, we obtain the following inverse function result for the class of locally Lipschitz functions.

Corollary 3.3.2 *Let $f : \mathbb{R}^n \to \mathbb{R}^n$ be locally Lipschitz at $x_0 \in \mathbb{R}^n$. If the matrices of $\partial^C f(x_0)$ are invertible, then f admits locally an inverse at x_0 which is locally Lipschitz at $f(x_0)$.*

Proof. This is immediate from Theorem 3.3.1 and the fact that the Clarke generalized Jacobian is a bounded, upper semicontinuous pseudo-Jacobian map.   □

The following example illustrates the generality of Theorem 3.3.1.

Example 3.3.3 Let $f(x, y) = (g(x) + y^2, \cos(x) + h(y))$ be a map from $\mathbb{R}^2$ to $\mathbb{R}^2$, where g and h are real functions that are differentiable with $\lim_{x \to 0} g'(x) = -\infty$ and $\lim_{y \to 0} h'(y) = \infty$. It can be seen that

$$\partial f_1(x, y) = \begin{cases} \{(g'(x), 2y)\} & \text{if } x \neq 0, \\ \{(\alpha, 2y) : \alpha \leq -1\} & \text{if } x = 0 \end{cases}$$

is a pseudo-Jacobian of $f_1(x, y) := g(x) + y^2$, which is upper semicontinuous at $(0, 0)$. Similarly

$$\partial f_2(x, y) = \begin{cases} \{(-\sin(x), h'(y))\} & \text{if } y \neq 0, \\ \{(-\sin(x), \beta) : \beta \geq 1\} & \text{if } y = 0 \end{cases}$$

is a pseudo-Jacobian of $f_2(x, y) := \cos(x) + h(y)$, which is also upper semicontinuous at $(0, 0)$. The recession cones of $\partial f_1(0, 0)$ and $\partial f_2(0, 0)$, respectively, are $\{(\alpha, 0) : \alpha \leq 0\}$ and $\{(0, \beta) : \beta \geq 0\}$. Hence all the conditions of the inverse function theorem are verified, and f has an inverse in a neighborhood of $(g(0), 1 + h(0))$.

We now apply the inverse function theorem to derive an implicit function theorem.

Theorem 3.3.4 *Let f be a continuous function of two variables $(y, z) \in \mathbb{R}^n \times \mathbb{R}^m$ with $f(y_0, z_0) = 0$. Assume that f admits a pseudo-Jacobian map ∂f which is upper semicontinuous at (y_0, z_0) and the matrices $p \in L(\mathbb{R}^m, \mathbb{R}^m)$ such that there exists $q \in L(\mathbb{R}^n, \mathbb{R}^m)$ with $[qp] \in \overline{co}(\partial f(y_0, z_0)) \cup co[(\partial f(y_0, z_0))_\infty \backslash \{0\}]$ are invertible. Then there exists a Lipschitz continuous function g from a neighborhood U of y_0 in $\mathbb{R}^n$ to $\mathbb{R}^m$ such that*

$$g(y_0) = z_0$$
$$f(y, g(y)) = 0 \quad for \ all \ y \in U.$$

Proof. Let us consider the function F from $\mathbb{R}^n \times \mathbb{R}^m$ to $\mathbb{R}^n \times \mathbb{R}^m$ defined as follows.

$$F(y, z) = (y, f(y, z)) \quad for \ (y, z) \in \mathbb{R}^n \times \mathbb{R}^m.$$

We wish to apply the inverse function theorem for $F = (f_1, f)$, where $f_1(y, z) = y$. We see that $\{(I, 0)\} \subset L(\mathbb{R}^{n+m}, \mathbb{R}^n)$, where I is the $n \times n$ identity matrix, is a bounded pseudo-Jacobian of f_1 which is upper semicontinuous at (y_0, z_0). This and the hypotheses of the theorem show that all conditions of the inverse function theorem are satisfied. So, we obtain an inverse function F^{-1} : for every $(y, 0)$ in a neighborhood of $(y_0, 0)$, one has

$$F^{-1}(y, 0) = (y, z)$$

for some $z \in \mathbb{R}^m$. By putting $g(y) = z$ (the last m components of $F^{-1}(y, 0)$), we see that $g(y)$ is Lipschitz continuous at y_0. Moreover, $f(y, g(y)) = 0$ and $g(y_0) = z_0$. The proof is complete. $\square$

The implicit function theorem for locally Lipschitz functions reads as follows.

Corollary 3.3.5 *Let f be a locally Lipschitz function of two variables $(y, z) \in \mathbb{R}^n \times \mathbb{R}^m$ with $f(y_0, z_0) = 0$. Assume that the matrices $p \in L(\mathbb{R}^m, \mathbb{R}^m)$ such that there exists $q \in L(\mathbb{R}^n, \mathbb{R}^m)$ with $[pq] \in \partial^C f(y_0, z_0)$ are invertible. Then there exists a Lipschitz continuous function g from a neighborhood U of y_0 in $\mathbb{R}^n$ to $\mathbb{R}^m$ such that*

$$g(y_0) = z_0$$
$$f(y, g(y)) = 0 \quad for \ all \ y \in U.$$

Proof. This is immediate from Theorem 3.3.4 and from the upper semicontinuity of the Clarke generalized Jacobian map. $\square$

Now we complete this section with an example which shows that in the inverse function theorem the invertibility condition of the matrices in the recession cones cannot be dropped.

Example 3.3.6 Let $f : \mathbb{R}^2 \to \mathbb{R}^2$ be defined by

$$f(x, y) = (-x + y^{1/3}, -x^3 + y).$$

Then a pseudo-Jacobian is given by

$$\partial f(0, 0) = \left\{ \begin{pmatrix} -1 & \alpha \\ 0 & 1 \end{pmatrix} : \alpha \geq 0 \right\}$$

and its recession cone is given by

$$(\partial f(0, 0))_\infty = \left\{ \begin{pmatrix} 0 & \alpha \\ 0 & 0 \end{pmatrix} : \alpha \geq 0 \right\}.$$

It is easy to see that $\overline{\text{co}}(\partial f(0, 0)) = \partial f(0, 0)$ and that every element of $\partial f(0, 0)$ is invertible. Now let $u = -x + y^{1/3}$ and $v = -x^3 + y$. Then it follows that

$$3ux^2 + 3u^2x + u^3 - v = 0.$$

For $v = 0$ and $u \neq 0$, we get that $x^2 + ux + (u^2/3) = 0$. Because this equation has no solution for x, the function f does not admit an inverse near 0. The condition on the invertibility of the matrices of the recession cones is violated.

3.4 Convex Interior Mapping Theorems

Let us state a special case of the standard minimax that is needed in the sequel.

Lemma 3.4.1 (Minimax theorem) *Let $v_0 \in \mathbb{R}^m$, let $D \subseteq \mathbb{R}^n$ be a nonempty convex compact set, and let $Q \subseteq L(\mathbb{R}^n, \mathbb{R}^m)$ be a nonempty convex set. Then we have*

$$\sup_{M \in Q} \inf_{u \in D} \langle v_0, M(u) \rangle = \inf_{u \in D} \sup_{M \in Q} \langle v_0, M(u) \rangle.$$

Proof. Let us denote by α and β the values of the left-hand side and the right-hand side, respectively, in the equality expressed in the lemma. It is plain that $\alpha \leq \beta$. So, the main chore is to show the inverse inequality. We do it first for the case when Q is bounded. Let us fix a positive ε and consider the function

$$h(M, u) := \langle v_0, M(u) \rangle + \varepsilon \|u\|.$$

We wish to prove that there are $u_\varepsilon \in D$ and $M_\varepsilon \in Q$ such that

$$h(M, u_\varepsilon) - \varepsilon \leq h(M_\varepsilon, u_\varepsilon) \leq h(M_\varepsilon, u) \qquad (3.19)$$

for every $u \in D$ and $M \in Q$. In fact, denote by $g(M) := \inf_{u \in D} h(M, u)$. For each $M \in Q$, there exists a unique element $e(M) \in D$ minimizing $h(M, \cdot)$ on D because h is strictly convex in u. Furthermore, there is some $M_\varepsilon \in Q$ such that

$$g(M_\varepsilon) \geq \sup_{M \in Q} g(M) - \varepsilon.$$

Denote by u_ε the element $e(M_\varepsilon)$ that minimizes $h(M_\varepsilon, \cdot)$ on D. It is clear that u_ε and M_ε satisfy the second inequality of relation (3.19). To prove the first inequality of the said relation, let $M \in Q$ be given. Then for each $\lambda \in (0, 1)$, the element $u_\lambda := e((1-\lambda)M_\varepsilon + \lambda M)$ minimizes $h((1-\lambda)M_\varepsilon + \lambda M, \cdot)$ on D. Because D is compact, one may assume that u_{λ_k} converges to some $\bar{u} \in D$ where $\{\lambda_k\}$ is a sequence of positives converging to 0. Then

$$h((1 - \lambda_k)M_\varepsilon + \lambda_k M, u) \geq h((1 - \lambda)M_\varepsilon + \lambda M, u_{\lambda_k})$$
$$\geq (1 - \lambda_k)h(M_\varepsilon, u_{\lambda_k}) + \lambda_k h(M, u_{\lambda_k}).$$

By the continuity of h, this implies $h(M_\varepsilon, u) \geq h(M_\varepsilon, \bar{u})$, and again, by the strict convexity of h in u one has $\bar{u} = u_\varepsilon$. In this way, for $M \in Q$ one obtains

$$g(M_\varepsilon) \geq g((1 - \lambda)M_\varepsilon + \lambda M) - \varepsilon$$
$$\geq h((1 - \lambda)M_\varepsilon + \lambda M, u_\lambda) - \varepsilon$$
$$\geq (1 - \lambda_k)h(M_\varepsilon, u_\varepsilon) + \lambda_k h(M, u_{\lambda_k}) - \varepsilon,$$

which yields

$$g(M_\varepsilon) = h(M_\varepsilon, u_\varepsilon) \geq L(M, u_\lambda) - \varepsilon.$$

When λ tends to 0, the latter inequality gives the first inequality of (3.19). By letting ε tend to 0 in (3.19), we derive $\alpha \geq \beta$ and hence the requested equality. For the case when Q is unbounded, it suffices to notice that α is the limit of $\alpha_k := \sup_{M \in Q \cap kB_{n \times m}} \inf_{u \in D} \langle v_0, M(u) \rangle$ and β is the limit of $\beta_k := \inf_{u \in D} \sup_{M \in Q \cap kB_{n \times m}} \langle v_0, M(u) \rangle$, and $\alpha_k = \beta_k$ according to the first part of the proof. $\qquad \square$

Theorem 3.4.2 *Let C be a nonempty convex set in $\mathbb{R}^n$ and let $f \colon \mathbb{R}^n \to \mathbb{R}^m$ be a continuous function. Assume that*

(i) $\partial f \colon \mathbb{R}^n \rightrightarrows L(\mathbb{R}^n, \mathbb{R}^m)$ is a pseudo-Jacobian map of f which is upper semicontinuous at $a \in \mathrm{cl}(C)$.

(ii)   Every matrix of the set $\overline{\mathrm{co}}(\partial f(a)) \cup \mathrm{co}[(\partial f(a))_\infty \backslash \{0\}]$ is surjective on C at a.

Then $f(a) \in \text{int}(f(C))$.

Proof. Without loss of generality we may assume that $a = 0$ and $f(a) = 0$. Moreover, by Proposition 3.1.11, we may also assume that C is closed. We obtain the conclusion by establishing the inclusion

$$\frac{\delta}{4k} B_m \subseteq f(\delta B_n \cap C).$$

Suppose the inclusion is false. Then we can find $\bar{y}$ with $\|\bar{y}\| \le \delta/4k$ such that

$$\bar{y} \notin f(\delta B_n \cap C).$$

We define a real function $\varphi : \mathbb{R}^n \to \mathbb{R}$ by

$$\varphi(x) := \|\bar{y} - f(x)\| + \frac{2}{\delta}\|\bar{y}\| \cdot \|x\|.$$

It is clear that φ is continuous. Hence it attains its minimum on the compact set $\delta B_n \cap C$ at some point $\bar{x} \in \delta B_n \cap C$. We claim that

$$\bar{x} \in \text{int}(\delta B_n) \cap C. \tag{3.20}$$

In fact, if $\|\bar{x}\| = \delta$, then

$$\varphi(\bar{x}) = \|\bar{y} - f(\bar{x})\| + 2\|\bar{y}\| > \varphi(0) = \|\bar{y}\|$$

because $\bar{x} \in C \cap \delta B_n$ and $\bar{y} \notin f((\delta B_n) \cap C)$, which is impossible for $\bar{x}$ being a minimum point.

It follows from (3.20) that

$$\text{cone}(C - \bar{x}) = \text{cone}[(B_n \cap C) - \bar{x}].$$

Consequently, if $\partial\varphi(\bar{x})$ is a pseudo-differential of φ at $\bar{x}$, then Theorem 2.1.16 yields

$$\sup_{\xi \in \partial\varphi(\bar{x})} \langle \xi, u \rangle \ge 0 \quad \text{for all } u \in C - \bar{x}. \tag{3.21}$$

Let us now find an appropriate pseudo-differential of φ at $\bar{x}$. To this purpose, note that $\bar{y} \ne f(\bar{x})$, therefore the function $y \to \|\bar{y} - y\|$ is Gâteaux differentiable at $y = f(\bar{x})$ and its derivative at this point equals $(f(\bar{x}) - \bar{y})/\|\bar{y} - f(\bar{x})\|$. Furthermore, for the function $x \to \|x\|$, the closed unit ball B_n is a pseudo-differential at any point. We now apply the sum rule and the chain rule to obtain the following pseudo-differential of φ at $\bar{x}$,

$$\partial\varphi(\bar{x}) := \left\{ \frac{f(\bar{x}) - \bar{y}}{\|\bar{y} - f(\bar{x})\|} M + \frac{2}{\delta}\|\bar{y}\|\xi : M \in Q, \xi \in B_n \right\},$$

where $Q := \text{co}(\partial f(\bar{x}) + (\partial f(\bar{x}))_\infty^\delta)$.

With this pseudo-differential, inequality (3.21) becomes

$$\sup_{M \in Q, \xi \in B_n} \langle \frac{f(\overline{x}) - \overline{y}}{\|\overline{y} - f(\overline{x})\|} M + \frac{2}{\delta} \|\overline{y}\| \xi, u \rangle \geq 0 \quad \text{for } u \in C - \overline{x}.$$

This implies

$$\frac{1}{2k} \geq - \sup_{M \in Q} \langle \frac{f(\overline{x}) - \overline{y}}{\|\overline{y} - f(\overline{x})\|}, M(u) \rangle \quad \text{for } u \in B_n \cap (C - \overline{x}),$$

or equivalently,

$$\frac{1}{2k} \geq \sup_{u \in B_n \cap (C - \overline{x})} \left(- \sup_{M \in Q} \langle \frac{f(\overline{x}) - \overline{y}}{\|\overline{y} - f(\overline{x})\|}, M(u) \rangle \right)$$

$$\geq - \inf_{u \in B_n \cap (C - \overline{x})} \sup_{M \in Q} \langle \frac{\overline{y} - f(\overline{x})}{\|\overline{y} - f(\overline{x})\|}, M(u) \rangle.$$

In virtue of Lemma 3.4.5, the last inequality gives

$$\frac{1}{2k} \geq - \sup_{M \in Q} \inf_{u \in B_n \cap (C - \overline{x})} \langle \frac{f(\overline{x}) - \overline{y}}{\|\overline{y} - f(\overline{x})\|}, M(u) \rangle. \tag{3.22}$$

According to Proposition 3.1.8, for each $M \in Q$ and for k large, we have the inclusion

$$B_m \subseteq kM[B_n \cap (C - \overline{x})].$$

In particular, there is $u \in B_n \cap (C - \overline{x})$ such that $M(u) = \frac{1}{4}((\overline{y} - f(\overline{x}))/\|\overline{y} - f(\overline{x})\|).$

Hence (3.22) implies

$$\frac{1}{2k} \geq \frac{1}{k}$$

which is impossible. This completes the proof.   □

Example 3.4.3 Let $f(x, y) = (g(x) + y^2, \cos(x) + h(y))$ be a map from $\mathbb{R}^2$ to $\mathbb{R}^2$, where g and h are real functions that are differentiable with $\lim_{x \to 0} g'(x) = -\infty$ and $\lim_{y \to 0} h'(y) = \infty$. It can be seen that

$$\partial f_1(x, y) = \begin{cases} \{(g'(x), 2y)\} & \text{if } x \neq 0, \\ \{(\alpha, 2y) : \alpha \leq -1\} & \text{if } x = 0 \end{cases}$$

is a pseudo-Jacobian of $f_1(x, y) := g(x) + y^2$, which is upper semicontinuous at $(0, 0)$. Similarly,

$$\partial f_2(x, y) = \begin{cases} \{(-\sin(x), h'(y))\} & \text{if } y \neq 0, \\ \{(-\sin(x), \beta) : \beta \geq 1\} & \text{if } y = 0 \end{cases}$$

is a pseudo-Jacobian of $f_2(x, y) := \cos(x) + h(y)$, which is also upper semi-continuous at $(0, 0)$. Then ∂f, defined by

$$\partial f(x, y) = (\partial f_1(x, y), \ \partial f_2(x, y)),$$

is a pseudo-Jacobian map of f and is upper semicontinuous at $(0, 0)$, where

$$\partial f(0, 0) = \left\{ \begin{pmatrix} \alpha & 0 \\ 0 & \beta \end{pmatrix} : \alpha \leq -1, \ \beta \geq 1 \right\}$$

and

$$(\partial f(0, 0))_\infty = \left\{ \begin{pmatrix} \alpha & 0 \\ 0 & \beta \end{pmatrix} : \alpha \leq 0, \ \beta \geq 0 \right\}.$$

Then all the conditions of Theorem 3.4.2 are satisfied and its conclusion holds.

When f is a locally Lipschitz function, Theorem 3.4.2 yields Pourciau's convex interior mapping theorem.

Corollary 3.4.4 *Suppose that* $f \colon \mathbb{R}^n \to \mathbb{R}^m$ *is locally Lipschitz and* C *is a convex set in* $\mathbb{R}^n$. *If every matrix of the Clarke generalized Jacobian* $\partial^C f(a)$ *of* f *at* $a \in \mathrm{cl}(C)$ *is surjective on* C *at* a, *then* $f(a) \in \mathrm{int}(f(C))$.

Proof. When f is locally Lipschitz, the Clarke generalized Jacobian map $x \to \partial^C f(x)$ is a pseudo-Jacobian map with bounded convex values that is upper semicontinuous. The corollary is then immediate from Theorem 3.4.2. □

A Convex Interior Mapping Theorem Using Partial Pseudo-Jacobians

For application purposes we derive a convex interior mapping theorem in which partial pseudo-Jacobians are involved.

Lemma 3.4.5 *Let* $F_i \colon \mathbb{R}^n \rightrightarrows \mathbb{R}^{k_i}$, $i = 1, 2$, *be set-valued maps with closed values that are upper semicontinuous at* $a \in \mathbb{R}^n$. *Then for every* $\delta \geq 0$, *the set-valued map* $F^\delta \colon \mathbb{R}^n \rightrightarrows \mathbb{R}^{k_1} \times \mathbb{R}^{k_2}$ *defined by*

$$F^\delta(x) = (F_1(x) + [F_1(x)]_\infty^\delta, \ F_2(x) + [F_2(x)]_\infty^\delta)$$

is upper semicontinuous at a.

Proof. Let $\varepsilon > 0$ be given. By the upper semicontinuity, there is some $\delta > 0$ such that for $i = 1, 2$,

$$F_i(x) \subseteq F_i(a) + \varepsilon B_{k_i},$$

whenever $x \in B_n$. Thus, for each $x \in B_n$,

$$[F_i(x)]_\infty \subseteq [F_i(a)]_\infty.$$

Consequently,

$$F^\delta(x) \subseteq (F_1(a) + [F_1(a)]_\infty^\delta + \varepsilon B_{k_1}, \ F_2(a) + [F_2(a)]_\infty^\delta + \varepsilon B_{k_2})$$
$$\subseteq F^\delta(a) + \varepsilon[B_{k_1} \times B_{k_2}]$$

which shows that F^δ is upper semicontinuous at a.   □

Theorem 3.4.6 *Let $C \subseteq \mathbb{R}^n = \mathbb{R}^{n_1} \times \mathbb{R}^{n_2}$ be a nonempty convex set and let $f \colon \mathbb{R}^{n_1} \times \mathbb{R}^{n_2} \to \mathbb{R}^m$ be a continuous function. Assume that*

(i) $\partial_x f$ and $\partial_y f$ are partial pseudo-Jacobian maps of f with respect to x and y, respectively, and are upper semicontinuous at $a \in \mathrm{cl}(C)$.

(ii) Every matrix (MN) where $M \in \overline{\mathrm{co}}(\partial_x f(a)) \cup \mathrm{co}[(\partial_x f(a))_\infty \backslash \{0\}]$ and $N \in \overline{\mathrm{co}}(\partial_y f(a)) \cup \mathrm{co}[(\partial_y f(a))_\infty \backslash \{0\}]$ is surjective on C at a.

Then $f(a) \in \mathrm{int}(f(C))$.

Proof. We proceed in a similar way as in the proof of Theorem 3.4.2. In view of Proposition 2.2.11, the set

$$Q := (\partial_x f(a) + (\partial_x f(a))_\infty^\delta, \ \partial_y f(a) + (\partial_y f(a))_\infty^\delta)$$

is a pseudo-Jacobian of f at a (formerly $\bar{x}$). Now Proposition 3.1.9 yields

$$B_m \subseteq k(MN)[B_n \cap (C - \bar{a})]$$

for every $(MN) \in Q$. By this the same contradiction is obtained.   □

Another particular case of the convex interior mapping theorem is obtained when C is the whole space.

The Interior Mapping Theorem

Corollary 3.4.7 *Let $f \colon \mathbb{R}^n \to \mathbb{R}^m$ be a continuous function. Assume that f admits a pseudo-Jacobian map ∂f which is upper semi-continuous at a. If every matrix of the set $\overline{\mathrm{co}}(\partial f(a)) \bigcup \mathrm{co}((\partial f(a))_\infty \backslash \{0\})$ is surjective, then for every open set $U \subset \mathbb{R}^n$ containing a, one has $f(a) \in \mathrm{int}(f(U))$.*

Proof. Concretize Theorem 3.4.2 to the case $C = \mathbb{R}^n$. $\square$

The Scalar Interior Mapping Theorem

A stronger form of the interior mapping theorem follows from Proposition 3.1.12 in the case where f is a real-valued function.

Theorem 3.4.8 *Let U be an open subset of $\mathbb{R}^n$ and $a \in U$. Let f be a continuous function from $\mathbb{R}^n$ into $\mathbb{R}$. Assume f admits a pseudo-Jacobian ∂f that is upper semicontinuous at a. If every matrix of the set $\tilde{\partial} f(a) :=$ $\overline{co}(\partial f(a)) \cup ((co(\partial f(a)))_\infty \backslash \{0\})$ is surjective, then*

$$f(a) \in int(f(U)).$$

Proof. The proof is similar to that of Theorem 3.4.2. The only difference is that we use Proposition 3.1.12 instead of Proposition 3.1.11 and by assuming $\overline{y} > f(x)$ for $x \in B_\delta(a)$, we define $\Phi(x)$ from U into $\mathbb{R}$ by

$$\Phi(x) = \overline{y} - f(x) + \frac{2}{\delta} |\overline{y} - f(a)| \, |x - a|.$$

Then one arrives at the formula

$$0 = -A + \frac{2}{\delta} |\overline{y} - f(a)| h,$$

for some $A \in \overline{co}(\partial f(\overline{x}))$ and some $h \in \mathbb{R}^n$ with $\|h\| \leq 1$. We then have

$$0 = -A(x - a) + \frac{2}{\delta} |\overline{y} - f(a)| h(x - a)$$

for any $x \in \mathbb{R}^n$. Then the rest of the proof is essentially the same as that of Theorem 3.4.2. $\square$

It is worth observing that in the proof of Theorem 3.4.8, Proposition 3.1.12 is directly applied without using any chain rule. Moreover, the convex hull of the set $(\partial f(a))_\infty \backslash \{0\}$ contains the set $co[(\partial f(a))_\infty] \backslash \{0\}$. They coincide whenever the convex hull of the recession cone $(\partial f(a))_\infty$ is a pointed cone.

A Convex Interior Mapping Theorem Using Fréchet Pseudo-Jacobians

Let $\Omega : \mathbb{R}^n \rightrightarrows \mathbb{R}^m$ be a set-valued map that is a bounded fan and let $K \subseteq \mathbb{R}^n$ be a closed and convex cone. The Banach constant of Ω with respect to K is given by

$$c(\Omega, K) := -\sup_{\|\xi\|=1, \xi \in R^m} \inf_{x \in K \cap B_n} s(\xi, x),$$

where

$$s(\xi, x) = \sup_{y \in \Omega(x)} \langle \xi, y \rangle$$

is the support function of the set $\Omega(x) \subseteq \mathbb{R}^m$. The next result is known as Ioffe's controllability theorem.

Lemma 3.4.9 *Suppose that $C \subseteq \mathbb{R}^n$ is a nonempty and convex set and $f : \mathbb{R}^n \to \mathbb{R}^m$ is continuous with a prederivative Ω at $x_0 \in \mathrm{cl}(C)$. If the Banach constant of Ω with respect to the tangent cone $T(C, x_0)$ to C at x_0 is strictly positive, then for every $\delta > 0$, one has*

$$f(x_0) \in \mathrm{int}(f(C \cap (x_0 + \delta B_n))).$$

Proof. Without loss of generality we may assume that $x_0 = 0$ and $f(x_0) = 0$. We first prove the lemma for the case when $C = K$. It follows from the definition that there is some positive $c > 0$ such that

$$\sup_{\|\xi\|=1, \xi \in \mathbb{R}^m} \inf_{x \in K \cap B_n} s(\xi, x) < -c.$$

Because Ω is a prederivative of f at $x_0 = 0$, one has

$$f(h) = f(h) - f(0) \in \Omega(h) + r(h)\|h\|B_m,$$

where $r(h) \to 0$ as $h \to 0$. Choose two small positive numbers $\varepsilon < c/2$ and $\lambda < \delta$ so that $|r(h)| < \varepsilon$ whenever $\|h\| < \lambda$. Consider an enlarged fan of Ω defined by

$$\Omega_0(h) = \Omega(h) + \varepsilon\|h\|B_m.$$

It is clear that

$$\inf_{x \in K \cap B_n} s_0(\xi, x) \le -\frac{c}{2} \quad \text{for each } \xi \in \mathbb{R}^m \text{ with } \|\xi\| = 1$$

$$f(h) \in \Omega_0(h) \quad \text{for every } h \in \mathbb{R}^n, \text{ with } \|h\| < \lambda,$$

where $s_0(\xi, x)$ is the support function of the set $\Omega_0(x) \subseteq \mathbb{R}^m$. As $\Omega_0(x)$ is a strictly convex and compact set, the support function $s_0(\xi, x)$ is strictly convex in x. Therefore, for every $\xi \in \mathbb{R}^m$ with $\|\xi\| = 1$, there exists a unique element $\phi(\xi) \in K \cap B_n$ with $\|\phi(\xi)\| = 1$ such that

$$s_0(\xi, \phi(\xi)) = \inf_{x \in K \cap B_n} s_0(\xi, x) \le -\frac{c}{2}.$$

Moreover, the function $\xi \to \phi(\xi)$ is continuous on the unit sphere of $\mathbb{R}^m$ and so it can be extended to all $\mathbb{R}^m$ by

$$\phi(\xi) = \begin{cases} 0 & \text{if } \xi = 0; \\ \|\xi\|\phi(\xi/\|\xi\|) & \text{otherwise.} \end{cases}$$

We consider the function $p \colon \mathbb{R}^m \to \mathbb{R}^m$ defined by

$$p(y) = f(\phi(y)) \quad \text{for } y \in \mathbb{R}^m$$

and show that

$$0 \in \text{int}(p(\lambda B_m)). \tag{3.23}$$

First observe that for each $y \in \mathbb{R}^m$,

$$\langle y, p(y) \rangle = \langle y, f(\phi(y)) \rangle \le s_0(y, \phi(y)) \le -\frac{c}{2}\|y\|^2. \tag{3.24}$$

We wish to find, for each $u \in (\lambda c/2)B_m$, an element $v \in \lambda B_m$ such that $p(v) = u$, which will yield (3.23). To this end, for $u \in R^m$ consider the function $q_u \colon R^m \to R^m$ given by

$$q_u(y) = \begin{cases} y + p(y) - u & \text{if } \|y + p(y) - u\| \le \lambda, \\ \frac{\lambda(y+p(y)-u)}{\|y+p(y)-u\|} & \text{otherwise.} \end{cases}$$

Then q_u is a continuous function from λB_m to itself. According to the Browder fixed point theorem (stating that every continuous function from a nonempty convex and compact set to itself possesses a fixed point), there is an element $v \in \lambda B_m$ such that $q_u(v) = v$.

If $r = \|v + p(v) - u\| \le \lambda$, then by the definition of q_u, we obtain $p(v) = u$ as requested.

If $r > \lambda$, then $\|v\| = \|q_u(v)\| = \lambda$ and

$$u_u(v) = \frac{\lambda}{r}(v + p(v) - u) = v.$$

By multiplying by v, one derives

$$(r - \lambda)\|v\|^2 = \lambda\langle v, p(v) \rangle - \lambda\langle v, u \rangle$$

which together with (3.24) yields

$$\left((r - \lambda) + \frac{\lambda c}{2}\right) \le (r - \lambda)\|v\|^2 - \lambda\langle v, p(v) \rangle$$

$$\le -\lambda\langle v, u \rangle$$

$$\le \lambda\|v\| \cdot \|u\| \le \lambda^2\|u\|.$$

Hence $0 < r - \lambda \le \|u\| - (\lambda c/2)$. This means that whenever $\|u\| < \lambda c/2$ we must have $r \le \lambda$ and consequently $p(v) = u$, establishing (3.23).

Furthermore, because $\phi(\lambda B_m) \subseteq K \cap (\lambda B_n)$, we have

$$p(\lambda B_m) = f(\phi(\lambda B_m)) \subseteq f(K \cap (\lambda B_n)) \subseteq f(K \cap (\delta B_n))$$

and by (3.23), $0 \in \mathrm{int}(f(K \cap (\delta B_n)))$.

To finish the proof we take up the general case in which C is not necessarily identical to K. For each $\varepsilon > 0$, define a convex cone

$$K_\varepsilon := \{y : y + \varepsilon\|y\|B_n \subseteq K\}.$$

It is obvious that K_ε possesses the following properties:

(a)   There is a positive $\delta' < \delta$ such that

$$K_\varepsilon \cap (\delta' B_n) \subseteq C \cap \delta' B_n.$$

(b)   The Hausdorff distance $h(K_\varepsilon \cap B_n,\ K \cap B_n)$ between $K_\varepsilon \cap B_n$ and $K \cap B_n$ tends to 0 as ε tends to 0.

It follows from (b) that $c(\Omega, K_\varepsilon)$ tends to $c(\Omega, K)$ when $\varepsilon \to 0$. Thus, for ε sufficiently small, $c(\Omega, K_\varepsilon) > 0$. In virtue of the first part,

$$0 \in \mathrm{int}(f(K_\varepsilon \cap (\delta' B_n)))$$

which together with (a) produces

$$0 \in \mathrm{int}(f(C \cap (\delta B_n))).$$

The proof is complete.   □

Corollary 3.4.10 *Suppose that $C \subseteq \mathbb{R}^n$ is a nonempty and convex set, and $f : \mathbb{R}^n \to \mathbb{R}^m$ is continuous and admits $\partial f(x)$ as a bounded Fréchet pseudo-Jacobian at $x \in \mathrm{cl}(C)$. If elements of $\partial f(x)$ are surjective on C at x, then for every positive $\delta > 0$ one has*

$$f(x) \in \mathrm{int}(f(C \cap (x + \delta B_n))).$$

Proof. Let Ω be a fan defined by the set $\partial f(x)$. In view of Proposition 1.7.10, this fan is a prederivative of f at x. The equi-surjectivity of $\partial f(x)$ on C at x implies that the Banach constant $c(\Omega, T(C, x)) > 0$. According to Ioffe's controllability theorem, we have $f(x) \in \mathrm{int}(f(C \cap (x + \delta B_n)))$ as requested.   □

Corollary 3.4.11 *Suppose that $C \subseteq \mathbb{R}^n$ is a nonempty and convex set and $f : \mathbb{R}^n \to \mathbb{R}^m$ is locally Lipschitz at $x \in \mathrm{cl}(C)$. If $\partial f(x)$ is a bounded pseudo-Jacobian of f at x such that $\mathrm{co}(\partial f(x))$ is equi-surjective on C at x, then for each $\delta > 0$ one has*

$$f(x) \in \mathrm{int}(f(C \cap (x + \delta B_n))).$$

Proof. Apply the previous corollary and Proposition 1.7.4.   □

3.5 Metric Regularity and Pseudo-Lipschitzian Property

The concepts of openness, metric regularity, and the pseudo-Lipschitzian property (or the Aubin property) are very closely to one another. A major development in the area of set-valued variational analysis in recent years has been the establishment of equivalences among these concepts and their characterizations by means of coderivatives [94, 107] or by slopes [45]. In this section, we see that pseudo-Jacobians provide us with a favorable apparatus to examine metric regularity and the pseudo-Lipschitzian property of a particular class of set-valued maps.

Equi-Surjectivity with Respect to a Set

Let $C \subseteq \mathbb{R}^n$ be a nonempty set, $K \subseteq \mathbb{R}^m$ a nonempty closed set with $0 \in K$ and let M be an $m \times n$-matrix. We say that M is *surjective* on C at $x \in \mathrm{cl}(C)$ with respect to K (or K-*surjective* for short) if

$$M(x) \in \mathrm{int}(M(C) + K). \tag{3.25}$$

Given a nonempty set $\Gamma \subseteq L(\mathbb{R}^n, \mathbb{R}^m)$, it is said to be *equi-surjective on C around $x \in \mathrm{cl}(C)$ with respect to K* (or *equi-K-surjective* for short) if there are positive numbers α and δ such that

$$\alpha B_m \subseteq M(C - x') + K \tag{3.26}$$

for every $x' \in \mathrm{cl}(C) \cap (x + \delta B_n)$ and for every $M \in \Gamma$.

We notice that when $K = \{0\}$ the above definition reduces to the one given in Section 3.1.

Proposition 3.5.1 *If C and K are convex sets, then a matrix M is K-surjective on C at $x_0 \in \mathrm{cl}(C)$ if and only if*

$$0 \in \mathrm{int}(M(T(C, x_0)) + K).$$

Consequently, M is K-surjective on C at $x_0 \in \mathrm{cl}(C)$ if and only if it is K-surjective on $C \cap (x_0 + B_n)$ at x_0.

Proof. When C is convex, one has $C - x_0 \subseteq T(C, x_0)$. Hence (3.25) implies

$$0 \in \mathrm{int}(M(C - x_0) + K) \subseteq \mathrm{int}(M(T(C, x_0)) + K).$$

Conversely, assume $0 \notin \mathrm{int}(M(C - x_0) + K)$. Because the set $M(C - x_0) + K$ is convex, by the separation theorem, one can find some $\xi \in \mathbb{R}^m \setminus \{0\}$ such that

$$0 \leq \langle \xi, M(x - x_0) + y \rangle \quad \text{for every } x \in C \text{ and } y \in K.$$

As $0 \in K$, it follows from the latter inequality that $0 \leq \langle \xi, M(x - x_0) \rangle$ for every $x \in C$ and $0 \leq \langle \xi, y \rangle$ for every $y \in K$. Hence, for every $v \in T(x_0, C)$, one also has

$$0 \leq \langle \xi, M(v) + y \rangle \text{ for every } y \in K.$$

Consequently, $0 \notin \text{int}(M(T(C, x_0)) + K)$. For the last assertion, it suffices to use the fact that $T(C, x_0) = T(C \cap (x_0 + B_n), x_0)$. $\square$

When C is closed and convex and K is not convex, the conclusion of the previous proposition is no longer true. This is seen in the next example.

Example 3.5.2 Let M be the identity 2×2-matrix, $K = \{(0,0), (0,-2)\}$ and

$$C = \{(x_1, x_2) \in \mathbb{R}^2 : (x_1)^2 + (x_2 - 1)^2 \leq 1\}.$$

For $x_0 = (0,0)$, we have

$$M(C - x_0) + K = C \cup \{C + (0, -2)\}$$
$$M(T(C, x_0)) + K = \{(x_1, x_2) \in \mathbb{R}^2 : x_2 \geq -2\}.$$

This shows that $0 \in \text{int}(M(T(C, x_0) + K))$, but $0 \notin \text{int}(M(C - x_0) + K))$.

The next proposition is an extension of Proposition 3.1.8.

Proposition 3.5.3 Let $C \subset \mathbb{R}^n$ be a nonempty convex set with $0 \in \text{cl}(C)$ and $K \subseteq \mathbb{R}^m$ a convex closed set with $0 \in K$. Let $F: \mathbb{R}^n \rightrightarrows L(\mathbb{R}^n, \mathbb{R}^m)$ be a set-valued map with closed values that is upper semicontinuous at 0. If every element of the set $\overline{co}(F(0)) \cup \text{co}((F(0))_\infty \setminus \{0\})$ is K-surjective on C at 0, then there exists some $\delta > 0$ such that the set

$$\bigcup_{y \in \delta B_n} \overline{co} \left[F(y) + (F(y))_\infty^\delta \right]$$

is equi-K-surjective on C around 0.

Proof. We follow the argument used in the proof of Proposition 3.1.8. Suppose to the contrary that the conclusion is not true. Thus, for each $k \geq 1$ and $\delta = 1/k$, there exist $x_k \in ((1/k)B_n) \cap \text{cl}(C)$, $v_k \in B_m$, and $M_k \in \bigcup_{y \in (1/k)B_n} \overline{co} \left[F(y) + (F(y))_\infty^\delta \right]$ such that

$$v_k \notin k(M_k(C - x_k) + K). \tag{3.27}$$

Without loss of generality we may assume, as in the proof of Proposition 3.1.8, that

$$\lim_{k\to\infty} v_k = v_0 \in B_m$$

and either

$$\lim_{k\to\infty} M_k = M_0 \in \overline{\mathrm{co}} F(0) \tag{3.28}$$

or

$$\lim_{k\to\infty} t_k M_k = M_* \in \mathrm{co}\left[(F(0))_\infty\backslash\{0\}\right], \tag{3.29}$$

where $\{t_k\}$ is some sequence of positive numbers converging to 0.

Let us see that (3.28) and (3.29) lead to a contradiction. First assume that (3.28) holds. By hypothesis, M_0 is K-surjective on C at 0, which means that

$$0 \in \mathrm{int}(M_0(C) + K).$$

In view of Proposition 3.5.1, there exist some $\varepsilon > 0$ and $k_0 \geq 1$ such that

$$v_0 + \varepsilon B_m \subseteq k_0(M_0(C \cap B_n) + K). \tag{3.30}$$

For this ε, choose $k_1 \geq k_0$ so that

$$\|M_k - M_0\| < \varepsilon/4 \quad \text{for } k \geq k_1. \tag{3.31}$$

We now show that there is $k_2 \geq k_1$ such that

$$v_0 + \frac{\varepsilon}{2} B_m \subseteq k_0(M_0(B_n \cap (C - x_k)) + K) \quad \text{for } k \geq k_2. \tag{3.32}$$

Indeed, if this is not the case, then one may assume that for each x_k there is some $b_k \in (\varepsilon/2)B_m$ satisfying

$$v_0 + b_k \notin k_0(M_0(B_n \cap (C - x_k)) + K).$$

Because the set on the right-hand side is convex, by the separation theorem, there exists some $\xi_k \in \mathbb{R}^m$ with $\|\xi_k\| = 1$ such that

$$\langle \xi_k, \, v_0 + b_k \rangle \; \leq \; \langle \xi_k, k_0(M_0(x) + y)\rangle \quad \text{for all} \quad x \in B_n \cap (C - x_k), y \in K.$$

Using subsequences if needed, one may again assume that

$$\lim_{k\to\infty} b_k = b_0 \in \frac{\varepsilon}{2} B_m,$$
$$\lim_{k\to\infty} \xi_k = \xi_0 \text{ with } \|\xi_0\| = 1.$$

It follows then

$$\langle \xi_0, v_0 + b_0 \rangle \leq \langle \xi_0, k_0(M_0(x) + y \rangle \quad \text{for all } x \in B_n \cap C, \, y \in K.$$

The point $v_0 + b_0$ being an interior point of the set $v_0 + \varepsilon B_m$, the obtained inequality contradicts (3.30). By this, (3.32) is true. It follows from (3.30), (3.31), and (3.32) for $k \geq k_2$ that

$$v_0 + \frac{\varepsilon}{2} B_m \subseteq k_0 \{ M_0[B_n \cap (C - x_k)] + K \}$$
$$\subseteq k_0 \{ M_k[B_n \cap (C - x_k)] + (M_0 - M_k)[B_n \cap (C - x_k)] + K \}$$
$$\subseteq k_0 \{ M_k[B_n \cap (C - x_k)] + \frac{\varepsilon}{4} B_m + K \}.$$

Because the set $M_k((C - x_k) \cap B_n) + K$ is convex, we deduce from the above inclusion that

$$v_0 + \frac{\varepsilon}{4} B_m \subseteq k_0(M_k(B_n \cap (C - x_k)) + K), \quad \text{for } k \geq k_2.$$

Now we choose $k \geq k_2$ so large that $v_k \in v_0 + (\varepsilon/4) B_m$ and obtain

$$v_k \in k(M_k(B_n \cap (C - x_k)) + K)$$

which contradicts (3.27). The case of (3.29) is proven by the same technique. $\square$

The next result is a generalization of Proposition 3.1.11.

Proposition 3.5.4 *Assume that the hypotheses of Proposition 3.5.3 hold. Then there is a closed convex set D containing 0 with $D\backslash\{0\} \subseteq C$ such that the set*

$$\bigcup_{y \in \delta B_n} \overline{\text{co}}[F(y) + (F(y))_\infty^\delta]$$

is equi-K-surjective on D around 0.

Proof. Let $\{D_k\}$ be an increasing sequence of closed convex sets that exists by Lemma 3.1.10; that is, D_k satisfy

$$0 \in D_k \subseteq C \cup \{0\} \quad \text{and} \quad C \subseteq \text{cl}[\cup_{k=1}^\infty D_k].$$

Our aim is to apply Proposition 3.5.3 to the sets D_k. We show that for k sufficiently large, every matrix of the set $\overline{\text{co}}(F(0)) \cup \text{co}\,[(F(0))_\infty\backslash\{0\}]$ is K-surjective on D_k at 0. Suppose to the contrary that for each $k = 1, 2, \ldots$ there is $M_k \in \overline{\text{co}}(F(0)) \cup \text{co}[(F(0))_\infty\backslash\{0\}]$ such that

$$0 \notin \text{int}(M_k(D_k \cap B_n) + K).$$

Because the set on the right-hand side is convex, by using the separation theorem, we find $\xi_k \in \mathbb{R}^m$ with $\|\xi_k\| = 1$ such that

$$0 \leq \langle \xi_k, M_k(x) + y \rangle \quad \text{for } x \in D_k \cap B_n \text{ and } y \in K. \qquad (3.33)$$

Without loss of generality we may assume that

$$\lim_{k \to \infty} \xi_k = \xi_0 \quad \text{with } \|\xi_0\| = 1$$

and either

$$\lim_{k\to\infty} M_k = M_0 \in \overline{co}(F(0)) \cup co[(F(0))_\infty\backslash\{0\}]$$

or there is a positive sequence $\{t_k\}$ such that

$$\lim_{k\to\infty} t_k M_k = M_0 \in co[(F(0))_\infty\backslash\{0\}].$$

In both cases (3.33) yields

$$0 \le \langle \xi_0, M_0(x) + y \rangle \quad \text{for } x \in C \cap B_n \text{ and } y \in K.$$

This contradicts the K-surjectivity of M_0 on C at 0. Thus, for k sufficiently large, Proposition 3.5.3 is applicable to the set $D = D_k$ and produces the desired result. $\square$

Generalized Inequality Systems

Let $f_0 \colon \mathbb{R}^n \to \mathbb{R}^m$ be a continuous function. Let $C \subset \mathbb{R}^n$ be a nonempty convex set and $K \subset \mathbb{R}^m$ a nonempty closed convex set containing the origin of the space. We consider the following generalized inequality system

$$0 \in f_0(x) + K, \ x \in C. \tag{3.34}$$

Given a parameter set $P \subset \mathbb{R}^r$ and a perturbation function $f : \mathbb{R}^n \times P \to \mathbb{R}^m$ with $f(x, p_0) = f_0(x)$, the parametric inequality system

$$0 \in f(x,p) + K, \ x \in C. \tag{3.35}$$

with $p \in P$ is called a perturbation of system (3.34). For each $p \in P$, the solution set

$$G(p) := \{x \in C : 0 \in f(x,p) + K\}$$

is sometimes called the *implicit set-valued map* defined by system (3.35).

In particular, when $K = \mathbb{R}_+^s \times \{0\}_{m-s}$ with $0 \le s \le m$, that is,

$$K = \{y = (y_1,\ldots,y_m) \in \mathbb{R}^m : y_1 \ge 0,\ldots, y_s \ge 0,\ y_{s+1} = \cdots = y_m = 0\},$$

system (3.34) becomes a system of s inequalities and $m - s$ equalities on the set C,

$$f_{0i}(x) \le 0, i = 1,\ldots,s$$
$$f_{0j}(x) = 0, j = s+1,\ldots,n$$
$$x \in C.$$

Below we present some sufficient conditions that guarantee the stability (the lower semicontinuity) of the implicit set-valued map G. The following variational principle of Ekeland is used.

Lemma 3.5.5 (Ekeland's variational principle) *Suppose that $A \subseteq \mathbb{R}^n$ is a nonempty and closed set, and $h : A \to \mathbb{R}$ is a lower semicontinuous function whose infimum $\inf_A h$ on the set A is finite. Suppose further that $x_0 \in A$ satisfies $h(x) \le \inf_A f + \varepsilon$ for some positive ε. Then for each λ there exists a point $\overline{x} \in A$ such that*

(i) $\|\overline{x} - x_0\| \le \lambda$.
(ii) $h(\overline{x}) \le h(x_0)$.
(iii) $\overline{x}$ is the unique minimizer of the function $x \mapsto h(x) + (\varepsilon/\lambda)\|x - \overline{x}\|$ on A.

Proof. We consider the function

$$g(x) := h(x) + \frac{\varepsilon}{\lambda}\|x - x_0\|$$

for $x \in A$. It is lower semicontinuous and the level set $\{x \in A : g(x) \le g(x_0)\}$ is nonempty (because it contains x_0) and closed. Moreover, as $\inf_A h$ is finite, that set is bounded, hence compact. Therefore, the set of minimizers of g, which is denoted A_0, is nonempty and compact. The function h being lower semicontinuous, admits a minimizer, say $\overline{x}$, on the set A_0. We show that $\overline{x}$ satisfies our requirements. Indeed, for $x \in A_0$ and $x \ne \overline{x}$ one has

$$h(\overline{x}) = h(\overline{x}) + \frac{\varepsilon}{\lambda}\|\overline{x} - \overline{x}\| \le h(x) < h(x) + \frac{\varepsilon}{\lambda}\|x - \overline{x}\|$$

and for $x \in A \setminus A_0$ one has $g(\overline{x}) < g(x)$; that is,

$$h(\overline{x}) + \frac{\varepsilon}{\lambda}\|\overline{x} - x_0\| < h(x) + \frac{\varepsilon}{\lambda}\|x - x_0\|,$$

which implies

$$h(\overline{x}) < h(x) + \frac{\varepsilon}{\lambda}\|x - \overline{x}\|.$$

By this (iii) follows. Setting $x = x_0$ in the above inequalities, we derive

$$h(\overline{x}) + \frac{\varepsilon}{\lambda}\|\overline{x} - x_0\| \le h(x_0) \le \inf_A h + \varepsilon \le h(\overline{x}) + \varepsilon,$$

which yields (i) and (ii).   □

Theorem 3.5.6 *Let $f_0 \colon \mathbb{R}^n \to \mathbb{R}^m$ be a continuous function, $f \colon \mathbb{R}^n \times P \to \mathbb{R}^m$ a perturbation of f_0, and x_0 a solution of system (3.34). Let $\partial_1 f$ be a pseudo-Jacobian map of f with respect to the variable x. Assume that*

(i) Each element of the set $\overline{\mathrm{co}}(\partial_1 f(x_0, p_0)) \cup \mathrm{co}((\partial_1 f(x_0, p_0))_\infty \setminus \{0\})$ is $(f_0(x_0) + K)$-surjective on C at x_0.
(ii) $\partial_1 f$ is upper semicontinuous in a neighborhood of (x_0, p_0).

Then there exist neighborhoods U of p_0 in P and V of x_0 in $\mathbb{R}^n$ such that

$$G(p) \cap V \neq \emptyset \quad \text{for each } p \in U$$

and the set-valued map $p \mapsto G(p) \cap V$ is lower semicontinuous on U.

Proof. Let us construct neighborhoods U of p_0 and V of x_0 such that $G(p) \cap V \neq \emptyset$ for each $p \in U$. By hypothesis we apply Proposition 3.5.1 and Proposition 3.5.3 to find two positives α and δ such that

$$2\alpha B_m \subset M(T(C,x)) + f_0(x_0) + K$$

for each $x \in (x_0 + \delta B_n) \cap C$ and for each matrix

$$M \in \Gamma := \bigcup_{x \in (x_0 + \delta B_n) \cap C,\, p \in (p_0 + \delta B_r) \cap P} \overline{co}(\partial_1 f(x,p) + (\partial_1 f(x,p))^\delta_\infty).$$

Because $f(x,p)$ is continuous, we may assume that $f(x,p) - f_0(x_0) \in \alpha B_m$ for $x \in (x_0 + \delta B_n) \cap C$ and $p \in (p_0 + \delta B_r) \cap P$. Therefore, for these x and p and for $M \in \Gamma$, one still has

$$\alpha B_m \subset M(B_n \cap T(C,x)) + f(x,p) + K. \tag{3.36}$$

Observe that if C is not closed, according to Proposition 3.5.4 we may assume that the latter inclusion remains true not only for C, but for some closed convex subset $C_0 \subseteq C$ containing x_0 too. Denote by

$$d(x,p) = \inf\{\|f(x,p) + v\| : v \in K\},$$

the distance from the origin of the space to the set $f(x,p) + K$. Because f is continuous, it is clear that this distance is a continuous function of (x,p). Moreover, as x_0 is a solution of system (3.34), $d(x_0, p_0) = 0$. Therefore, for the positives α and δ above, there is $\delta_1 \in (0, \delta)$ such that

$$d(x,p) \leq \alpha\delta/4 \text{ for all } x \in (x_0 + \delta_1 B_n) \cap C, p \in (p_0 + \delta_1 B_r) \cap P.$$

We set

$$U := (p_0 + \delta_1 B_r) \cap P$$
$$V := \text{int}(x_0 + \delta B_n)$$

and prove that these are the neighborhoods requested. We may also assume that C is closed, otherwise C_0 is used instead of C in the reasoning that follows. Let $p \in U$ be fixed and consider the function $d(.,p)$ on the set $(x_0 + \delta B_n) \cap C$. Because $d(x,p) \geq 0$ for every x and $d(x_0, p) \leq \alpha\delta/4$, in view of Ekeland's variational principle (Lemma 3.5.5), there exists $\overline{x} \in (x_0 + \delta B_n) \cap C$ such that

$$d(\overline{x}, p) \le d(x_0, p)$$

$$\|\overline{x} - x_0\| \le \delta/2$$

$$d(\overline{x}, p) \le d(x, p) + (\alpha/2)\|x - \overline{x}\| \quad \text{for all } x \in (x_0 + \delta B_n) \cap C. \ (3.37)$$

It follows that $\overline{x} \in \text{int}(x_0 + \delta B_n)$. Now we prove that $d(\overline{x}, p) = 0$ which means that $0 \in f(\overline{x}, p) + K$, and hence $G(p) \cap V \ne \emptyset$. Indeed, assume to the contrary that $d(\overline{x}, p) \ne 0$. Let $\overline{y} \in f(\overline{x}, p) + K$ realize the distance $d(\overline{x}, p)$; that is,

$$\|\overline{y}\| = d(\overline{x}, p) = \inf\{\|f(\overline{x}, p) + y\| : y \in K\}.$$

This $\overline{y}$ exists and is unique because the set $f(\overline{x}, p) + K$ is a closed convex set. It is clear that the unit vector $-\overline{v} := -\overline{y}/\|\overline{y}\|$ belongs to the normal cone to the set $f(\overline{x}, p) + K$ at $\overline{y}$:

$$-\overline{v} \in N(f(\overline{x}, p) + K, \overline{y}).$$

In particular, $\overline{v}$ belongs to the positive polar cone to the set $f(\overline{x}, p) + K$. Furthermore, set $\overline{w} = \overline{y} - f(\overline{x}, p) \in K$. Then

$$d(x, p) \le \|f(x, p) + \overline{w}\| \quad \text{for every } x \in \mathbb{R}^n.$$

Define

$$\varphi(x) = \|f(x, p) + \overline{w}\| + (\alpha/2)\|x - \overline{x}\|$$

for every $x \in \mathbb{R}^n$. It follows from (3.37) that

$$\varphi(\overline{x}) \le \varphi(x) \quad \text{for all } x \in (x_0 + \delta B_n) \cap C.$$

This and the fact that $\overline{x} \in \text{int}(x_0 + \delta B_n)$ imply that $\overline{x}$ is a local minimum point of φ on C. By Theorem 2.1.16, one has

$$\sup_{\xi \in \partial f(\overline{x})} \langle \xi, u \rangle \ge 0 \quad \text{for all } u \in T(C, \overline{x}), \tag{3.38}$$

where $\partial \varphi(\overline{x})$ is any pseudo-differential of φ at $\overline{x}$. Let us compute a pseudo-differential of φ. Because $\overline{y} \ne 0$, the function norm $y \mapsto \|y\|$ is continuously differentiable at $\overline{y}$ and its derivative is $\overline{v}$. By the chain rule stated in Corollary 2.4.5, for every $\varepsilon \in (0, \delta)$, the closure of the set

$$\overline{v} \circ [\partial_1 f(\overline{x}, p) + (\partial_1 f(\overline{x}, p))_\infty^\varepsilon]$$

is a pseudo-differential of the function $x \mapsto \|f(x, p) + \overline{w}\|$ at $\overline{x}$. Moreover, the set $(\alpha/2)B_n$ is also a pseudo-differential of the function $x \mapsto \|x - \overline{x}\|$ at $\overline{x}$. By the sum rule, Theorem 2.1.1, the closure of the set

$$\overline{v} \circ [\partial_1 f(\overline{x}, p) + (\partial_1 f(\overline{x}, p))_\infty^\varepsilon] + (\alpha/2)B_n$$

as well as the set

$$\partial\varphi(\overline{x}) := \mathrm{cl}\{\overline{v} \circ \mathrm{co}[\partial_1 f(\overline{x},p) + (\partial_1 f(\overline{x},p))_\infty^\varepsilon] + (\alpha/2)B_n\} \qquad (3.39)$$

is a pseudo-differential of φ at $\overline{x}$. Denote by

$$Q = \overline{\mathrm{co}}\,(\partial_1 f(\overline{x},p) + (\partial_1 f(\overline{x},p))_\infty^\varepsilon),$$
$$D = T(C,\overline{x}) \cap B_n.$$

We now show that

$$\sup_{M \in Q} \inf_{v \in D} \langle \overline{v}, M(v)\rangle \le -\alpha \qquad (3.40)$$

$$\inf_{v \in D} \sup_{M \in Q} \langle \overline{v}, M(v)\rangle \ge -\alpha/2. \qquad (3.41)$$

If these inequalities are true, then, in view of the minimax theorem (Lemma 3.4.5), we arrive at a contradiction: $-\alpha/2 \le -\alpha$. By this $d(\overline{x},p) = 0$ and $G(p) \cap V \ne \emptyset$. Our aim at the moment is to prove (3.40) and (3.41). Indeed, because $Q \subseteq \Gamma$, for every $M \in Q$, in view of (3.36) there exist $v \in T(C,\overline{x}) \cap B_n$ and $w \in f(\overline{x},p) + K \cap B_m$ such that

$$-\alpha\overline{v} = M(v) + w.$$

Then

$$-1 = -\langle \overline{v}, \overline{v}\rangle = (1/\alpha)\langle \overline{v}, M(v) + w\rangle.$$

Because $\overline{v}$ is positive on the set $f(\overline{x},p)+K$, one has $\langle \overline{v}, w\rangle \ge 0$ and therefore

$$\langle \overline{v}, M(v)\rangle \le -\alpha.$$

This yields

$$\inf_{v \in D} \langle \overline{v}, M(v)\rangle \le -\alpha$$

and (3.40) is obtained.

For relation (3.41), let $v \in D$ be arbitrarily given. It follows from (3.38) and (3.39) that for each $\varepsilon_1 > 0$, one can find $M \in Q$ and $\xi \in B_n$ such that

$$\overline{v} \circ M(v) + (\alpha/2)\langle \xi, v\rangle \ge -\varepsilon_1.$$

Consequently,

$$\langle \overline{v}, M(v)\rangle \ge -(\alpha/2)\langle \xi, v\rangle - \varepsilon_1 \ge -\alpha/2 - \varepsilon_1.$$

Hence

$$\sup_{M \in Q} \langle \overline{v}, M(v)\rangle \ge -\alpha/2 - \varepsilon_1.$$

This being true for every $\varepsilon_1 > 0$, we deduce that

$$\sup_{M \in Q} \langle \overline{v}, M(v) \rangle \geq -\alpha/2$$

which implies (3.41).

To complete the proof it remains to show that the set-valued map $p \mapsto G(p) \cap V$ is lower semicontinuous on U. In fact, let $p \in U$ and $x \in G(p) \cap V$ be given. Let $\varepsilon > 0$. Choose $\tau \in (0, \varepsilon)$ so that $(x + \tau B_n) \cap C \subset V$. Using the same technique as above with (x, p) instead of (x_0, p_0), we can find a neighborhood U' of p in P such that for every $p' \in U'$ there is some $x' \in (x + \tau B_n) \cap C$ satisfying

$$0 \in f(x', p') + K.$$

Thus, $x' \in G(p') \cap (x + \tau B_n) \subset G(p') \cap V$. By this the lower semicontinuity is established.   □

Using the above theorem we can derive an open mapping theorem with respect to a given set.

Corollary 3.5.7 *Let $C \subset \mathbb{R}^n$ be a nonempty convex set and $K \subset \mathbb{R}^m$ be a nonempty closed convex set. Let $f_0 : \mathbb{R}^n \to \mathbb{R}^m$ be continuous and $x_0 \in \mathrm{cl}(C)$. Assume that f_0 admits a pseudo-Jacobian mapping ∂f_0 which is upper semicontinuous on a neighborhood of x_0, and each element of the set $\overline{\mathrm{co}}(\partial f_0(x_0)) \cup \mathrm{co}((\partial f_0(x_0))_\infty \setminus \{0\})$ is $(f_0(x_0) + K)$-surjective on C at x_0. Then*

$$0 \in \mathrm{int}(f_0(C) + K).$$

Proof. Let $P = \mathbb{R}^m$, $p_0 = 0$, and $f(x, p) = f_0(x) - p$ for $x \in \mathbb{R}^n$. It is clear that x_0 is a solution of the generalized inequality system (3.34) and $f(x, p)$ is a perturbation of f_0. It is easy to see that all the hypotheses of Theorem 3.5.6 are satisfied, by which there exist a neighborhood U of $p_0 = 0$ and a neighborhood V of x_0 such that $G(p) := \{x \in C : p \in f(x) + K\} \cap V$ is nonempty for all $p \in U$. This implies that $U \subset f(C \cap V) + K$, and completes the proof.   □

When K reduces to the origin of the space, Corollary 3.5.7 presents a convex interior mapping result (see Theorem 3.4.2).

Metric Regularity

Let us consider the parametric inequality system (3.35) by assuming additionally that C is closed. The implicit set-valued map

$$p \mapsto G(p) = \{x \in C : 0 \in f(x, p) + K\}$$

is said to be *metrically regular* at (x_0, p_0) if there exist a positive μ, a neighborhood U_1 of p_0 in P, and a neighborhood V_1 of x_0 such that

$$\rho(x, G(p)) \leq \mu\rho(0, f(x,p) + K) \quad \text{for every } p \in U_1 \text{ and } x \in V_1 \cap C. \quad (3.42)$$

Here $\rho(\cdot, \cdot)$ denotes the distance. Below we give a sufficient condition for the metric regularity of the map G.

Theorem 3.5.8 *Under the hypotheses of Theorem 3.5.6 the implicit set-valued map G is metrically regular at (x_0, p_0).*

Proof. Let δ, α, U, and V be defined as in the proof of Theorem 3.5.6. Because $0 \in f(x_0, p_0) + K$ and the function

$$d(x, p) := \inf_{y \in f(x,p)+K} \|y\|$$

is continuous, one can find a neighborhood $U_1 \subseteq U$ of p_0 and a neighborhood $V_1 \subseteq (x_0 + (\delta/2)B_n)$ of x_0 such that

$$d(x, p) < \frac{\alpha\delta}{2} \quad \text{for every } x \in V_1 \quad \text{and} \quad p \in U_1.$$

We wish to show that inequality (3.42) is satisfied for $\mu = 1/\alpha$. To this end, let $x \in V_1 \cap C$ and $p \in U_1$ be given. We have

$$\frac{2}{\delta}d(x, p) < \alpha.$$

Pick up two positives $\tau \in ((2/\delta)d(x,p), \alpha)$ and $\tau' \in (\tau, \alpha)$. Then one obtains

$$d(x, p) < \frac{\tau'}{\tau}d(x, p).$$

By applying Ekeland's variational principle to the function $d(\cdot, p)$, one can find $\bar{x} \in (x_0 + \delta B_n) \cap C$ such that

$$\|\bar{x} - x\| \leq d(x, p)/\tau$$
$$d(\bar{x}, p) \leq d(x', p) + \tau\|x' - \bar{x}\| \text{ for each } x' \in (x_0 + \delta B_n) \cap C.$$

We deduce that

$$\|\bar{x} - x_0\| \leq \|\bar{x} - x\| + \|x - x_0\| < d(x, p)/\tau + \delta/2 \leq \delta.$$

Thus $\bar{x} \in \text{int}(x_0 + \delta B_n)$. The same argument as in the proof of Theorem 3.5.6 yields the equality $d(\bar{x}, p) = 0$ or equivalently $0 \in f(\bar{x}, p) + K$. Consequently, $\bar{x} \in G(p)$ and

$$\rho(x, G(p)) \le \|x - \overline{x}\| \le d(x, p)/\tau.$$

By letting τ tend to α in the above inequalities, we derive

$$\rho(x, G(x)) \le \frac{1}{\alpha}\rho(0, f(x, p) + K)$$

for every $p \in U_1$ and $x \in V_1 \cap C$. The proof is complete. $\square$

Pseudo-Lipschitz Property

We still assume that C is closed. The map G is said to be *pseudo-Lipschitz* around (x_0, p_0) with modulus $\ell > 0$ if there exist neighborhoods $\overline{U}$ of p_0 in P and $\overline{V}$ of x_0 such that

$$G(p) \cap \overline{V} \subseteq G(p') + \ell\|p' - p\|B_n \tag{3.43}$$

for any p and $p' \in \overline{U}$.

Theorem 3.5.9 *Assume that in addition to the hypotheses of Theorem 3.5.6 there are a positive constant κ and neighborhoods U_0 of p_0 in P and V_0 of x_0 such that*

$$\|f(x, p') - f(x, p)\| \le \kappa\|p' - p\|$$

for all $p, p' \in U_0$ and $x \in V_0$. Then the implicit set-valued map G is pseudo-Lipschitz around (x_0, p_0).

Proof. Let δ, α, U, and V be defined as in the proof of Theorem 3.5.6. Choose $\theta > 0$ so small that

$$x_0 + \theta\kappa B_n \subseteq V \cap V_0$$
$$(p_0 + \alpha\theta B_r) \cap P \subseteq U \cap U_0.$$

Set $\ell = 2\kappa/\alpha$ and

$$\overline{U} = P \cap \text{int}(p_0 + (\alpha\theta/8)B_r)$$
$$\overline{V} = \text{int}(x_0 + (\theta\kappa/2)B_n).$$

We claim that (3.43) holds true. It suffices to prove that given $p, p' \in \overline{U}$ and $x \in G(p) \cap \overline{V}$, one has

$$\rho(x, G(p')) \le \ell\|p - p'\|. \tag{3.44}$$

Indeed, because $\|p - p'\| < \alpha\theta/4$ we can choose a positive ε verifying

$$\frac{2}{\theta}\|p - p'\| < \varepsilon < \frac{\alpha}{2}. \tag{3.45}$$

Consider the function ϕ on $\mathbb{R}^n$ defined by

$$\phi(x') = d(x', p') + \varepsilon \|x' - x\|.$$

It follows from the hypothesis of the theorem that for $w \in K$ with $d(x, p) = \|f(x, p) + w\| = 0$, one has

$$\begin{aligned}
\phi(x) = d(x, p') &= d(x, p') - d(x, p) \\
&\leq \|f(x, p') + w\| - \|f(x, p) + w\| \\
&\leq \kappa \|p - p'\|.
\end{aligned}$$

In view of (3.45), we deduce

$$\phi(x) \leq \varepsilon \kappa \theta / 2.$$

By applying Ekeland's variational principle to the function ϕ on the set $(x_0 + \theta \kappa B_n) \cap C$, we can find some $\bar{x} \in (x_0 + \theta \kappa B_n) \cap C$ such that

$$\|\bar{x} - x\| \leq \theta \kappa / 2$$
$$\phi(\bar{x}) \leq \phi(x') + \varepsilon \|x' - \bar{x}\|$$

for each $x' \in (x_0 + \theta \kappa B_n) \cap C$. This yields

$$\begin{aligned}
d(\bar{x}, p') + \varepsilon \|\bar{x} - x\| &\leq d(x, p') \qquad\qquad (3.46) \\
d(\bar{x}, p') &\leq d(x', p') + 2\varepsilon \|x' - \bar{x}\|
\end{aligned}$$

for each $x' \in (x_0 + \theta \kappa B_n) \cap C$. Because $x \in \mathrm{int}(x_0 + (\theta \kappa / 2) B_n)$, it follows that $\bar{x}$ is an interior point of the set $x_0 + \theta \kappa B_n$. Moreover, because $0 < 2\varepsilon < \alpha$ the argument of the proof of Theorem 3.5.6 can be applied to show that $d(\bar{x}, p') = 0$, or equivalently, $\bar{x} \in G(p')$. Inequality (3.46) yields

$$\|\bar{x} - x\| \leq d(x, p')/\varepsilon \leq (\kappa/\varepsilon) \|p - p'\|.$$

Consequently,

$$\rho(x, G(p')) \leq (\kappa/\varepsilon) \|p - p'\|.$$

By letting ε tend to $\alpha/2$ in the latter inequality, we deduce (3.44). This completes the proof. □

Corollary 3.5.10 *Let $C \subset \mathbb{R}^n$ be a nonempty convex set and $K \subset \mathbb{R}^m$ be a nonempty closed convex set. Let $f : \mathbb{R}^n \to \mathbb{R}^m$ be continuous and $x_0 \in \bar{C}$. Assume that f admits a pseudo-Jacobian mapping ∂f which is upper semicontinuous on a neighborhood of x_0, and each element of the set $\overline{co}(\partial f(x_0)) \cup co((\partial f(x_0))_\infty \setminus \{0\})$ is $(f(x_0) + K)$-surjective on C at x_0. Then the implicit set-valued map*

$$p \mapsto G(p) := \{x \in C : p \in f(x) + K\}$$

is pseudo-Lipschitz around $(x_0, 0)$, *and there exist a positive* μ, *a neighborhood of* 0 *in* $\mathbb{R}^m$, *and a neighborhood* V *of* x_0 *such that*

$$\rho(x, G(p)) \leq \mu\rho(p, f(x) + K)$$

for all $p \in U$ *and* $x \in V$.

Proof. Consider the system (3.35) with $P = \mathbb{R}^m, p_0 = 0, f_0(x) = f(x)$, and $f(x, p) = f(x) - p$ for $x \in \mathbb{R}^n, p \in \mathbb{R}^m$. Apply Theorem 3.5.8 and Theorem 3.5.9 to this system to obtain the result. $\qquad\square$

Let us now consider a simple example showing that, in general, the metric regularity of implicit set-valued maps does not imply the pseudo-Lipschitz property.

Example 3.5.11 Let $n = m = r = 1$, $C = \mathbb{R}$, $K = \{0\}$, $f(x, p) = x(p + 1) - p^{1/3}$ for all $x, p \in \mathbb{R}$. Let $p_0 = 0$ and $x_0 = 0$. Then the map $p \mapsto G(p)$, where $G(p) = \{x \in C : 0 \in f(x, p) + K\}$, is metrically regular at (p_0, x_0), but it is not pseudo-Lipschitz around this point. It is easily verified that the assumptions of Theorem 3.5.8 are satisfied, whereas the assumptions of Theorem 3.5.9 are not.

Here is another example showing that for implicit set-valued maps the pseudo-Lipschitz property does not imply the metric regularity.

Example 3.5.12 Let $n = m = r = 1$, $C = \mathbb{R}$, $K = \{0\}$, $f(x, p) = x^3 - p^3$, $p_0 = 0$, and $x_0 = 0$. Because $G(p) = \{x \in C : 0 \in f(x, p) + K\} = \{p\}$ for every p, $G(\cdot)$ is pseudo-Lipschitz at (p_0, x_0). However, there does not exist any $\mu > 0$ such that

$$d(x, G(p)) \leq \mu d(0, f(x, p) + K)$$

for all (x, p) in a neighborhood of (x_0, p_0). Indeed, because

$$d(x, G(p)) = |x - p| \quad \text{and} \quad d(0, f(x, p) + K) = |x^3 - p^3|,$$

such a constant μ cannot exist.

We conclude this section with an example in which coderivatives cannot be used to obtain the pseudo-Lipschitz property of a map, whereas a suitably chosen pseudo-Jacobian may help to produce the desired result.

Example 3.5.13 Let $f_0(x) = x^{1/3}$ for every $x \in \mathbb{R}$ and $f(x, p) = (p + 1)x^{1/3} - p$ for every $(x, p) \in \mathbb{R} \times \mathbb{R}$. Let $P = \mathbb{R}$, $C = \mathbb{R}$, $K = \{0\}$, $p_0 = 0$, and $x_0 = 0$. For every $p \in (-1, 1)$, the solution set $G(p)$ of system (3.35) is given by the formula $G(p) = \{p^3/(p+1)^3\}$. It is clear that

$$\partial_1 f(x, p) = \begin{cases} [\alpha, +\infty) & \text{if } x = 0, \\ \{\frac{1}{3}(p+1)x^{-2/3}\} & \text{if } x \neq 0, \end{cases}$$

where $\alpha > 0$ is chosen arbitrarily, is a pseudo-Jacobian map of $f(\cdot, p)$. It can be seen that the hypotheses of Theorem 3.5.6 are satisfied. Hence there exist neighborhoods U of p_0 and V of x_0 such that $G(p) \cap V$ is nonempty for every $p \in U$, and the set-valued map $p \mapsto G(p) \cap V$ is lower semicontinuous on U. By Theorem 3.5.8, $G(\cdot)$ is metrically regular at (p_0, x_0), that is, there exist constant $\mu > 0$ and neighborhoods U_1 of p_0 and V_1 of x_0 such that (3.42) is valid. Because the condition of Theorem 3.5.9 is satisfied for $\kappa = 2$, $U_0 = \mathbb{R}$, and $V_0 = (-1, 1)$, the map $G(\cdot)$ is pseudo-Lipschitz around (p_0, x_0). Notice that the coderivative of the function $f(\cdot, p)$ is empty at $x = 0$, so it tells us nothing about the pseudo-Lipschitzian property of G. However, it should be noted that the coderivative of the inverse set-valued map G^{-1} does yield the pseudo-Lipschitzian property of G. Moreover, it gives a precise estimate for the Lipschitz modulus [94].

4

Nonsmooth Mathematical Programming Problems

In this chapter we present first- and second-order optimality conditions for nonsmooth mathematical programming problems. Conditions that are necessary or sufficient for optimality of various classes of mathematical programming problems are given. They cover composite programming problems as well as multiobjective programming problems.

4.1 First-Order Optimality Conditions

Problems with Equality Constraints

Let U be an open subset of $\mathbb{R}^n$; let $f, h_1, \ldots, h_m : U \to \mathbb{R}$ be real-valued functions. We consider the following mathematical programming problem with m equality constraints,

$$(PE) \qquad \begin{aligned} &\text{minimize} \ \ f(x) \\ &\text{subject to } h_i(x) = 0, \, i = 1, \ldots, m. \end{aligned}$$

The vector function whose components are $h_1, \ldots, h_m$ is denoted h and the feasible solution set, or the constraint set, is denoted C; that is

$$C := \{x \in \mathbb{R}^n : h_i(x) = 0, \quad i = 1, 2, \ldots, m\}.$$

We also use the notation

$$\tilde{\partial} h(x) := \overline{\text{co}}\,(\partial h(x)) \cup \text{co}((\partial h(x))_\infty \backslash \{0\})$$

if $\partial h(x)$ is a pseudo-Jacobian of h at x. The following theorem gives us a necessary condition for local optimal solutions of the problem (PE).

Theorem 4.1.1 *For the problem (PE), assume that f and h are continuous on U. Assume also that $F = (f, h)$ admits a pseudo-Jacobian map ∂F*

which is upper semicontinuous at $\overline{x} \in U$ and that (PE) has a local optimal solution $\overline{x}$. Then there are numbers $\lambda_0 \geq 0, \lambda_1, \ldots, \lambda_m$ not all zero such that

$$0 \in \lambda \circ (\overline{co}(\partial F(\overline{x})) \cup co((\partial F(\overline{x}))_\infty \backslash \{0\})),$$

where $\lambda = (\lambda_0, \ldots, \lambda_m)$.

Proof. We first note that the set $\tilde{\partial} F(\overline{x})$ must contain an element from the space $L(\mathbb{R}^n, \mathbb{R}^{m+1})$ which is not surjective. This is obvious in the case where $n < m + 1$, because $m + 1$ of n-dimensional vectors are linearly dependent. If each $A \in \tilde{\partial} F(\overline{x})$ is surjective, then $f(\overline{x})$ would lie in the interior of $F(U)$ by the interior mapping theorem (Corollary 3.4.7). This would ensure the existence of a positive $\epsilon > 0$ and a point $y \in U$ such that

$$F(y) = (f(\overline{x}) - \epsilon, 0, \ldots, 0),$$

contradicting the optimality of $\overline{x} \in C$. Let $M \in \tilde{\partial} F(\overline{x})$ not be surjective. Then M can be written as $M = (M_0, \ldots, M_m)$, where $M_0, \ldots, M_m$ are linearly dependent. Thus,

$$\lambda_0 M_0 + \cdots + \lambda_m M_m = 0$$

for some nonzero element $(\lambda_0, \ldots, \lambda_m)$ of $\mathbb{R}^{m+1}$. One may choose λ_0 to be nonnegative. $\qquad\square$

The inclusion stated in Theorem 4.1.1 is called a general Lagrange multiplier rule. When F is continuously differentiable, the classical Jacobian matrix $\nabla F(\overline{x})$ can be used as a pseudo-Jacobian of F at $\overline{x}$. The multiplier rule is then written in the form

$$\lambda_0 \nabla f(\overline{x}) + \lambda_1 \nabla h_1(\overline{x}) + \cdots + \lambda_m \nabla h_m(\overline{x}) = 0,$$

and called the Fritz John optimality condition. If λ_0 is strictly positive, by dividing the above equality by λ_0, one obtains a multiplier rule, called the Kuhn–Tucker optimality condition, in which the coefficient corresponding to the objective function f is equal to 1.

Now assume that f and each h_i, $i = 1, \ldots, m$, admit pseudo-Jacobian maps ∂f and ∂h_i which are upper semicontinuous at $\overline{x}$. If $\overline{x}$ is a solution to (PE), then there are numbers $\lambda_0 \geq 0, \lambda_1, \ldots, \lambda_m$ not all zero such that

$$0 \in \lambda \circ G(\overline{x}),$$

where $\lambda = (\lambda_0, \ldots, \lambda_m)$, and the map G is defined by

$$G(x) := \overline{co}\,(\partial f(x)) \times \overline{co}(\partial h_1(x)) \times \cdots \times \overline{co}\,(\partial h_m(x)) \cup$$
$$\cup co(((\partial f(x))_\infty \times (\partial h_1(x))_\infty \times \cdots \times (\partial f_m(x))_\infty) \backslash \{0\}).$$

To see this, define for each $x \in \mathbb{R}^n$,

$$\partial F(x) := \partial f(x) \times \partial h_1(x) \times \cdots \times \partial h_m(x).$$

Then ∂F is a pseudo-Jacobian of F that is upper semicontinuous at $\bar{x}$, and

$$\overline{\mathrm{co}}\,(\partial F(x)) \subseteq \overline{\mathrm{co}}\,(\partial f(x)) \times \overline{\mathrm{co}}\,(\partial h_1(x)) \times \cdots \times \overline{\mathrm{co}}\,(\partial h_m(x)).$$

Moreover,

$$(\partial F(x))_\infty \subseteq (\partial f(x))_\infty \times (\partial h_1(x))_\infty \times \cdots \times (\partial h_m(x))_\infty.$$

Hence

$$\tilde{\partial} F(x) = \overline{\mathrm{co}}\,\partial F(x) \cup \mathrm{co}((\partial F(x))_\infty \backslash \{0\}) \subseteq G(x).$$

It is worth noting that the set $G(x)$, in general, is distinct from the set

$$\overline{\mathrm{co}}(\partial f(x)) \times \cdots \times \overline{\mathrm{co}}(\partial h_m(x)) \cup (\mathrm{co}((\partial f(x))_\infty \backslash \{0\}) \times \cdots \times \mathrm{co}((\partial h_m(x))_\infty \backslash \{0\})).$$

See Example 4.1.4 for details.

Corollary 4.1.2 *For the problem (PE), let $F = (f, h)$ be locally Lipschitz at $\bar{x} \in U$. If $\bar{x}$ is a minimizer of (PE), then there are numbers $\lambda_0 \geq 0, \lambda_1, \ldots, \lambda_m$ not all zero such that*

$$0 \in \partial^C (\lambda \circ F)(\bar{x})$$

where $\lambda = (\lambda_0, \ldots, \lambda_m)$.

Proof. Because $\partial^C F$ is upper semicontinuous at $\bar{x}$ and bounded, the conclusion follows from Theorem 4.1.1 by noting that $\lambda \circ \partial^C F(\bar{x}) = \partial^C (\lambda \circ F)(\bar{x})$. $\qquad \square$

In Section 4.3, we present a Lagrange multiplier rule, which is fairly sharper than the condition in Corollary 4.1.2 for locally Lipschitz problems. A multiplier rule in which the first component λ_0 is zero has very little interest because it does not contain any information on the objective function f. Here is one of regularity conditions, called constraint qualification, which guarantees that $\lambda_0 \neq 0$:

(CQ_1) All matrices formed by the last m rows of elements of the set $\overline{\mathrm{co}}\,(\partial F(x)) \cup \mathrm{co}((\partial F(x))_\infty \backslash \{0\})$ are of maximal rank.

Corollary 4.1.3 *Under the hypothesis of Theorem 4.1.1, if the constraint qualification (CQ_1) holds, then there are numbers $\lambda_1, \ldots, \lambda_m$ such that $0 \in \lambda \circ \tilde{\partial} F(\bar{x})$, where $\lambda = (1, \lambda_1, \ldots, \lambda_m)$.*

Proof. It follows directly from Theorem 4.1.1 that there exist numbers $\lambda_0 \geq 0, \lambda_1, \ldots, \lambda_m$ not all zero such that

$$0 \in (\lambda_0, \ldots, \lambda_m) \circ \tilde{\partial} F(\bar{x}).$$

Let $a_0, a_1, \ldots, a_m$ be the rows of the matrix $M \in \tilde{\partial} F(\bar{x})$ for which $0 = (\lambda_0, \ldots, \lambda_m) \circ M$. If $\lambda_0 = 0$, then $\lambda_1 a_1 + \cdots + \lambda_m a_m = 0$ and the maximal rank condition would be violated. Thus $\lambda_0 \neq 0$ and one may set it equal to 1.   □

We provide a numerical example to illustrate the fact that the recession cone component in the Lagrange multiplier condition cannot, in general, be dropped for optimization problems involving (non-Lipschitz) continuous functions.

Example 4.1.4 Consider the following problem,

$$\text{minimize } x_3 + x_4^2$$
$$\text{subject to } 2x_1^{2/3}\text{sign}(x_1) + x_2^4 - 2x_3 = 0$$
$$2x_1^{1/3} + x_2^2 - \sqrt{2}x_4 = 0.$$

Let $F = (f, h_1, h_2)$ where

$$f(x_1, x_2, x_3, x_4) = x_3 + x_4^2,$$
$$h_1(x_1, x_2, x_3, x_4) = 2x_1^{2/3}\text{sign}(x_1) + x_2^4 - 2x_3,$$
$$h_2(x_1, x_2, x_3, x_4) = 2x_1^{1/3} + x_2^2 - \sqrt{2}x_4.$$

We are interested in the point $x = 0$, at which F evidently is continuous but not Lipschitz. A pseudo-Jacobian of F at 0 and its recession cone are given, respectively, by

$$\partial F(0) = \left\{ \begin{pmatrix} 0 & 0 & 1 & 0 \\ 2\alpha & 0 & -2 & 0 \\ 2\alpha^2 & 0 & 0 & -\sqrt{2} \end{pmatrix} : \alpha \geq 1 \right\},$$

$$(\partial F(0))_\infty = \left\{ \begin{pmatrix} 0 & 0 & 0 & 0 \\ 0 & 0 & 0 & 0 \\ \beta & 0 & 0 & 0 \end{pmatrix} : \beta \geq 0 \right\}.$$

Hence

$$\tilde{\partial} F(0) = \overline{\text{co}} \left\{ \begin{pmatrix} 0 & 0 & 1 & 0 \\ 2\alpha & 0 & -2 & 0 \\ 2\alpha^2 & 0 & 0 & -\sqrt{2} \end{pmatrix} : \alpha \geq 1 \right\} \cup \left\{ \begin{pmatrix} 0 & 0 & 0 & 0 \\ 0 & 0 & 0 & 0 \\ \beta & 0 & 0 & 0 \end{pmatrix} : \beta > 0 \right\}.$$

Clearly, each $M \in \overline{co}(\partial F(0))$ is of maximal rank. So, $(\lambda_0, \lambda_1, \lambda_2) \circ M \neq 0$ for any $(\lambda_0, \lambda_1, \lambda_2) \neq 0$. But for any matrix $N \in (\partial F(0))_\infty$, $(1, 1, 0) \circ N = 0$. Hence the conclusion of Theorem 4.1.1 holds. By this the point $x = 0$ is susceptible to be a local optimal solution of the problem. Direct calculation confirms that it is.

Problems with Mixed Constraints

In this section we study mathematical programming problems with mixed (equality and inequality) constraints. Let $f, g_i, h_j : \mathbb{R}^n \to \mathbb{R}$, $i = 1, \ldots, p, j = 1, \ldots, q$ be real-valued functions. We consider the following problem,

$$(P) \qquad \begin{array}{l} \text{minimize } f(x) \\ \text{subject to } g_i(x) \leq 0, \qquad i = 1, \ldots, p, \\ \qquad\qquad\ \ h_j(x) = 0, \qquad j = 1, \ldots, q. \end{array}$$

We denote by $g = (g_1, \ldots, g_p)$, $h = (h_1, \ldots, h_q)$, and $F = (f, g, h)$. Below is a multiplier rule for the problem (P). The proof of this rule is based on the convex interior mapping theorem (Theorem 3.4.2).

Theorem 4.1.5 *Assume that F is continuous and admits a pseudo-Jacobian map ∂F which is upper semicontinuous at $\overline{x} \in \mathbb{R}^n$. If $\overline{x}$ is a local optimal solution of (P), then there exists a nonzero vector $(\alpha, \beta, \gamma) \in \mathbb{R} \times \mathbb{R}^p \times \mathbb{R}^q$ with $\alpha \geq 0$, $\beta = (\beta_1, \ldots, \beta_p)$ with $\beta_i \geq 0$ such that*

$$\beta_i g_i(\overline{x}) = 0, i = 1, \ldots, p,$$

$$0 \in (\alpha, \beta, \gamma) \circ (\overline{co}(\partial F(\overline{x})) \cup co[(\partial F(\overline{x}))_\infty \backslash \{0\}]).$$

Proof. Let $\varepsilon > 0$ be given so that $f(x) \geq f(\overline{x})$ for every feasible $x \in \overline{x} + \varepsilon B_n$. Without loss of generality we may assume $\overline{x} = 0$ and $F(\overline{x}) = 0$. Let us denote

$$W = \{(t, a, 0) \in \mathbb{R} \times \mathbb{R}^p \times \mathbb{R}^q : t < 0, \ a_i < 0, i = 1, \ldots, p\},$$
$$C = (\varepsilon B_n) \times W \subseteq \mathbb{R}^n \times \mathbb{R}^{1+p+q}.$$

Let us also define a vector function $\phi : \mathbb{R}^n \times \mathbb{R}^{1+p+q} \to \mathbb{R}^{1+p+q}$ by

$$\phi(x, w) = F(x) - w.$$

By denoting by I the identity $(1 + p + q) \times (1 + p + q)$-matrix, we see that

$$(x, w) \mapsto \partial_x \phi(x, w) = \partial F(x)$$
$$(x, w) \mapsto \partial_w \phi(x, w) = \{I\}$$

are partial pseudo-Jacobian maps of ϕ which are upper semicontinuous at $(0, 0)$. Moreover,

$$(\partial_x \phi(x, w))_\infty = (\partial F(x))_\infty, \quad (\partial_w \phi(x, w))_\infty = \{0\}.$$

Furthermore, we observe that

$$\phi(0, 0) \notin \phi((\varepsilon B_n) \times W),$$

otherwise we can find some $x \in \varepsilon B_n$ and $w \in W$ such that

$$0 = \phi(0, 0) = F(x) - w$$

which shows that x is feasible for (P) and $f(x) < f(\overline{x})$ and contradicts the hypothesis. It follows that

$$\phi(0, 0) \notin \text{int}(\phi((\varepsilon B_n) \times W)).$$

In view of the convex interior mapping theorem (Theorem 3.4.2), there exists a matrix from the set

$$(\overline{\text{co}}(\partial F(0)) \cup \text{co}[(\partial F(0))_\infty \backslash \{0\}], -I),$$

of the form $(M, -I)$ such that

$$(M, -I)(0, 0) \notin \text{int}((M, -I)((\varepsilon B_n) \times W)).$$

Because the set on the right-hand side is convex, we apply the separation theorem to find a nonzero vector $(\alpha, \beta, \gamma) \in \mathbb{R}^{1+p+q}$ such that

$$\langle (\alpha, \beta, \gamma), (M, -I)(x, w) \rangle \geq 0 \quad \text{for all } (x, w) \in (\varepsilon B_n) \times W.$$

This is equivalent to

$$\langle (\alpha, \beta, \gamma), M(x) \rangle \geq \langle (\alpha, \beta, \gamma), w \rangle \quad \text{for all } x \in \mathbb{R}^n, \ w \in W.$$

Because the scalar product is continuous, the latter inequality remains true for all $x \in \mathbb{R}^n$ and $w \in \text{cl}(W)$. One deduces $\alpha \geq 0$ when setting $x = 0, w = (t, a, 0)$ with $t = -1, a = 0$, and $\beta_i \geq 0$ when setting $x = 0, t = 0$, and $a_i = -1, a_j = 0$ for $j \neq i$. The condition $\beta_i g_i(\overline{x}) = 0$ is evident because $g_i(\overline{x}) = 0$. Furthermore, with $w = 0$, the above inequality yields

$$\langle (\alpha, \beta, \gamma), M(x) \rangle \geq 0, \text{ for all } x \in \mathbb{R}^n$$

which implies $(\alpha, \beta, \gamma) \circ M = 0$. $\square$

The condition $\beta_i g_i(\overline{x}) = 0, i = 1, \ldots, p$ is called the complementarity condition. It says that if the constraint $g_i(x) \leq 0$ is not active at $\overline{x}$ (i.e., $g_i(\overline{x}) < 0$), then the corresponding multiplier β_i must be zero.

When f, g, and h are locally Lipschitz, Theorem 4.1.5 gives the classical multiplier rule for Lipschitz problems.

Corollary 4.1.6 *Assume that F is locally Lipschitz and $\bar{x}$ is a local optimal solution of (P). Then there exists a nonzero vector $(\alpha, \beta, \gamma) \in \mathbb{R}^{1+p+q}$ with $\alpha \geq 0, \beta_i \geq 0$ such that*

$$\beta_i g_i(\bar{x}) = 0, \ i = 1, \ldots, p,$$

$$0 \in (\alpha, \beta, \gamma) \circ \partial^C F(\bar{x}).$$

Proof. We use the Clarke generalized Jacobian $\partial^C F$ as an upper semicontinuous pseudo-Jacobian of F and apply Theorem 4.1.5 to produce the desired result.   □

A Kuhn–Tucker condition for the problem (P) can be obtained similarly to the problem (PE). To this purpose we introduce a new constraint qualification:

(CQ_2) All matrices formed by the last q rows of elements of the set $\overline{\text{co}}\,(\partial F(x)) \cup \text{co}((\partial F(x))_\infty \setminus \{0\})$ are of maximal rank; and for each element M whose rows are $M_0, M_1, \ldots, M_{p+q}$ of that set, there exists a vector $v \in \mathbb{R}^n$ such that

$$\langle M_i, v \rangle < 0 \quad \text{if } g_i(\bar{x}) = 0, i \in \{1, \ldots, p\},$$

$$\langle M_j, v \rangle = 0 \quad \text{for } j = p+1, \ldots, p+q.$$

Corollary 4.1.7 *Assume that F is continuous and $\bar{x}$ is a local optimal solution of (P). Under the hypothesis of Theorem 4.1.5 and the constraint qualification (CQ_2), there exists a vector $(\beta, \gamma) \in \mathbb{R}^P \times \mathbb{R}^q$, where $\beta = (\beta_1, \ldots, \beta_p)$ with $\beta_i \geq 0$, such that $\beta_i g_i(\bar{x}) = 0, i = 1, \ldots, p,$ and*

$$0 \in (1, \beta, \gamma) \circ \{\overline{\text{co}}(\partial F(\bar{x})) \cup \text{co}[(\partial F(\bar{x}))_\infty \setminus \{0\}]\}.$$

Proof. By Theorem 4.1.5, we can find a nonzero vector $(\alpha, \beta, \gamma) \in \mathbb{R} \times \mathbb{R}^p \times \mathbb{R}^q$ satisfying the conclusion of that theorem. Let M be a $(1+p+q) \times n$ matrix of the set $\tilde{\partial} F(\bar{x})$ such that

$$0 = (\alpha, \beta, \gamma) \circ M.$$

Assume to the contrary that $\alpha = 0$. By multiplying both sides of the above vector equality by the vector v and by taking into account the complementarity condition, we obtain the sum

$$\sum_{i \in \{1, \ldots, p\}, g_i(\bar{x}) = 0} \beta_i \langle M_i, v \rangle + \sum_{j=1}^q \gamma_j \langle M_{p+j}, v \rangle = 0.$$

In view of (ii), we deduce $\beta_i = 0$ for $i = 1, \ldots, p$. The multiplier rule now becomes

$$O = (0, O_p, \gamma) \circ M,$$

where O_p denotes the null vector of $\mathbb{R}^p$. This contradicts the hypothesis (i). Thus $\alpha \neq 0$ and one may set $\alpha = 1$.   □

Locally Lipschitz Programming

We now study a mathematical programming problem of the form:

$$(PL) \qquad \begin{aligned} &\text{minimize} \;\; f(x) \\ &\text{subject to} \;\; g_i(x) \leq 0, \, i = 1, \ldots, p, \\ &\qquad\qquad\quad h_j(x) = 0, \;\; j = 1, \ldots, q, \\ &\qquad\qquad\quad x \in Q, \end{aligned}$$

where $f, g_i, h_j : \mathbb{R}^n \to \mathbb{R}$, $i = 1, \ldots, p, j = 1, \ldots, q$ are (not necessarily differentiable) locally Lipschitz functions and Q is a closed convex subset of $\mathbb{R}^n$. For this case, a multiplier rule can be established without upper semicontinuity of the pseudo-Jacobian map.

Theorem 4.1.8 *Assume that $F = (f, g, h)$ is locally Lipschitz and that it admits a bounded pseudo-Jacobian $\partial F(\bar{x})$ at $\bar{x}$. If $\bar{x}$ is a local minimizer of (PL), then there exist Lagrange multipliers $\lambda_0 \geq 0, \ldots, \lambda_p \geq 0, \lambda_{p+1}, \ldots, \lambda_{p+q}$, not all zero, such that*

$$\lambda_i g_i(\bar{x}) = 0, \qquad i = 1, \ldots, m$$

$$0 \in \lambda \circ \mathrm{co}(\partial F(\bar{x})) + N(Q, \bar{x}),$$

where $\lambda = (\lambda_0, \ldots, \lambda_m)$.

Proof. Assume for simplicity that $f(\bar{x}) = 0$ and $g(\bar{x}) = 0$. We denote

$$Z := \mathbb{R}^n \times \mathbb{R}^{p+1}$$
$$S := Q \times \mathbb{R}_+^{p+1} = \{z = (x, a) \in Z : x \in Q, a_i \geq 0, i = 0, \ldots, p\}.$$

Clearly, S is a closed convex set and the tangent cone to S at $\bar{z} = (\bar{x}, 0)$ is given by

$$T(S, \bar{z}) = T(Q, \bar{x}) \times \mathbb{R}_+^{p+1},$$

where $T(Q, \bar{x})$ is the tangent cone to Q at $\bar{x}$ and $\mathbb{R}_+^{p+1}$ is the nonnegative octant of $\mathbb{R}^{p+1}$. Let $Y = \mathbb{R}^{p+q+1}$ and let $G: Z \to Y$ be a map defined as follows.

$$(G(x, a))_i = \begin{cases} f(x) + a_0 & i = 0, \\ g_i(x) + a_i & i = 1, \ldots, p, \\ h_{i-p}(x) & i = p+1, \ldots, p+q. \end{cases}$$

Then G is locally Lipschitz and the set

$$\partial G(z) = \{(M, I) : M \in \partial F(x)\}$$

is a bounded pseudo-Jacobian of G at z, where $I \in L(\mathbb{R}^{p+1}, \mathbb{R}^{p+q+1})$ is defined by

$$I = [e_1, \ldots, e_{p+1}],$$

with $e_i = (0, \ldots, 0, 1, 0, \ldots, 0)^{tr}$.

Because $\bar{x}$ is a minimizer of (PE), $G(\bar{z}) = (f(\bar{x}), g(\bar{x}), h(\bar{x}))$ cannot be in the interior of $G(S \cap (\bar{z} + \lambda B_Z))$ for any $\lambda > 0$. Otherwise, there would exist some point $y \in S \cap (\bar{z} + \lambda_0 B_Z)$ for some $\lambda_0 > 0$ such that

$$\begin{aligned}
f(y) &< f(\bar{x}) \\
g_i(y) &= g_i(\bar{x}), \quad i = 1, \ldots, p, \\
h_j(y) &= h_j(\bar{x}), \quad j = 1, \ldots, q,
\end{aligned}$$

which implies that y is a feasible point and hence contradicts the hypothesis that $\bar{x}$ is a minimizer. In view of Corollary 3.4.11, the set $\mathrm{co}(\partial G(\bar{z}))$ is not equi-surjective on S at $\bar{z}$. Hence there exists an element $M \in \mathrm{co}(\partial F(\bar{x}))$ such that the matrix (M, I) is not surjective on S at $\bar{z}$; that is, $0 \notin \mathrm{int}((M, I)(S - \bar{z}))$. The separation theorem gives us the existence of a nonzero vector $\lambda = (\lambda_0, \ldots, \lambda_{p+q}) \in \mathbb{R}^{p+q+1}$ such that

$$\langle \lambda, (M, I)(x - \bar{x}, a) \rangle \geq 0$$

for every $(x, a) \in S$. By setting $x = \bar{x}$, we deduce that $\lambda_i \geq 0$ for $i = 0, \ldots, p$. By setting $a = 0$, we have

$$\langle \lambda, M(x - \bar{x}) \rangle \geq 0$$

for every $x \in Q$. Hence $\lambda \circ M \in N(Q, \bar{x})$, and so the conclusion follows. $\square$

Corollary 4.1.9 *Let $\bar{x}$ be a local optimal solution to (PL). Assume that the functions f, g, and h are locally Lipschitz and admit bounded pseudo-differentials $\partial f(\bar{x}), \partial g_i(\bar{x})$, and $\partial h_j(\bar{x})$ at $\bar{x}$. Then there exist Lagrange multipliers $\lambda_0 \geq 0, \ldots, \lambda_p \geq 0, \lambda_{p+1}, \ldots, \lambda_{p+q}$, not all zero, such that*

$$\lambda_i g_i(\bar{x}) = 0, \qquad i = 1, \ldots, p$$

$$0 \in \lambda_0 \mathrm{co}(\partial f(\bar{x})) + \sum_{i=1}^{p} \lambda_i \mathrm{co}(\partial g_i(\bar{x})) + \sum_{j=1}^{q} \lambda_{j+p} \mathrm{co}(\partial h_j(\bar{x})) + N(Q, \bar{x}).$$

Proof. Because $\partial F(x) = \partial f_0(x) \times \cdots \times \partial f_m(x)$ is a bounded pseudo-Jacobian of F at x, the conclusion follows from Theorem 4.1.8. $\square$

The standard form of the Lagrange multiplier rule for the Michel-Penot subdifferentials follows easily from Corollary 4.1.9.

Corollary 4.1.10 *If $\bar{x}$ is a solution to (PL), then there exist multipliers $\lambda_0 \geq 0, \ldots, \lambda_p \geq 0, \lambda_{p+1}, \ldots, \lambda_{p+q}$, not all zero, such that*

$$\lambda_i g_i(\bar{x}) = 0, \qquad i = 1, \ldots, p$$

$$0 \in \lambda_0 \partial^{MP} f(\bar{x}) + \sum_{i=1}^{p} \lambda_i \partial^{MP} g_i(\bar{x}) + \sum_{j=1}^{q} \lambda_{i+p} \partial^{MP} h_j(\bar{x}) + N(Q, \bar{x}).$$

Proof. Choose the Michel-Penot subdifferential ∂^{MP} as a pseudo-differential and apply Corollary 4.1.9. $\qquad\square$

A version of the Lagrange multiplier rule for the Clarke subdifferential follows from Theorem 4.1.8.

Corollary 4.1.11 *For the problem (PL), let $F = (f, g, h)$. If $\bar{x}$ is a solution to (PL), then there exist multipliers $\lambda_0 \geq 0, \ldots, \lambda_p \geq 0, \lambda_{p+1}, \ldots, \lambda_{p+q}$, not all zero, such that*

$$\lambda_i g_i(\bar{x}) = 0, \qquad i = 1, \ldots, p$$

$$0 \in \lambda \circ \partial^C F(\bar{x}) + N(Q, \bar{x}),$$

where $\lambda = (\lambda_0, \ldots, \lambda_m)$.

Proof. Let $\partial F(x) = \partial^C F(x)$. Then the conclusion follows directly from Theorem 4.1.8. $\qquad\square$

The following example illustrates that the multiplier rule of Theorem 4.1.8 is sharper than the one given in Corollary 4.1.10.

Example 4.1.12 Consider the problem

$$\text{minimize} \quad (x_1 + 1)^2 + x_2^2$$
$$\text{subject to} \ \ 2x_1 + |x_1| - |x_2| = 0.$$

Clearly, $(0, 0)$ is the minimum point of the above problem. Let f_0 denote the objective function $(x_1 + 1)^2 + x_2^2$ and let f_1 denote the constraint function $2x_1 + |x_1| - |x_2|$. Then f_0 is continuously differentiable, and therefore we can take its gradient at $(0, 0)$ as a pseudo-differential at this point. Thus,

$$\overline{co}(\partial f_0(0, 0)) = \partial^{MP} f_0(0, 0) = \partial^C f_0(0, 0) = \{(2, 1)\}.$$

The constraint function f_1 is not differentiable at $(0, 0)$, but locally Lipschitz at this point. It is clear that its Michel-Penot subdifferential coincides with the Clarke subdifferential and is given by

$$\partial^{MP} f_1(0,0) = \partial^C f_1(0,0) = \overline{co}\{(3,-1);(1,1);(1,-1);(3,1)\}.$$

It is easy to see that the set

$$\partial f_1(0,0) = \{(3,-1);(1,1)\}$$

is a pseudo-differential of f_1 at $(0,0)$. Moreover, for $\lambda_0 = 1$ and $\lambda_1 = -1$, one has

$$(0,0) \in \lambda_0 \overline{co}(\partial f_0(0,0)) + \lambda_1 \overline{co}(\partial f_1(0,0)).$$

The set in the right hand side of the latter inclusion is strictly contained in the Michel-Penot subdifferential of the function $\lambda_0 f_0 + \lambda_1 f_1$ at $(0,0)$, which is given by

$$\partial^{MP}(\lambda_0 f_1 + \lambda_1 f_1)(0,0) = \overline{co}\{(1,-1);(-1,1);(-1,-1);(1,1)\}.$$

A Kuhn–Tucker-type necessary optimality condition can be obtained under a constraint qualification. For instance, if we choose $\partial f(\overline{x}) = \partial^{MP} f(\overline{x})$ and $\partial F_1(\overline{x}) = \partial^{MP} g_1(\overline{x}) \times \cdots \times \partial^{MP} h_q(\overline{x})$, then a constraint qualification for (PL) can be stated as

(i)   For every element M of the set $(\partial^{MP} h_1(\overline{x})^{tr}, \ldots, \partial^{MP} h_q(\overline{x})^{tr})$ the system

$$M^{tr}(u) \in N(Q, \overline{x}), \quad u \in \mathbb{R}^q$$

has only one solution $u = 0$.

(ii)   There exists a vector v from the tangent cone $T(Q, \overline{x})$ such that

$$\begin{aligned} \langle \partial^{MP} g_i(\overline{x})^{tr}, v \rangle &< 0, &&\text{if } g_i(\overline{x}) = 0, i \in \{1, \ldots, p\} \\ \langle \partial^{MP} h_i(\overline{x})^{tr}, v \rangle &= 0, &&i = p+1, \ldots, m. \end{aligned}$$

We notice that when $\overline{x}$ is an interior point of Q, the normal cone $N(Q, \overline{x})$ collapses to $\{0\}$, and the first condition of the above constraint qualification is given in a familiar form: the matrices of the set

$$(\partial^{MP} h_1(\overline{x})^{tr}, \ldots, \partial^{MP} h_q(\overline{x})^{tr})$$

have maximal rank.

Corollary 4.1.13 *If $\overline{x} \in \mathbb{R}^n$ is a solution to (PL) and the above constraint qualification for problem (PL) is satisfied at $\overline{x}$, then there exist multipliers $\lambda_1, \ldots, \lambda_m$ such that*

$$\lambda_i g_i(\overline{x}) = 0, \ i = 1, \ldots, p$$

$$0 \in \partial^{MP} f(\overline{x}) + \sum_{i=1}^{p} \lambda_i \partial^{MP} g_i(\overline{x}) + \sum_{j=1}^{q} \lambda_{j+p} \partial^{MP} h_j(\overline{x}) + N(Q, \overline{x}).$$

Proof. By applying Theorem 4.1.8 and using the Michel–Penot subdifferential, we can find numbers $\lambda_0, \ldots, \lambda_{p+q}$ with $\lambda_i \geq 0, i = 0, \ldots, p$ such that

$$0 \in \lambda_0 \partial^{MP} f(\overline{x}) + \sum_{i=1}^{p} \lambda_i \partial^{MP} g_i(\overline{x}) + \sum_{j=1}^{q} \lambda_{j+p} \partial^{MP} h_j(\overline{x}) + N(Q, \overline{x}).$$

Notice that in the second term on the right-hand side the multipliers λ_i corresponding to $g_i(\overline{x}) \neq 0$ are all zero because of the complementarity condition. If $\lambda_0 = 0$, then multiplying both sides of the above inclusion by the vector $v \in T(Q, \overline{x})$ and using (ii) of the constraint qualification, we conclude that the multipliers λ_i corresponding to $g_i(\overline{x}) = 0$ are equal to zero. Then the above inclusion becomes

$$0 \in \sum_{j=1}^{q} \lambda_{j+p} \partial^{MP} h_j(\overline{x}) + N(Q, \overline{x}).$$

But this contradicts the hypothesis (i) of the constraint qualification. $\quad\square$

Example 4.1.14 Consider the following minimax problem,

$$(CP) \qquad \min_{x \in \mathbb{R}^n} \max_{1 \leq k \leq s} f_0^k(x)$$
$$\text{subject to} \quad f_i(x) \leq 0, \ i = 1, \ldots, p,$$
$$f_i(x) = 0, \ i = p+1, \ldots, m,$$
$$x \in Q,$$

where $f_0^1, \ldots, f_0^s, f_1, \ldots, f_m : \mathbb{R}^n \to \mathbb{R}$ are locally Lipschitz functions and Q is a closed convex subset of $\mathbb{R}^n$ containing $\overline{x}$. The function f_0, defined by

$$f_0(x) = \max\{f_0^k : k = 1, \ldots, s\},$$

is easily seen to be Lipschitz near $\overline{x}$. For any x, $I(x)$ denotes the set of indices j for which $f_0^j(x) = f_0(x)$.

In the following we deduce the optimality conditions for the above minimax problem.

Corollary 4.1.15 *Assume that* $f_0^1, \ldots, f_0^s, f_1, \ldots, f_m$ *are locally Lipschitz. Suppose that* $F_1 = (f_1, \ldots, f_m)$ *admits a bounded pseudo-Jacobian* $\partial F_1(\overline{x})$ *at* $\overline{x}$. *If* $\overline{x} \in \mathbb{R}^n$ *is a solution of (CP), then there exist multipliers* $\lambda_0 \geq 0, \ldots, \lambda_p \geq 0, \lambda_{p+1}, \ldots, \lambda_m$ *not all zero such that*

$$\lambda_i f_i(\overline{x}) = 0, \qquad i = 1, \ldots, m$$

$$0 \in \lambda_0 \mathrm{co}\Big(\bigcup_{j \in I(\overline{x})} \partial f_0^j(\overline{x}) \Big) + \lambda \circ \mathrm{co}(\partial F_1(\overline{x})) + N(Q, \overline{x}).$$

Proof. By Corollary 4.1.9 there exist multipliers $\lambda_0 \geq 0, \ldots, \lambda_p \geq 0, \lambda_{p+1}, \ldots, \lambda_m$, not all zero, such that

$$\lambda_i f_i(\overline{x}) = 0, \qquad i = 1, \ldots, m$$

$$0 \in \lambda_0 \mathrm{co}(\partial f_0(\overline{x})) + \lambda \circ \mathrm{co}(\partial F_1(\overline{x})) + N(Q, \overline{x}).$$

The direct calculation of $\partial f_0(\overline{x})$ shows that $\partial f_0(\overline{x}) := \bigcup_{j \in I(\overline{x})} \partial f_0^j(\overline{x})$ is a pseudo-differential of f_0 at $\overline{x}$ (see also Theorem 2.1.9). Indeed, for each $h \in \mathbb{R}^n$,

$$f_0^+(\overline{x}; h) = \max_{j \in I(\overline{x})} (f_0^j)^+(\overline{x}; h) \leq \max_{j \in I(\overline{x})} \max_{\xi^j \in \partial f_0^j(\overline{x})} \langle \xi^j, h \rangle = \max_{\xi \in \bigcup_{j \in I(\overline{x})} \partial f_0^j(\overline{x})} \langle \xi, h \rangle$$

and

$$f_0^-(\overline{x}; h) \geq \max_{j \in I(\overline{x})} (f_0^j)^-(\overline{x}; h) \geq \max_{j \in I(\overline{x})} \min_{\xi^j \in \partial f_0^j(\overline{x})} \langle \xi^j, h \rangle \geq \min_{\xi \in \bigcup_{j \in I(\overline{x})} \partial f_0^j(\overline{x})} \langle \xi, h \rangle.$$

Hence the condition holds. $\qquad\qquad\qquad\qquad\qquad\qquad\qquad\qquad\qquad\qquad\square$

We conclude by noting that in Corollary 4.1.15 if we further assume that $f_0^k, k = 1, \ldots, s$, are also Gâteaux differentiable at $\overline{x}$, then there exist multipliers $\lambda_0 \geq 0, \ldots, \lambda_p \geq 0, \lambda_{p+1}, \ldots, \lambda_m$ not all zero such that

$$\lambda_i f_i(\overline{x}) = 0, \qquad i = 1, \ldots, m$$

$$0 \in \lambda_0 \mathrm{co}\Big(\bigcup_{j \in I(\overline{x})} \nabla f_0^j(\overline{x}) \Big) + \lambda \circ \mathrm{co}(\partial F_1(\overline{x})) + N(Q, \overline{x}).$$

Moreover, by imposing a constraint qualification similar to that for problem (PL) (Corollary 4.1.13) one can obtain the optimality condition in which the first multiplier λ_0 is equal to one.

4.2 Second-Order Conditions

Necessary Conditions

Let $f: \mathbb{R}^n \to \mathbb{R}$, $g: \mathbb{R}^n \to \mathbb{R}^p$, and $h: \mathbb{R}^n \to \mathbb{R}^q$ be continuous functions. We consider the constrained mathematical programming problem (P) again:

$$\begin{aligned}
(P) \qquad\qquad \text{minimize } & f(x) \\
\text{subject to } & g(x) \leq 0 \\
& h(x) = 0.
\end{aligned}$$

We know from the previous section (Theorem 4.1.5) that if f, g, and h are continuously differentiable and x_0 is a local solution of problem (P), then there exists a nonzero vector $(\lambda_0, \lambda, \mu) \in \mathbb{R} \times \mathbb{R}^p \times \mathbb{R}^q$ such that

$$\lambda_0 \nabla f(x_0) + \langle \lambda, \nabla g(x_0) \rangle + \langle \mu, \nabla h(x_0) \rangle = 0,$$

$$\lambda_0 \geq 0, \lambda_i \geq 0 \text{ and } \lambda_i g_i(x_0) = 0, \ i = 1, \ldots, p.$$

Similarly to the case of problems with equality constraints, we say that the Kuhn–Tucker condition is satisfied at x_0 if the above rule holds with $\lambda_0 = 1$. Now we develop second-order optimality conditions for problem (P) by assuming that the data f, g, and h are differentiable and that the Kuhn–Tucker condition with a multiplier $(\lambda, \mu) \in \mathbb{R}^p \times \mathbb{R}^q$ is satisfied. Denote

$$\begin{aligned}
L(x) \quad &:= f(x) + \langle \lambda, g(x) \rangle + \langle \mu, h(x) \rangle. \\
X \quad &:= \{x \in \mathbb{R}^n : g(x) \leq 0, \langle \lambda, g(x) \rangle = 0 \text{ and } h(x) = 0\}. \\
T(X, x_0) \quad &:= \{v \in \mathbb{R}^n : v = \lim t_i(x_i - x_0), x_i \in X, x_i \to x_0, t_i > 0\}. \\
T_0(X, x_0) \quad &:= \{v \in \mathbb{R}^n : \text{there is } \delta > 0 \text{ such that } x_0 + tv \in X \text{ for } t \in [0, \delta]\}.
\end{aligned}$$

The function L is the Lagrangian associated with the multiplier (λ, μ); the set X is the set of feasible solutions x satisfying $\lambda_i g_i(x) = 0, i = 1, \ldots, k$; the set $T(X, x_0)$ is the contingent cone of X at x_0, which coincides with the tangent cone defined in Chapter 2 when the set is convex, and $T_0(X, x_0)$ is the set of feasible directions of X. We wish now to establish second-order optimality conditions for problem (P) where the data f, g, and h are of class C^1. We express these conditions by using pseudo-Hessian matrices and recession matrices.

Theorem 4.2.1 *Assume that the following conditions hold*

(i)   *The functions f, g, and h are continuously differentiable and x_0 is a local minimizer of the problem (P).*
(ii)   *The Kuhn–Tucker condition is satisfied at x_0, for some vector $(\lambda, \mu) \in \mathbb{R}^k \times \mathbb{R}^\ell$.*
(iii)   *$\partial^2 L(x_0)$ is a pseudo-Hessian of L at x_0.*

Then for each $u \in T_0(X, x_0)$, there is $M \in \partial^2 L(x_0) \cup ([\partial^2 L(x_0)]_\infty \setminus \{0\})$ such that

$$\langle u, M(u) \rangle \geq 0.$$

If in addition, L has a pseudo-Hessian map $\partial^2 L$ that is upper semicontinuous at x_0, then the conclusion is true for each $u \in T(X, x_0)$.

Proof. Let $u \in T_0(X, x_0)$. There is $\delta > 0$ such that $[x_0, x_0 + \delta u] \subset X$. Because x_0 is a local solution, there is $i_0 \geq 1$ such that $\delta > 1/i_0$ and

$$L(x_0 + u/i) - L(x_0) = f(x_0 + u/i) - f(x_0) \geq 0, \text{ for } i \geq i_0.$$

In view of the classic mean value theorem, there is $t_i \in (0, \delta)$ such that

$$L(x_0 + u/i) - L(x_0) = \nabla L(x_0 + t_i u)(u/i), \quad \text{for } i \geq i_0.$$

Then

$$\langle u, \nabla L(x_0 + t_i u)\rangle \geq 0, \quad \text{for } i \geq i_0$$

which together with (ii) implies

$$\limsup_{t \downarrow 0} \frac{\langle u, \nabla L(x_0 + tu) - \nabla L(x_0)\rangle}{t} \geq 0.$$

By the definition of pseudo-Hessian we derive

$$0 \leq (u \circ \nabla L)^+(x_0, u) \leq \sup_{M \in \partial^2 L(x_0)} \langle u, M(u)\rangle.$$

Then there exists a sequence of pseudo-Hessian matrices $\{M_i\} \subset \partial^2 L(x_0)$ such that

$$\lim_{i \to \infty} \langle u, M_i(u)\rangle \geq 0.$$

If the sequence $\{M_i\}$ is bounded, then we may assume that it converges to some $M \in \partial^2 L(x_0)$ because the latter set is closed, and obtain

$$\langle u, M(u)\rangle \geq 0.$$

If the sequence $\{M_i\}$ is unbounded, we may assume that

$$\lim_{i \to \infty} \|M_i\| = \infty \quad \text{and} \quad \lim_{i \to \infty} \frac{M_i}{\|M_i\|} = M_0 \in (\partial^2 L(x_0))_\infty \setminus \{0\},$$

and obtain

$$\langle u, M_0(u)\rangle \geq 0.$$

Suppose now that $\partial^2 L$ is a pseudo-Hessian map of L which is upper semicontinuous at x_0. Let $u \in T(X, x_0)$. Because the case $u = 0$ is trivial, we may assume that there is a sequence $\{x_i\} \subset X$ converging to x_0 such that

$$u = \lim_{i \to \infty} \frac{x_i - x_0}{\|x_i - x_0\|}.$$

Furthermore, as x_0 is a local minimizer, there is some $i_0 \geq 1$ such that

$$L(x_i) - L(x_0) = f(x_i) - f(x_0) \geq 0, \quad \text{for } i \geq i_0.$$

In view of the Taylor expansion, we have

$$L(x_i) - L(x_0) - \nabla L(x_0)(x_i - x_0) \in \frac{1}{2}\overline{\text{co}}(\langle x_i - x_0, \partial^2 L(y_i)(x_i - x_0)\rangle),$$

for some $y_i \in (x_0, x_i)$. This and the Kuhn–Tucker condition yield the existence of a matrix $M_i \in \partial^2 L(y_i)$ such that

$$\langle x_i - x_0, M_i(x_i - x_0) \rangle \geq -\frac{\|x_i - x_0\|^2}{i}, \quad \text{for } i \geq i_0.$$

As in the first part of the proof, if the sequence $\{M_i\}$ is bounded, then we may assume that it converges to some $M \in \partial^2 L(x_0)$, due to the upper semicontinuity of the pseudo-Hessian map $\partial^2 L$. The latter inequality implies

$$\langle u, M(u) \rangle = \lim_{i \to \infty} \langle \frac{x_i - x_0}{\|x_i - x_0\|}, M_i(\frac{x_i - x_0}{\|x_i - x_0\|}) \rangle \geq \lim_{i \to \infty} \left(-\frac{1}{i} \right) = 0.$$

If the sequence $\{M_i\}$ is unbounded, then due to the upper semicontinuity of the pseudo-Hessian map $\partial^2 L$, we may assume that

$$\lim_{i \to \infty} \|M_i\| = \infty \quad \text{and} \quad \lim_{i \to \infty} \frac{M_i}{\|M_i\|} = M_0 \in (\partial^2 L(x_0))_\infty \setminus \{0\}.$$

We deduce

$$\langle u, M_0(u) \rangle = \lim_{i \to \infty} \langle \frac{x_i - x_0}{\|x_i - x_0\|}, \frac{M_i}{\|M_i\|}(\frac{x_i - x_0}{\|x_i - x_0\|}) \rangle \geq \lim_{i \to \infty} \left(-\frac{1}{i\|M_i\|} \right) = 0.$$

This completes the proof. $\square$

The second part of Theorem 4.2.1 can be improved by requiring a certain regularity condition of $\partial^2 L$ instead of upper semicontinuity when ∇L is locally Lipschitz. Let S be a nonempty subset of $\mathbb{R}^n$; let $f : \mathbb{R}^n \to \mathbb{R}$ be C^1 and let $a \in S$. We say that the pseudo-Hessian set-valued map $\partial^2 f : \mathbb{R}^n \rightrightarrows L(\mathbb{R}^n, \mathbb{R}^n)$ is *regular* at a with respect to S if for each $u \in S$

$$\limsup_{\substack{A' \in \partial^2 f(a+tu') \\ u' \to u, \, t \downarrow 0}} \langle A'(u'), u' \rangle \leq \max_{A \in \partial^2 f(a)} \langle A(u), u \rangle. \tag{4.1}$$

This condition means that for each $u \in S$ and for each sequence $u_k \to u$, $t_k \downarrow 0$, and $A_k \in \partial^2 f(a + t_k u_k)$,

$$\limsup_{k \to \infty} \langle A_k(u_k), u_k \rangle \leq \max_{A \in \partial^2 f(a)} \langle A(u), u \rangle.$$

It is easy to see from the definition that if the map $\partial^2 f$ is locally bounded at a then

$$\limsup_{\substack{A' \in \partial^2 f(a+tu') \\ u' \to u, \, t \downarrow 0}} \langle A'(u'), u' \rangle$$

is finite. We now see that upper semicontinuity of the map $\partial^2 f$ at a guarantees regularity at a.

Lemma 4.2.2 *Let f be a C^1-function; let $\partial^2 f(x)$ be a pseudo-Hessian of f for each $x \in \mathbb{R}^n$ and let $a \in S \subset \mathbb{R}^n$. If the set-valued map $\partial^2 f$ is upper semicontinuous at a, then $\partial^2 f$ is regular at a with respect to S.*

Proof. Let $u \in S$ and let the sequences $u_k \to u$, $t_k \downarrow 0$, and $A_k \in \partial^2 f(a + t_k u_k)$. Because $\partial^2 f$ is locally bounded,

$$l := \limsup_{\substack{A' \in \partial^2 f(a+tu') \\ u' \to u,\, t \downarrow 0}} \langle A'(u'), u' \rangle$$

is finite. Suppose that

$$l > \max_{A \in \partial^2 f(a)} \langle A(u), u \rangle = \langle A_o u, u \rangle,$$

where $A_0 \in \partial^2 f(a)$. Define $\varepsilon = l - \langle A_0(u), u \rangle > 0$. Then there exists a subsequence, again denoted by $\langle A_k(u_k), u_k \rangle$, such that

$$\langle A_0(u), u \rangle = l - \varepsilon < \lim_{k \to \infty} \langle A_k(u_k), u_k \rangle.$$

Because $\partial^2 f$ is upper semicontinuous at a, we can find a subsequence $A_{i_k} \in \partial^2 f(a + t_{i_k} u_{i_k})$, such that $A_{i_k} \to \bar{A} \in \partial^2 f(a)$ as $k \to \infty$. Hence

$$\langle A_0(u), u \rangle < \lim_{k \to \infty} \langle A_k(u_k), u_k \rangle$$
$$= \langle \bar{A}u, u \rangle \leq \langle A_0(u), u \rangle,$$

which is a contradiction and so

$$l \leq \max_{A \in \partial^2 f(a)} \langle A(u), u \rangle. \qquad \square$$

Clearly if f is twice continuously differentiable then $\partial^2 f(\cdot) = \{\nabla^2 f(\cdot)\}$ is regular at x with respect to each subset S of $\mathbb{R}^n$. If f is $C^{1,1}$ then $\partial^2 f := \partial_H^2 f$ is regular at each point. In other words, condition (4.1) is satisfied for a $C^{1,1}$-function by $\partial^2 f = \partial_H^2 f$. The following example shows that a pseudo-Hessian set-valued map of a $C^{1,1}$-function, which is not upper semi-continuous, satisfies the regularity condition (4.1).

Example 4.2.3 Let $h : \mathbb{R} \to \mathbb{R}$ be an odd function that is defined for $x \geq 0$ by

$$h(x) = \begin{cases} 2x - 1 & x \geq \frac{1}{2}; \\ -x + \frac{1}{2^{2n-1}} & x \in [\frac{1}{2^{2n}}, \frac{1}{2^{2n-1}}],\ n = 1, 2, \ldots, \\ 2x - \frac{1}{2^{2n}} & x \in [\frac{1}{2^{2n+1}}, \frac{1}{2^{2n}}],\ n = 1, 2, \ldots, \\ 0 & x = 0. \end{cases}$$

Define $f : \mathbb{R}^2 \to \mathbb{R}$ by

$$f(x_1, x_2) = \int_0^{|x_1|} h(t)dt + \frac{x_2^2}{2}.$$

Then f is a $C^{1,1}$-function because $\nabla f(x_1, x_2) = (h(x_1), x_2)$ is a locally Lipschitz function. A pseudo-Hessian set-valued map $\partial^2 f$ is given by

$$\partial^2 f(x_1, x_2) = \begin{cases} \left\{ \begin{pmatrix} -1 & 0 \\ 0 & 1 \end{pmatrix}, \begin{pmatrix} 2 & 0 \\ 0 & 1 \end{pmatrix} \right\} & x_1 = \pm\frac{1}{2^n},\ n = 1, 2, \ldots, \\ \left\{ \begin{pmatrix} 0 & 0 \\ 0 & 1 \end{pmatrix}, \begin{pmatrix} 2 & 0 \\ 0 & 1 \end{pmatrix} \right\} & x_1 = 0, \\ \left\{ \begin{pmatrix} h'(x_1) & 0 \\ 0 & 1 \end{pmatrix} \right\} & \text{otherwise.} \end{cases}$$

It is easy to verify that $\partial^2 f$ is regular at $(0,0)$ and locally bounded at $(0,0)$. However, it is not upper semicontinuous at $(0,0)$ because

$$\begin{pmatrix} -1 & 0 \\ 0 & 1 \end{pmatrix} \in \partial^2 f((\frac{1}{2^n}, 0)) \quad \text{but} \quad \begin{pmatrix} -1 & 0 \\ 0 & 1 \end{pmatrix} \notin \partial^2 f((0,0)).$$

It is also worth noting that

$$\partial_H^2 f((0,0)) = \left\{ \begin{pmatrix} \alpha & 0 \\ 0 & 1 \end{pmatrix} \mid \alpha \in [-1,\ 2] \right\}$$

and that $\mathrm{co}(\partial^2 f((0,0))) \subset \partial_H^2 f((0,0))$.

Theorem 4.2.4 *Assume that the problem (P) has a local optimal solution a. Let the Kuhn–Tucker condition be satisfied at a by (λ, μ). Suppose that for each $x \in \mathbb{R}^n$, $\partial^2 L(x)$ is a pseudo-Hessian of $L(\cdot)$ at x. If the set-valued map $\partial^2 L(\cdot)$ is locally bounded at a and regular at a with respect to $T(X, a)$, then for every $u \in T(X, a)$ one can find some $M \in \partial^2 L(a)$ such that*

$$\langle M(u), u \rangle \geq 0.$$

Proof. Let $u \in T(X, a)$. Then there exist sequences $t_k \downarrow 0$ and $u_k \to u$ as $k \to \infty$ such that, for every k, $a + t_k u_k \in X$. So,

$$L(a + t_k u_k) = f(a + t_k u_k).$$

Now it follows from the Taylor expansion (Theorem 2.2.20) that

$$L(a + t_k u_k) \leq L(a, \lambda, \mu) + t_k \langle \nabla L(a), u_k \rangle + \frac{t_k^2}{2} \langle N_k(u_k), u_k \rangle$$

where $N_k \in \partial^2 L(a + \bar{t}_k u_k)$ and $0 < \bar{t}_k < t_k$. Noting that a is a local minimum of the problem (P), we get

$$L(a) = f(a),$$
$$\nabla L(a) = 0,$$
$$f(a + t_k u_k) \geq f(a),$$

for sufficiently large k. Thus, for sufficiently large k,

$$\langle N_k(u_k), u_k \rangle \geq 0.$$

Because the set-valued map $\partial^2 L$ is locally bounded at a, the sequence $\{N_k\}$ is bounded. Hence this sequence has a subsequence, again denoted $\{N_k\}$, which converges to a matrix N. As $k \to \infty$, the sequence $a + \bar{t}_k u_k$ converges to a. Then it follows that

$$\langle N(u), u \rangle = \lim_{k \to \infty} \langle N_k(u_k), u_k \rangle \geq 0.$$

Hence

$$\limsup_{\substack{A' \in \partial^2 f(a + tu') \\ u' \to u, \, t \downarrow 0}} \langle A'(u'), u' \rangle \geq \langle N(u), u \rangle \geq 0,$$

and so, by the regularity assumption, we get that $\max_{A \in \partial^2 f(a)} \langle A(u), u \rangle \geq 0$ as requested. $\qquad\square$

Corollary 4.2.5 *Assume that functions f, g_i, and h_j, for each i, j in problem (P) are $C^{1,1}$ and that the problem (P) has a local optimal solution a. If the constraint qualification (CQ_2) holds at a, then there exist $\lambda_i \geq 0$ satisfying $\lambda_i g_i(a) = 0$, for $i = 1, 2, \ldots, p$, and $\mu \in \mathbb{R}^q$ such that $\nabla L(a) = 0$, and for every $u \in T(X, a)$ there exists some $M \in \partial_H^2 L(a)$ satisfying*

$$\langle M(u), u \rangle \geq 0.$$

Proof. Choose $\partial_H^2 L(a)$ as a pseudo-Hessian of $L(\cdot)$ at a. The result then follows from Theorem 4.2.4 because the map $\partial_H^2 L(\cdot)$ is upper semicontinuous at a. $\qquad\square$

The following example shows that Theorem 4.2.4 provides sharper optimality conditions than the conditions of Corollary 4.2.5.

Example 4.2.6 Consider the problem

$$\text{minimize} \int_0^{|x_1|} h(t)dt + \frac{x_2^2}{2}$$
$$\text{subject to } x_1 \geq 0, \quad x_2 \geq 0,$$

where $f(x_1, x_2) = \int_0^{|x_1|} h(t)dt + x_2^2/2$, $g_1(x_1, x_2) = x_1$, $g_2(x_1, x_2) = x_2$, and h is given as in Example 4.2.3. Then f is a $C^{1,1}$ function. The point $(0,0)$ is a solution of the problem. The Kuhn–Tucker condition is satisfied at $(0,0)$ by $\lambda = (\lambda_1, \lambda_2) = (0,0)$ and the condition of Theorem 4.2.4 is verified by the matrix

$$\begin{pmatrix} 0 & 0 \\ 0 & 1 \end{pmatrix} \in \partial^2 L((0,0)) = \partial^2 f((0,0)) \subset \partial_H^2 f((0,0)),$$

for each vector $(u_1,\ u_2)$ from the tangent cone to X at $(0,0)$ which is given by

$$T(X,(0,0)) = \{(x_1,\ x_2) \in \mathbb{R}^2 \mid x_1 \geq 0,\ x_2 \geq 0\}.$$

It can be seen that under certain conditions elements of the tangent cone $T(X, a)$ can be obtained explicitly in terms of the gradients of the functions g_i and h_j. Namely, if the vectors $\nabla g_i(a),\ i \in I(a),\ \nabla h_j(a),\ j = 1, 2, \ldots, q$ are linearly independent, where $I(a)$ is the set of active indices (i.e., $i \in I(a)$ if and only if $g_i(a) = 0$), then $u \in T(X, a)$ if and only if u is a solution to the linear system

$$\begin{aligned} \langle \nabla g_i(a), u \rangle &= 0 \quad \text{for } i \text{ such that } \lambda_i > 0, \\ \langle \nabla g_i(a), u \rangle &\leq 0 \quad \text{for } i \text{ such that } \lambda_i = 0 \text{ and } g_i(a) = 0, \\ \langle \nabla h_j(a), u \rangle &= 0 \quad \text{for } j = 1, 2, \ldots, q. \end{aligned}$$

Here $I(a)$ is the active index set at a; that is, $i \in I(a)$ if and only if $g_i(a) = 0$.

Sufficient Conditions

In this section we derive second-order sufficient conditions for local solutions of problem (P). The feasible set of this problem is denoted S, and the contingent cone to S at $x \in S$ is denoted $T(S, x)$.

Theorem 4.2.7 *Assume that the following conditions hold*

(i)   *The functions f, g, and h are continuously differentiable.*
(ii)   *The Kuhn–Tucker condition is satisfied at x_0, for some $(\lambda, \mu) \in \mathbb{R}^p \times \mathbb{R}^q$.*
(iii)   *There is a pseudo-Hessian map $\partial^2 L$ of L that is upper semicontinuous at x_0 such that for every $u \in T(S, x_0) \setminus \{0\}$ and $M \in \partial^2 L(x_0) \cup ([\partial^2 L(x_0)]_\infty \setminus \{0\})$, one has*

$$\langle u, M(u) \rangle > 0.$$

Then x_0 is a locally unique solution of the problem (P).

Proof. Suppose to the contrary that there is $x_i \in S$ such that $\lim_{i \to \infty} x_i = x_0$ and $f(x_i) \leq f(x_0)$. We may assume that

$$\lim_{i \to \infty} \frac{x_i - x_0}{\|x_i - x_0\|} = u \in T(S, x_0).$$

It follows that

$$L(x_i) - L(x_0) = f(x_i) - f(x_0) + \langle \lambda, g(x_i) \rangle \leq 0.$$

Using the Taylor expansion (Theorem 2.2.20), we express

$$L(x_i) - L(x_0) - \nabla L(x_0)(x_i - x_0) \in \frac{1}{2} \overline{co} \langle x_i - x_0, \partial^2 L(y_i)(x_i - x_0) \rangle,$$

for some $y_i \in (x_0, x_i)$. This and the Kuhn-Tucker condition yield the existence of a matrix $M_i \in \partial^2 L(y_i)$ such that

$$\langle x_i - x_0, M_i(x_i - x_0) \rangle \leq \frac{\|x_i - x_0\|^2}{i}.$$

If the sequence $\{M_i\}$ is bounded, then we may assume that it converges to some $M \in \partial^2 L(x_0)$, due to the upper semicontinuity of the pseudo-Hessian map $\partial^2 L$. The latter inequality implies

$$\langle u, M(u) \rangle = \lim_{i \to \infty} \langle \frac{x_i - x_0}{\|x_i - x_0\|}, M_i(\frac{x_i - x_0}{\|x_i - x_0\|}) \rangle \leq 0,$$

which contradicts the hypothesis. If the sequence $\{M_i\}$ is unbounded, then due to the upper semicontinuity of the pseudo-Hessian map $\partial^2 L$, we may assume that

$$\lim_{i \to \infty} \|M_i\| = \infty \quad \text{and} \quad \lim_{i \to \infty} \frac{M_i}{\|M_i\|} = M_0 \in (\partial^2 L(x_0))_\infty \setminus \{0\}.$$

We deduce

$$\langle u, M_0(u) \rangle = \lim_{i \to \infty} \langle \frac{x_i - x_0}{\|x_i - x_0\|}, \frac{M_i}{\|M_i\|}(\frac{x_i - x_0}{\|x_i - x_0\|}) \rangle \leq 0,$$

which again contradicts the hypothesis. This completes the proof.   □

The upper semicontinuity of $\partial^2 L$ is unnecessary when ∇L admits a Fréchet pseudo-Jacobian. We say then $\partial^2 L$ is a Fréchet pseudo-Hessian of L.

Theorem 4.2.8 *Assume that the following conditions hold*

(i)   The functions f, g, and h are continuously differentiable.

(ii)   The Kuhn–Tucker condition is satisfied at x_0, for some $(\lambda, \mu) \in \mathbb{R}^p \times \mathbb{R}^q$.

(iii) There is a Fréchet pseudo-Hessian $\partial^2 L$ of L at x_0 such that for every $u \in T(S, x_0) \setminus \{0\}$ and $M \in \partial^2 L(x_0) \cup ([\partial^2 L(x_0)]_\infty \setminus \{0\})$, one has

$$\langle u, M(u) \rangle > 0.$$

Then x_0 is a locally unique solution of the problem (P).

Proof. We follow the proof of the previous theorem. The expression for $L(x_i) - L(x_0) - \nabla L(x_0)(x_i - x_0)$ can now be written as

$$L(x_i) - L(x_0) - \nabla L(x_0)(x_i - x_0) = \langle M_i(t_i(x_i - x_0)) + o(t_i \| x_i - x_0 \|), x_i - x_0 \rangle$$

for some $M_i \in \partial^2 L(x_0)$ and some $t_i \in (0, 1)$ with $o(t_i \| x_i - x_0 \|)/\| t_i(x_i - x_0) \|$ tending to 0 as $i \to \infty$. The rest of the proof remains without change. $\square$

Next we give more sufficient conditions in the case when a pseudo-Hessian of L in a neighborhood of a is known. Let $J = \{i \in I(a) \ : \ \lambda_i > 0\}$. Define

$$Y = \{y \in B_n \ : \ \langle y, \nabla g_i(a) \rangle = 0, \ i \in J, \ \langle y, \nabla h_j(a) \rangle = 0, \ j = 1, 2, \ldots, q\}$$

and for $\varepsilon > 0$ and $\delta > 0$ define

$$Z(\varepsilon, \delta) = \{u \in B_n \ : \ \| u - y \| < \varepsilon, \ \text{for some } y \in Y,$$
$$\text{and } a + \bar{\delta}(u)u \in C, \ \text{for some } 0 < \bar{\delta}(u) < \delta\}.$$

Theorem 4.2.9 *Let a be a feasible point for (P). Suppose that the Kuhn–Tucker condition is satisfied at a by $(\lambda, \mu) \in \mathbb{R}^p \times \mathbb{R}^q$. Assume that for each x in a neighborhood of a, $\partial^2 L(x)$ is a pseudo-Hessian of L at a. If there exist $\varepsilon > 0$ and $\delta > 0$ such that for each $u \in Z(\varepsilon, \delta)$ and for each $0 < \alpha < 1$,*

$$\langle M(u), u \rangle \geq 0$$

for every $M \in \partial^2 L(a + \alpha u)$, then a is a local minimizer of the problem (P).

Proof. If a is not a local minimizer, then there exists a sequence $\{x_k\}$ such that x_k is feasible for (P), $x_k \to a$ as $k \to +\infty$, and $f(x_k) < f(a)$ for each k.

Let $x_k = a + \delta_k u_k$, where $\| u_k \| = 1$, $\delta_k > 0$, $\delta_k \to 0$ as $k \to +\infty$. Because $\| u_k \| = 1$, the sequence $\{u_k\}$ has a convergent subsequence. Without loss of generality, we assume that $u_k \to y$ as $k \to +\infty$, with $\| y \| = 1$.

By the mean value theorem, we have

$$0 > f(x_k) - f(a) = \delta_k u_k \nabla f(a + \eta_{0k}\delta_k u_k), \ 0 < \eta_{0k} < 1,$$
$$0 \geq g_i(x_k) - g_i(a) = \delta_k u_k \nabla g_i(a + \eta_{ik}\delta_k u_k), \ 0 < \eta_{ik} < 1, \ \forall i \in I(a),$$
$$0 = h_j(x_k) - h_j(a) = \delta_k u_k \nabla h_j(a + \xi_{jk}\delta_k u_k), \ 0 < \xi_{jk} < 1, \ \forall j = 1, \ldots, q.$$

Dividing the above inequalities and the equality by δ_k and taking limits as $k \to +\infty$, we obtain

$$\langle y, \nabla f(a) \rangle \leq 0, \ \langle y, \nabla g_i(a) \rangle \leq 0, \ \forall i \in I(a), \ \langle y, \nabla h_j(a) \rangle = 0, \ \forall j.$$

Suppose that $\langle y, \nabla g_i(a) \rangle < 0$ for at least one $i \in J$. Then we get

$$0 \geq \langle y, \nabla f(a) \rangle = -\sum_{i \in J} \lambda_i \langle y, \nabla g_i(a) \rangle - \sum_{j=1}^{q} \mu_j \langle y, \nabla h_j(a) \rangle > 0.$$

This is a contradiction. Thus $\langle y, \nabla g_i(a) \rangle = 0$ for all $i \in J$ or $J = \phi$. Then $y \in Y$. Because the Kuhn–Tucker conditions are satisfied at a by λ_i, μ_j, we have

$$\lambda_i \geq 0, \ \lambda_i g_i(a) = 0, \ i = 1, \ldots, \ p,$$

$$\nabla L(a) = \nabla f(a) + \sum_{i \in I(a)} \lambda_i \nabla g_i(a) + \sum_{j=1}^{q} \mu_j \nabla h_j(a) = 0.$$

Because $f(a) > f(x_k)$, it follows from the latter inequalities and from the Taylor expansion for $L(x)$ at a (Theorem 2.2.20) that

$$f(a) > f(x_k)$$

$$\geq f(x_k) + \sum_{i \in I(a)} \lambda_i g_i(x_k) + \sum_{j=1}^{q} \mu_j h_j(x_k)$$

$$\geq f(a) + \sum_{i \in I(a)} \lambda_i g_i(a) + \sum_{j=1}^{q} \mu_j h_j(a)$$

$$+ \delta_k u_k^{tr} \left(\nabla f(a) + \sum_{i \in I(a)} \lambda_i \nabla g_i(a) + \sum_{j=1}^{q} \mu_j \nabla h_j(a) \right)$$

$$+ \frac{1}{2} \min_{M_k \in co(\partial^2 L(a + \theta_k \delta_k u_k))} \langle M_k(\delta_k u_k), \ \delta_k u_k \rangle$$

$$= f(a) + \frac{1}{2} \delta_k^2 \min_{M_k \in \partial^2 L(a + \theta_k \delta_k u_k)} \langle M_k(u_k), \ u_k \rangle$$

$$= f(a) + \frac{1}{2} \delta_k^2 \langle M_k^0(u_k), u_k \rangle$$

for some $M_k^0 \in \partial^2 L(a + \theta_k \delta_k u_k)$ and $0 < \theta_k < 1$. Hence for any k, one has

$$0 > \langle M_k^0 u_k, u_k \rangle. \tag{4.2}$$

By construction, $\|u_k\| = 1$, $u_k \to y \in Y$, $\delta_k \to 0$ as $k \to +\infty$, $0 < \theta_k \delta_k < 1$ when k is large, and $a + \delta_k u_k$ is feasible for every k. Hence for k large, $u_k \in Z(\varepsilon, \delta)$ and by assumption

$$\langle M_k^0(u_k), u_k \rangle \geq 0.$$

This is a contradiction with (4.2). Then a is a local minimizer of (P). $\quad\square$

Theorem 4.2.10 *Let a be a feasible solution for (P). Suppose that the Kuhn–Tucker condition is satisfied at a by $(\lambda, \mu) \in \mathbb{R}^p \times \mathbb{R}^q$. Assume that for each x in a neighborhood of a, $\partial^2 L(x)$ is a pseudo-Hessian of $L(\cdot)$ at a. If there exist $\varepsilon > 0$ and $\delta > 0$ such that for each $u \in Z(\varepsilon, \delta)$ and for each $0 < \alpha < 1$, one has*

$$\langle M(u), u \rangle > 0 \quad \text{for all } M \in \partial^2 L(a + \alpha u),$$

then a is a strict local minimizer of the problem (P).

Proof. The proof is only a slight modification of that of Theorem 4.2.9 and so it is omitted. $\quad\square$

Example 4.2.11 *(Necessary condition)* Consider the following problem:

$$\text{minimize } x^{4/3} - y^4$$
$$\text{subject to } -x^2 + y^4 \leq 0.$$

It is clear that $(0,0)$ is a local optimal solution of this problem. By setting $\lambda = 1$, we see that the Kuhn–Tucker condition is verified at this solution. The Lagrangian function L is given by

$$L(x) = x^{4/3} - y^4 - x^2 + y^4 = x^{4/3} - x^2.$$

The gradient map of L is given by

$$\nabla L(x, y) = \left(\frac{4}{3} x^{1/3} - 2x, 0 \right).$$

Because this gradient map is not locally Lipschitz at $(0,0)$, the Clarke generalized Hessian of L does not exist. Let us define

$$\partial^2 L(x, y) := \left\{ \begin{pmatrix} \frac{4}{9} x^{-2/3} - 2 & 0 \\ 0 & 0 \end{pmatrix} \right\}, \text{ for } x \neq 0,$$

and

$$\partial^2 L(0,y) := \left\{ \begin{pmatrix} \alpha & 0 \\ 0 & -1/\alpha \end{pmatrix} : \alpha \geq 2 \right\}.$$

A simple calculation confirms that this is a pseudo-Hessian map of L which is upper semicontinuous at $(0,0)$. In this example, the set X mentioned before Theorem 4.2.1 is given by

$$X := \{(x,y) \in \mathbb{R}^2 : x^2 = y^4\}.$$

In particular, $u = (0,1) \in T(X,(0,0))$. For each $M \in \partial^2 L(0,0)$, we have

$$\langle u, M(u) \rangle = -\frac{1}{\alpha} < 0$$

as $\alpha \geq 2$. The recession cone of $\partial^2 L(0,0)$ is given by

$$(\partial^2 L(0,0))_\infty = \left\{ \begin{pmatrix} \alpha & 0 \\ 0 & 0 \end{pmatrix} : \alpha \geq 0 \right\}.$$

By choosing

$$M = \begin{pmatrix} 1 & 0 \\ 0 & 0 \end{pmatrix} \in (\partial^2 L(0,0))_\infty \setminus \{0\},$$

we do have $\langle u, M(u) \rangle \geq 0$, as desired.

Example 4.2.12 *(Sufficient condition)* Consider the following problem;

$$\min \quad -x^{4/3} - y^4$$
$$\text{subject to } y^4 - x^2 = 0.$$

As in the previous example, by setting $\mu = 1$ we see that the Kuhn–Tucker condition is satisfied at $(0,0)$. The Lagrangian function L is given by

$$L(x,y) = -x^{4/3} - x^2,$$

and its gradient map is given by

$$\nabla L(x,y) = (-\frac{4}{3}x^{1/3} - 2x, 0).$$

This gradient map is not locally Lipschitz at $(0,0)$. Let us define

$$\partial^2 L(x,y) := \left\{ \begin{pmatrix} -\frac{4}{9}x^{-2/3} - 2 & 0 \\ 0 & 0 \end{pmatrix} \right\}, \text{ for } x \neq 0,$$

and

$$\partial^2 L(0,y) := \left\{ \begin{pmatrix} -\alpha & 0 \\ 0 & 1/\alpha \end{pmatrix} : \alpha \geq 2 \right\}.$$

It is not hard to see that this is a pseudo-Hessian map of L which is upper semicontinuous at $(0,0)$. The feasible set S of this problem coincides with the set X of Example 4.2.11, so the contingent cone to this set at $(0,0)$ is given by

$$T(S,(0,0)) = \{(0,\beta) \in \mathbb{R}^2 : \beta \in R\}.$$

For each $u = (0,\beta)$ with $\beta \neq 0$ and for each $M \in \partial^2 L(0,0)$, we have

$$\langle u, M(u) \rangle = \frac{\beta^2}{\alpha} > 0$$

as $\alpha \geq 2$. Despite this, the point $(0,0)$ is not a local optimal solution of the problem. Let us look at the recession condition of Theorem 4.2.8. The recession cone of $\partial^2 L(0,0)$ is given by

$$(\partial^2 L(0,0))_\infty = \left\{ \begin{pmatrix} -\alpha & 0 \\ 0 & 0 \end{pmatrix} : \alpha \geq 0 \right\}.$$

By choosing

$$M = \begin{pmatrix} -1 & 0 \\ 0 & 0 \end{pmatrix} \in (\partial^2 L(0,0))_\infty \setminus \{0\},$$

we derive

$$\langle u, M(u) \rangle = 0,$$

and see that the sufficient condition on the recession Hessian matrices is violated.

4.3 Composite Programming Necessary Optimality Conditions

Consider the following convex composite minimization problem,

(CCP)   minimize $(g \circ F)(x)$
subject to $x \in C$, $f_i(x) \leq 0$, $i = 1, 2, \ldots, m,$

where $F : \mathbb{R}^n \to \mathbb{R}^m$ is a continuous nonsmooth map, $g : \mathbb{R}^m \to \mathbb{R}$ is a convex function, $C \subset \mathbb{R}^n$ is a closed convex set, and for each i, $f_i : \mathbb{R}^n \to \mathbb{R}$ is continuous. These kinds of problems are found in engineering applications. For instance, the min-max model with max-min constraints

minimize $\max_{i \in I} F_i(x)$
subject to $\max_{1 \leq k \leq r} \min_{1 \leq j \leq q_k} f_k^j(x) \leq 0$

can equivalently be written as the following inequality constrained problem of the form (CCP),

$$\min_{(x,\mu_1,\ldots,\mu_r)} (g \circ F)(x)$$

$$\text{subject to} \quad \sum_{j \in \mathbf{q}_k} \mu_k^j = 1, \ \mu_k^j \geq 0,$$

$$\sum_{j \in \mathbf{q}_k} \mu_k^j f_k^j(x) \leq 0, \ k \in \mathbf{r}, j \in \mathbf{q}_k,$$

where $I := \{1, 2, \ldots, m\}$, $\mathbf{r} := \{1, 2, \ldots, r\}$, $\mathbf{q}_k := \{1, 2, \ldots, q_k\}$, $g(x) = \max_{i \in I} x_i$, and $F(x) = (F_1(x), \ldots, F_m(x))$.

Models involving max-min constraints arise in the design of electronic circuits subject to manufacturing tolerances and postmanufacturing tuning, and in optimal steering of mobile robots in the presence of obstacles. The composite structure of the problem (CCP) is used in a variety of applications. For instance, to solve nonlinear equations $F_i(x) = 0$, $i = 1, 2, \ldots, m$, one minimizes the norm $||(F_1(x), \ldots, F_m(x))||$ which is a composite function of the norm function and the vector function $(F_1, \ldots, F_m)$. Similar problems of finding a feasible point of a system of continuous nonlinear inequalities $F_i(x) \leq 0$, $i = 1, 2, \ldots, m$, can be approached by minimizing $||F(x)^+||$ where $F_i^+ = \max(F_i, 0)$. Composite functions $g \circ F$ also appear in the form of an exact penalty function when solving a nonlinear programming problem. All these examples can be cast within the structure of (CCP). A variant of the nonsmooth composite model function $g \circ F$, where g is differentiable and F is continuous, also comes to light in the optimization reformulation of complementarity problems which we deal with in the next chapter. Also, continuous composite functions play an important role in the study of spectral functions such as the spectral abscissa and spectral radius that are continuous but are not locally Lipschitz. Variational analysis of such composite functions is of great interest in control theory and related areas.

Theorem 4.3.1 *For the problem (CCP), let $x \in \mathbb{R}^n$. Let $F : \mathbb{R}^n \to \mathbb{R}^m$ be a continuous map, $g : \mathbb{R}^m \to \mathbb{R}$ a convex function, and let $f_i : \mathbb{R}^n \to \mathbb{R}$ be continuous for each $i = 1, 2, \ldots, m$. Assume that F admits a pseudo-Jacobian map ∂F which is upper semicontinuous at x and that f_i admits a bounded pseudo-Jacobian $\partial f_i(x)$ at x, for each $i = 1, 2, \ldots, m$. If x is a local minimizer of the problem (CCP), then there exist nonnegative numbers $\lambda^0, \lambda^1, \ldots, \lambda^m$ with $\lambda^0 + \cdots + \lambda^m = 1$ such that $\lambda^i f_i(x) = 0$, $i = 1, 2, \ldots, m$,*

$$0 \in \left(\lambda^0 \overline{\text{co}}(\partial g(F(x)) \circ \partial F(x)) + \sum_{i=1}^{m} \lambda^i \partial f_i(x) \right.$$

$$\left. \cup \lambda^0 \text{co}\{\partial g(F(x)) \circ ((\partial F(x))_\infty \backslash \{0\})\} \right) - (C - x)^*.$$

Proof. Put $I(x) := \{i : f_i(x) = 0\}$, the active index set at x. Consider the system

$$y \in (C - x), \ (g \circ F)^+(x; y) < 0, \ f_i^+(x; y) < 0, \ i \in I(x). \tag{4.3}$$

We claim that this system has no solution. Otherwise, it follows from the definitions of the upper Dini derivative and the continuity of f_i that we can find a real number $\alpha > 0$ such that

$$x + \alpha y \in C, (g \circ F)(x + \alpha y) < (g \circ F)(x), \ f_i(x + \alpha y) < 0, \ i = 1, 2, \ldots, m$$

which contradicts local minimality at x. For $\varepsilon > 0$, define

$$A_\varepsilon := [\partial g(F(x)) + \varepsilon B_m]^{tr} \partial F(x),$$
$$P_\varepsilon := A_\varepsilon \cup \Big(\bigcup_{i \in I(x)} \partial f_i(x) \Big).$$

Because (4.3) has no solution, by the definition of pseudo-Jacobian, the following system also has no solution;

$$y \in (C - x), \ \sup_{v \in P_\varepsilon} \langle v, y \rangle < 0.$$

So, the separation theorem yields

$$0 \in \text{cl}(\text{co}(P_\varepsilon) - (C - x)^*).$$

Take $\varepsilon = 1/k$, $k \geq 1$. Then, by Caratheodory's theorem, we can represent 0 as

$$0 = \lambda_k^0 \sum_{j=1}^{n+1} \mu_k^j \Big(a_{jk} + \frac{1}{k} b_{jk}\Big)^{tr} c_{jk} + \sum_{i \in I(x)} \lambda_k^i d_{ik} - e_k + \frac{1}{k} l_k', \tag{4.4}$$

where

$$\lambda_k^0, \ \lambda_k^i \geq 0, \ \lambda_k^0 + \sum_{i \in I(x)} \lambda_k^i = 1, \ \mu_k^j \geq 0, \ \sum_{j=1}^{n+1} \mu_k^j = 1,$$
$$a_{jk} \in \partial g(F(x)), \ b_{jk} \in B_m, \ c_{jk} \in \partial F(x), \ j = 1, \ldots, n+1,$$
$$d_{ik} \in \text{co}(\partial l_i(x)), \ i \in I(x), \ e_k \in (C - x)^*, \ l_k' \in B_m.$$

Let

$$J := \{1, 2, \ldots, n+1\}, \ J_1 := \{j \in J : \{c_{jk}\}_{k \geq 1} \text{ is bounded}\}$$

and $J_2 := J \setminus J_1$. Then (4.4) can be rewritten as

$$0 = \lambda_k^0 \Big(\sum_{j \in J_1} \mu_k^j \Big(a_{jk} + \frac{1}{k} b_{jk}\Big)^{tr} c_{jk} + \sum_{j \in J_2} \mu_k^j \Big(a_{jk} + \frac{1}{k} b_{jk}\Big)^{tr} c_{jk} \Big)$$
$$+ \sum_{i \in I(x)} \lambda_k^i d_{ik} - e_k + \frac{1}{k} l_k'. \tag{4.5}$$

We may now assume, without loss of generality, the following sequences converge when k tends to ∞.

$$\lambda_k^0 \to \lambda^0 \in [0,1], \ \lambda_k^i \to \lambda^i \in [0,1] \quad \text{and} \quad \lambda^0 + \sum_{i \in I(x)} \lambda^i = 1,$$

$$\mu_k^j \to \mu^j \in [0,1] \quad \text{and} \quad \sum_{j=1}^{n+1} \mu^j = 1,$$

$$a_{jk} \to a_j \in \partial g(F(x)), \ b_{jk} \to b_j \in B(0,1), \ j = 1, \dots, n+1,$$

$$c_{jk} \to c_j \in \partial F(x), \ j \in J_1,$$

$$d_{ik} \to d_i \in \text{co}(\partial l_i(x)), \ i \in I(x), \quad \text{and}$$

$$l_k \to l' \in B_m.$$

Case 1: $J_2 = \phi$. In this case, we may assume $e_k \to e$ for some $e \in (C - x)^*$. Letting $k \to \infty$, (4.5) yields

$$0 = \lambda^0 \sum_{j=1}^{n+1} \mu^j a_j^{tr} c_j + \sum_{i \in I(x)} \lambda^i d_i - e$$

$$\in \lambda^0 \text{co}(\partial g(F(x))^{tr} \partial F(x)) + \sum_{i \in I(x)} \lambda^i \text{co}(\partial f_i(x)) - (C - x)^*.$$

Case 2: $J_2 \neq \phi$. If $\{\mu_k^j c_{jk}\}_{k \geq 1}$ is bounded for every $j \in J_2$, then $\mu^j = 0$ for all $j \in J_2$. Hence $\sum_{j \in J_1} \mu^j = 1$. So, we may assume that

$$\mu_k^j c_{jk} \to c_j \in (\partial F(x))_\infty, \ j \in J_2 \quad \text{and} \quad e_k \to e \in (C - x)^*.$$

Passing (4.5) to the limit, we get

$$0 \in \lambda^0 \Big(\sum_{j \in J_1} \mu^j a_j^{tr} c_j + \sum_{j \in J_2} a_j^{tr} c_j \Big) + \sum_{i \in I(x)} \lambda^i d_i - e$$

$$\in \lambda^0 (\text{co}(\partial g(F(x)) \circ \partial F(x)) + \text{co}(\partial g(F(x)) \circ (\partial F(x))_\infty)) +$$

$$+ \sum_{i \in I(x)} \lambda^i \text{co}(\partial f_i(x)) - (C - x)^*$$

$$\subset \lambda^0 \overline{\text{co}}(\partial g(F(x)) \circ \partial F(x)) + \sum_{i \in I(x)} \lambda^i \text{co}(\partial f_i(x)) - (C - x)^*,$$

because $\text{co}(\partial g(F(x)) \circ \partial F(x)) + \text{co}(\partial g(F(x)) \circ (\partial F(x))_\infty) \subset \overline{\text{co}}(\partial g(F(x)) \circ \partial F(x))$. This inclusion follows from the fact that

$$\partial g(F(x)) \circ (\partial F(x))_\infty \subset (\partial g(F(x)) \circ \partial F(x))_\infty \subset (\text{co}(\partial g(F(x))) \circ \partial F(x))_\infty$$

and that

$$\text{co}(\partial g(F(x))) \circ \partial F(x) + \text{co}(\partial g(F(x)) \circ (\partial F(x))_\infty)$$

$$\subset \overline{\text{co}}(\partial g(F(x))) \circ \partial F(x) + (\overline{\text{co}}(\partial g(F(x))) \circ \partial F(x))_\infty$$

$$= \overline{\text{co}}(\partial g(F(x))) \circ \partial F(x).$$

If there exists $j \in J_2$ such that $\{\mu_k^j c_{jk}\}_{k \geq 1}$ is unbounded, then by taking subsequences instead we may assume there exists $j_0 \in J_2$ such that

$$\|\mu_k^{j_0} c_{jk}\| \geq \|\mu_k^j c_{jk}\|, \ \forall j \in J_2, \ k \geq 1.$$

Then $\mu_k^j c_{jk}/\|\mu_k^{j_0} c_{j_0 K}\| \to c_j \in (\partial F(x))_\infty$, $j \in J_2$, and from (4.5), we may assume $e_k/\|\mu_k^{j_0} c_{j_0 k}\| \to e \in (C - x)^*$, because $(C - x)_\infty^* \subset (C - x)^*$. Put $J_3 := \{j \in J_2 : c_j \neq 0\}$. Then $J_3 \neq \phi$ because $j_0 \in J_3$. Now, by dividing (4.5) by $\|\mu_k^{j_0} c_{j_0 k}\|$ and passing to the limit with $k \to \infty$, we obtain

$$0 = \lambda^0 \sum_{j \in J_3} a_j^{tr} c_j - e \in \lambda^0 \mathrm{co}\, (\partial g(F(x)) \circ ((\partial F(x))_\infty \backslash \{0\})) - (C - x)^*.$$

Thus

$$0 \in \left(\lambda^0 \overline{\mathrm{co}}(\partial g(F(x))) \circ \partial F(x) + \sum_{i \in I(x)} \lambda^i \mathrm{co}(\partial f_i(x)) \right) \cup$$

$$\cup \lambda^0 \mathrm{co}\, (\partial g(F(x)) \circ ((\partial F(x))_\infty \backslash \{0\})) - (C - x)^*.$$

By choosing $\lambda^i = 0$ whenever $f_i(x) = 0$, we obtain the conclusion. $\qquad\square$

The conclusion of the preceding theorem does not ensure that the Lagrange multiplier $\lambda^0 \neq 0$. A suitable constraint qualification will ensure that $\lambda^0 \neq 0$ as we saw for a general constrained problem in the previous section.

Now consider the composite problem with max-min constraints

$$\text{(P)} \qquad\qquad \text{minimize} \ \min(g \circ F)(x)$$
$$\text{subject to} \ \max_{1 \leq k \leq r} \min_{1 \leq j \leq q_k} f_k^j(x) \leq 0 \, ,$$

where $F : \mathbb{R}^n \to \mathbb{R}^m$ and $f_k^j : \mathbb{R}^n \to \mathbb{R}$ are continuous, and $g : \mathbb{R}^m \to \mathbb{R}$ is convex.

Given an integer q, let Δ_q denote the q-simplex; that is,

$$\Delta_q := \left\{ \mu \in \mathbb{R}^q \ \Big| \ \sum_{j=1}^q \mu^j = 1, \ \mu^j \geq 0, \ j = 1, 2, \ldots, q \right\}.$$

Denote by $O_{l,s}$ the zero element of $L(\mathbb{R}^l, \mathbb{R}^s)$ and by O_l the zero element of $\mathbb{R}^l$, for $l, s \in \mathbb{N}$. For the sets $A \subset L(\mathbb{R}^l, \mathbb{R}^s)$ and $B \subset L(\mathbb{R}^q, \mathbb{R}^s)$, the product set $A \times B$ is given by

$$A \times B := \left\{ (a, b) \in L(\mathbb{R}^{l+q}, \mathbb{R}^s) \ | \ a \in A, \ b \in B \right\}.$$

Corollary 4.3.2 *For the problem (P), assume that F admits a pseudo-Jacobian map ∂F which is upper semicontinuous at $\overline{x}$ and f_k^j admits a bounded pseudo-Jacobian $\partial f_k^j(\overline{x})$, for each k and j. If $\overline{x}$ is a local minimizer* *for the problem (P), then there exist $\overline{\mu}_0 := (\overline{\mu}_0^0, \overline{\mu}_0^1, \ldots, \overline{\mu}_0^r) \in \Delta_{r+1}$, and $\overline{\mu}_k := (\overline{\mu}_k^1, \ldots, \overline{\mu}_k^{q_k}) \in \Delta_{q_k}$, such that*

$$\sum_{j=1}^{q_k} \overline{\mu}_0^k \overline{\mu}_k^j f_k^j(\overline{x}) = 0, \ k = 1, 2, \ldots, r$$

$$0 \in [\overline{\mu}_0^0 \overline{\mathrm{co}}(\partial g(F(\overline{x}))) \circ \partial F(\overline{x}) + \sum_{k=1}^{r} \sum_{j=1}^{q_k} \overline{\mu}_0^k \overline{\mu}_k^j \mathrm{co}(\partial f_k^j(\overline{x}))]$$

$$\cup [\overline{\mu}_0^0 \mathrm{co}(\partial g(F(\overline{x}))) \circ ((\partial F(\overline{x}))_\infty \backslash \{0\})].$$

Proof. Observe first that if $\overline{x}$ is a local minimizer for the problem (P), then there exist $\overline{\mu}_k \in \Delta_{q_k}$, $k = 1, 2, \ldots, r$, such that $(\overline{x}, \overline{\mu}_1, \ldots, \overline{\mu}_r)$ is a local minimizer for the following problem, denoted (P'),

$$\text{minimize}_{(x, \mu_1, \ldots, \mu_r)} \ (g \circ F)(x)$$
$$\text{subject to} \quad (x, \mu) \in \mathbb{R}^n \times \prod_{k=1}^{r} \Delta_{q_k},$$
$$\sum_{j=1}^{q_k} \mu_k^j f_k^j(x) \leq 0, \ k = 1, 2, \ldots, r,$$

where $\mu = (\mu_1, \ldots, \mu_r)$. Define $\tilde{F} : \mathbb{R}^n \times \prod_{k=1}^{r} \mathbb{R}^{q_k} \to \mathbb{R}^m$ by $\tilde{F}(x, \mu) = F(x)$ and $f_k : \mathbb{R}^n \times \prod_{k=1}^{r} \mathbb{R}^{q_k} \to \mathbb{R}$, $k = 1, 2, \ldots, r$, by

$$f_k(x, \mu) = \sum_{j=1}^{q_k} \mu_k^j f_k^j(x).$$

Put $C = \mathbb{R}^n \times \prod_{k=1}^{r} \Delta_{q_k}$. Rewrite (P') as (P''):

$$\text{minimize}_{(x, \mu)} \ (g \circ \tilde{F})(x, \mu)$$
$$\text{subject to} \quad (x, \mu) \in C,$$
$$f_k(x, \mu) \leq 0, \ k = 1, 2, \ldots, r.$$

It can be verified that the set

$$\partial \tilde{F}(x, \mu) := \partial F(x) \times \{O_{\ell, m}\}$$

is a pseudo-Jacobian of $\partial \tilde{F}$ at (x, μ), where $\ell = \sum_{k=1}^{r} q_k$. The upper semicontinuity of $\partial \tilde{F}$ at $(\overline{x}, \overline{\mu})$ follows from the upper semicontinuity of ∂F at $\overline{x}$. Now the set

$$\partial f_k(\overline{x}, \overline{\mu}) := \sum_{j=1}^{q_k} \overline{\mu}_k^j \partial f_k^j(\overline{x}) \times \{O_{\ell, 1}\} + \sum_{j=1}^{q_k} f_k^j(\overline{x}) e_k^j$$

is a bounded pseudo-Jacobian of f_k at $(\overline{x}, \overline{\mu})$, where

$$e_k^j := (O_n, O_{q_1}, \ldots, O_{q_{k-1}}, e_{j,k}, O_{q_{k+1}}, \ldots, O_{q_r})$$

and $e_{j,k}$ is the jth unit vector of $\mathbb{R}^{q_k}$. By Theorem 4.3.1, there exists $\bar{\mu}_0 \in \Delta_{r+1}$ such that

$$\bar{\mu}_0^k f_k(\bar{x}, \bar{\mu}) = 0, \ k = 1, 2, \ldots, r \tag{4.6}$$

and

$$O_{n+\ell} \in \left[\bar{\mu}_0^0 \overline{\mathrm{co}}(\partial g(F(\bar{x}))) \circ \partial \tilde{F}(\bar{x}, \bar{\mu}) + \sum_{k=1}^r \bar{\mu}_0^k \mathrm{co}(\partial f_k(\bar{x}, \bar{\mu})) \right]$$

$$\cup \left[\bar{\mu}_0^0 \mathrm{co}\, \partial g(F(\bar{x})) \circ ((\partial \tilde{F}(\bar{x}, \bar{\mu}))_\infty \backslash \{0_{n+\ell,m}\}) \right] - (C - (\bar{x}, \bar{\mu}))^*. \tag{4.7}$$

Now (4.6) can be rewritten as

$$\sum_{j=1}^{q_k} \bar{\mu}_0^k \bar{\mu}_k^j f_k^j(\bar{x}) = 0, \ k = 1, 2, \ldots, r.$$

It can be verified that

$\overline{\mathrm{co}}(\partial g(F(x))) \circ \partial \tilde{F}(\bar{x}, \bar{\mu}) = \overline{\mathrm{co}}(\partial g(F(x))) \circ \partial F(x) \times \{O_{\ell,m}\},$
$\overline{\mathrm{co}}(\partial f_k(\bar{x}, \bar{\mu})) \subset \sum_{j=1}^{q_k} \bar{\mu}_k^j \mathrm{co}(\partial f_k^j(\bar{x})) \times \{O_{\ell,1}\} + \sum_{j=1}^{q_k} f_k^j(\bar{x}) e_k^j,$
$\mathrm{co}(\partial g(F(\bar{x}))) \circ ((\partial \tilde{F}(\bar{x}, \bar{\mu}))_\infty \backslash \{0\}) = \mathrm{co}(\partial g(F(x))) \circ ((\partial F(x))_\infty \backslash \{O\}),$
$(C - (\bar{x}, \bar{\mu}))^* = \{O_n\} \times ((\prod_{k=1}^r \Delta_{q_k}) - (\bar{\mu}))^*.$

¿From these relations and (4.7), we get

$$O_{n+\ell} \in \left[\bar{\mu}_0^0 \overline{\mathrm{co}}(\partial g(F(\bar{x}))) \circ (\partial F(\bar{x}) \times \{O_{\ell,m}\}) \right.$$

$$\left. + \sum_{k=1}^r \sum_{j=1}^{q_k} \bar{\mu}_0^k \bar{\mu}_k^j (\mathrm{co}(\partial f_k^j(\bar{x})) \times \{O_{\ell,1}\}) + \sum_{k=1}^r \sum_{j=1}^{q_k} \bar{\mu}_0^k f_k^j(\bar{x}) e_k^j \right]$$

$$\bigcup \bar{\mu}_0^0 \overline{\mathrm{co}}(\partial g(F(\bar{x}))) \circ (((\partial F(\bar{x}))_\infty \backslash \{O_{n,m}\}) \times \{O_{\ell,m}\})$$

$$- \{O_n\} \times \left(\left(\prod_{k=1}^r \Delta_{q_k} \right) - (\bar{\mu}) \right)^*.$$

This implies that

$$O_n \in \left[\bar{\mu}_0^0 \overline{\mathrm{co}}(\partial g(F(\bar{x}))) \circ \partial F(\bar{x}) + \sum_{k=1}^r \sum_{j=1}^{q_k} \bar{\mu}_0^k \bar{\mu}_k^j \mathrm{co}(\partial f_k^j(\bar{x})) \right]$$

$$\cup \bar{\mu}_0^0 \mathrm{co}(\partial g(F(\bar{x}))) \circ ((\partial F(\bar{x}))_\infty \backslash \{O_{n,m}\}).$$

$\square$

Corollary 4.3.3 *Let $F : \mathbb{R}^n \to \mathbb{R}^m$ be a continuous map, let $g : \mathbb{R}^m \to \mathbb{R}$ be a convex function, and let $C \subset \mathbb{R}^m$ be a closed convex set. Assume that F admits a pseudo-Jacobian map ∂F which is upper semicontinuous at $x \in C$. If x is a local minimizer of the composite problem*

$$minimize \ (g \circ F)(x)$$
$$subject \ to \ \ x \in C,$$

then

$$0 \in \overline{co}(\partial g(F(x))) \circ \partial F(x) \cup co\,(\partial g(F(x)) \circ ((\partial F(x))_\infty \backslash \{0\})) - (C - x)^*.$$

Proof. The conclusion follows from the preceding theorem by taking for each i, $f_i(x) = -1$, for all x. In this case, $\lambda^i = 0$, for $i = 1, 2, \ldots, m$, and so $\lambda^0 = 1$. $\qquad\square$

The following example shows that the necessary condition in Corollary 4.3.3 is, in general, not valid without a recession cone condition.

Example 4.3.4 Let $F : \mathbb{R}^2 \to \mathbb{R}^2$ and $g : \mathbb{R}^2 \to \mathbb{R}$ be defined by

$$F(x, y) = \left(x^{2/3}\mathrm{sign}(x) + \frac{y^4}{2}, \sqrt{2}x^{1/3} + \frac{y^2}{\sqrt{2}}\right),$$

$g(u, v) = u + v^2$, and $C = \{(x, y) \in \mathbb{R}^2 \mid x \leq 0, \ y \leq 0\}$. Then F is continuous, but not Lipschitz, g is convex, and the composite function $g \circ F$ is given by

$$(g \circ F)(x, y) = x^{2/3}(\mathrm{sign}(x) + 2) + y^4 + 2x^{1/3}y^2.$$

The function $g \circ F$ attains its local minimum at $(0, 0)$. A pseudo-Jacobian of F at $(0, 0)$ and its recession cone are given, respectively, by

$$\partial F(0, 0) = \left\{\begin{pmatrix} \alpha & 0 \\ \alpha^2 & 0 \end{pmatrix} : \alpha \geq 1\right\}$$

$$\partial F(0, 0)_\infty = \left\{\begin{pmatrix} 0 & 0 \\ \beta & 0 \end{pmatrix} : \beta \geq 0\right\}.$$

Clearly, $0 \notin \overline{co}(\partial g(F(0, 0))) \circ \partial F(0, 0) - (C - (0, 0))^*$. However,

$$0 \in co\,(\partial g(F(0, 0)) \circ ((\partial F(0, 0))_\infty \backslash \{0\})) - (C - (0, 0))^*.$$

Sufficient Conditions

We now establish conditions which ensure that a feasible point is a local or strict local minimizer of $g \circ F$ over a closed convex set C. The next result presents a test for local optimality of the continuous convex composite function $g \circ F$.

Theorem 4.3.5 *Let $F : \mathbb{R}^n \to \mathbb{R}^m$ be a continuous map; let $g : \mathbb{R}^m \to \mathbb{R}$ be a convex function; let C be a closed convex subset of $\mathbb{R}^m$ and let $a \in C$. If there exists a neighborhood U of a such that F admits a pseudo-Jacobian map ∂F which is upper semicontinuous on U and if*

$$\langle w, x - a \rangle > 0,$$

for each $x \in C \cap U \backslash \{a\}$ and for each

$$w \in (\overline{co}(\partial g(F(x))) \circ \partial F(x)) \cup co\{\partial g(F(x)) \circ ((\partial F(x))_\infty \backslash \{0\})\},$$

then a is a local minimizer of $g \circ F$ over C.

Proof. Suppose that a is not a local minimizer of $g \circ F$ over C. Then there exists $y \in U \cap C$ such that $(g \circ F)(a) > (g \circ F)(y)$. By the continuity of $g \circ F$, we can find $b = y + \alpha(a - y)$ for some $\alpha \in (0, 1)$ with

$$(g \circ F)(b) > (g \circ F)(y).$$

Let $\varepsilon > 0$. Put $A_\varepsilon(x) := (\partial g(F(x)) + \varepsilon B_m^{tr}) \circ \partial F(x)$. Corollary 2.3.4 gives us for each $\varepsilon > 0$, $cl(A_\varepsilon(x))$ is a pseudo-Jacobian of $g \circ F$ at each $x \in U \cap C$. Take $\varepsilon = 1/k$, $k \in \mathbb{N}$. Because $(g \circ F)(b) - (g \circ F)(y) > 0$, in view of the mean value theorem, there exist $z_k = y + \alpha_k(b - y)$, and $\alpha_k \in (0, 1)$, such that $w_k^{tr}(b - y) > 0$, for some $w_k \in \overline{co}(A_{1/k})$. So, we can find $p_k \in co(A_{1/k}(z_k))$ satisfying

$$\langle p_k, (b - y) \rangle > 0.$$

By Caratheodory's theorem, p_k can be represented as

$$p_k = \sum_{i=1}^{n+1} \lambda_{ik} \langle u_{ik} + \frac{1}{k} a_{ik}, v_{ik} \rangle,$$

for some $u_{ik} \in \partial g(F(z_k))$, $a_{ik} \in B_m$, $v_{ik} \in \partial F(z_k)$, $\lambda_{ik} \geq 0$ with $\sum_{i=1}^{n+1} \lambda_{ik} = 1$. Now

$$\sum_{i=1}^{n+1} \lambda_{ik} \langle u_{ik} + \frac{1}{k} a_{ik}, v_{ik}(b - y) \rangle > 0. \tag{4.8}$$

Let

$$I := \{1, 2, \ldots, n+1\}, \ I_1 = \{i \in I : \{v_{ik}\}_{k \geq 1} \text{ is bounded}\}, \text{ and } I_2 := I \backslash I_1.$$

Then we may assume, without loss of generality, that $\lambda_{ik} \to \lambda_i$, $\sum_{i=1}^{n+1} \lambda_i = 1$, $z_k \to z \in [b, y]$, Clearly, $z \neq a$. By the continuity of F and the property of the subdifferential of convex functions, we may assume that $u_{ik} \to u_i \in \partial g(F(z))$. We may also assume that for each $i \in I_1$, $v_{ik} \to v_i$ for some v_i. The upper semicontinuity of ∂F at z implies $v_i \in \partial(z)$. Represent (4.8) as

$$\Big\langle \sum_{i\in I_1} \lambda_{ik}(u_{ik} + \frac{1}{k}a_{ik}) \circ v_{ik} + \sum_{i\in I_2}(\lambda_{ik}u_{ik} + \frac{1}{k}a_{ik}) \circ v_{ik}, (b-y)\Big\rangle > 0.$$

Employing the same method of proof as in the proof of Theorem 4.3.1, we find an element

$$w \in \overline{co}\, \partial g(F(z)) \circ \partial F(z) \cup co\,(\partial g(F(z)) \circ ((\partial F(z))_\infty \backslash \{0\}))$$

such that $\langle w, (b-y)\rangle \geq 0$. Because $z \in [b, y]$, there exists $\beta > 0$ such that $z - a = \beta(y - b)$. Hence $\langle w, z - a\rangle \leq 0$, which contradicts the hypothesis and so the proof is completed. $\qquad\square$

Theorem 4.3.6 *Let $F : \mathbb{R}^n \to \mathbb{R}^m$ be a continuous map; let $g : \mathbb{R}^m \to \mathbb{R}$ be a convex function and let $C \subset \mathbb{R}^n$ be a closed convex set. Assume that F admits a pseudo-Jacobian map ∂F which is upper semicontinuous on a neighborhood of $a \in C$ and that*

$$\langle w, y\rangle > 0$$

for all $w \in (\overline{co}(\partial g(F(a))) \circ \partial F(a) \cup (co(\partial g(F(a))) \circ ((\partial F(a))_\infty \backslash \{0\}))$, and for all $y \in T(C, a)$, where $T(C, a)$ is the contingent cone to C at a. Then a is a strict local minimizer of $g \circ F$ over C.

Proof. Suppose to the contrary that a is not a strict local minimizer of $g \circ F$ over C. Then there is $a_i \to a$, $a_i \in C \backslash \{a\}$ such that $(g \circ F)(a_i) - (g \circ F)(a) \leq 0$. We may assume that $a_i - a/\|a_i - a\| \to y \in T(C, a)$. We use the mean value theorem to infer that there exist some $c_i \in (a_i, a)$ and $\beta_i \in \overline{co}\,[\partial g(F(c_i)) + (1/i)B_m^{tr}\partial F(c_i)(a_i - a)]$ such that

$$\beta_i = (g \circ F)(a_i) - (g \circ F)(a) \leq 0.$$

Hence, for each i, we can find $p_i \in co\,(\partial g(F(c_i)) + (1/i)B_m) \circ \partial F(c_i)$ satisfying

$$\langle p_i, a_i - a\rangle - \frac{\|a - a_i\|}{i} \leq 0. \tag{4.9}$$

By Caratheodory's theorem, we can represent p_i as

$$p_i = \sum_{j=1}^{n+1} \lambda_{ji}\big(u_{ji} + \frac{1}{i}b_{ji}\big) \circ v_{ji},$$

where

$$\lambda_{ji} \geq 0,\ \sum_{j=1}^{n+1} \lambda_{ji} = 1,\ u_{ji} \in \partial g(F(c_i)),\ b_{ji} \in B_m,\ v_{ji} \in \partial F(c_i).$$

Let $J := \{1, 2, \ldots, n+1\}$, $J_1 := \{j \in J : \{v_{ji}\}_{i \geq 1} \text{ is bounded}\}$, and $J_2 := J \setminus J_1$. Divide (4.9) by $\|a_i - a\|$ to get

$$\langle \sum_{j \in J_1} \lambda_{ji}(u_{ji} + \frac{1}{i}b_{ji}) \circ v_{ji}, \frac{(a_i - a)}{\|a_i - a\|} \rangle$$

$$+ \langle \sum_{j \in J_2} \lambda_{ji}(u_{ji} + \frac{1}{i}b_{ji}) \circ v_{ji}, \frac{(a_i - a)}{\|a_i - a\|} \rangle - \frac{1}{i} \leq 0.$$

As in the proof of the preceding theorem, by passing to the limit in the latter inequality when i tends to ∞, we can find

$$w \in \overline{\mathrm{co}}(\partial g(F(a))) \circ \partial F(a) \cup \mathrm{co}\left((\partial g(F(a)) \circ ((\partial F(a))_\infty \setminus \{0\})\right)$$

satisfying $\langle w, y \rangle \leq 0$, which contradicts the hypothesis and so the proof is completed. $\square$

Second-Order Conditions

In this section, we prove second-order results for the following convex composite problem,

(CP) minimize $(g \circ F)(x)$

                           subject to $x \in C$,

where $g : \mathbb{R}^m \to \mathbb{R}$ is convex and $F : \mathbb{R}^n \to \mathbb{R}^m$ is Gâteaux differentiable. In order to introduce a new Lagrangian for this problem we define the conjugate (or the Fenchel transform) of the convex function g by

$$g^*(\xi) := \sup\{\langle \xi, x \rangle - g(x) : x \in \mathbb{R}^m\}, \quad \text{for } \xi \in \mathbb{R}^m.$$

This function takes values in $\mathbb{R} \cup \{+\infty\}$. We state some of the properties of conjugate functions needed in the sequel. Recall that ∂^{ca} denotes the subdifferential in the sense of convex analysis (see Section 1.4).

Lemma 4.3.7 *Let g be a convex function on $\mathbb{R}^m$. Then g is a convex function and the following assertions are equivalent for every vector x and ξ of the effective domains of g and g^**

(i) $g^*(\xi) + g(x) = \langle \xi, x \rangle$,
(ii) $\xi \in \partial^{ca}g(x)$.

Proof. Because for every fixed $x \in \mathbb{R}^n$, the function $\xi \mapsto \langle \xi, x \rangle - g(x)$ is affine, hence convex, the conjugate function being a supremum of convex functions is convex. For the equivalence of (i) and (ii), let $\xi \in \partial^{ca}(x)$. Then by definition one has

$$\langle \xi, x \rangle - g(x) \geq \langle \xi, y \rangle - g(y)$$

for every $y \in \mathbb{R}^m$, and so

$$\langle \xi, x \rangle - g(x) \geq \sup\{\langle \xi, y \rangle - g(y) : y \in \mathbb{R}^m\} = g^*(\xi).$$

On the other hand, by the definition of conjugate functions,

$$g^*(\xi) \geq \langle \xi, x \rangle - g(x).$$

Therefore, equality (i) is obtained.

Conversely, equality in (i) shows that

$$\sup_{y \in \mathbb{R}^m} (\langle \xi, y \rangle - g(y)) = \langle \xi, x \rangle - g(x).$$

Therefore, for every $y \in \mathbb{R}^m$ one has

$$\langle \xi, y \rangle - g(y) \leq \langle \xi, x \rangle - g(x),$$

which implies

$$g(y) - g(x) \geq \langle \xi, y - x \rangle.$$

According to Proposition 1.4.3, ξ is an element of $\partial^{ca} g(x)$. $\qquad \square$

Now we define the Lagrangian of the problem (CP) by

$$L(x, y^*) = \langle y^*, F(x) \rangle - g^*(y^*) \quad \text{for } x \in \mathbb{R}^n, \ y^* \in \mathbb{R}^m,$$

where g^* is the conjugate function of g. We define the ε-subdifferential of g at y by

$$\partial_\varepsilon g(y) = \{y^* \in \mathbb{R}^m : g(z) \geq g(y) + \langle y^*, z - y \rangle \text{ for all } z \in \mathbb{R}^m\}.$$

Let $h : \mathbb{R}^n \to \mathbb{R}$. A real-valued function $\phi(x, u)$ defined on $\mathbb{R}^n \times \mathbb{R}^n$ is said to be an *LMO-approximation* for h at z in the sense of Ioffe if $\phi(x, 0) = h(x)$ for any x in a neighborhood of z, if the function $u \to \phi(x, u)$ is convex and continuous, and if

$$\liminf_{y \to z, u \to 0} \|u\|^{-1}(\phi(y, u) - h(y + u)) \geq 0.$$

Lemma 4.3.8 *Let $\varepsilon > 0$ be given and let $\phi(x, u)$ be an LMO-approximation for a locally Lipschitz function h at z. Then the function*

$$\phi_\varepsilon(x, u) := \sup\{\langle u^*, u \rangle - \phi(x, u^*) : u^* \in \partial_\varepsilon \phi(x, 0)\}$$

is an LMO-approximation for h at z.

Proof. Let k be a Lipschitz rank for h and let $0 < \eta < k$ be given. Choose a positive $\delta \leq \varepsilon/(2k)$ such that

$$\phi(x, u) + \eta \|u\| \geq h(x + u) \quad \text{for } x \in z + \delta B_n, \ u \in \delta B_n. \tag{4.10}$$

We show that (4.10) remains valid when ϕ is replaced by ϕ_ε which will complete the proof. To this end, let us fix arbitrary elements x and u satisfying (4.10). It is clear that

$$\phi_\varepsilon(x, u) \leq \phi(x, u).$$

So, if equality holds, we are done. Hence we assume that $\phi_\varepsilon(x, u) < \phi(x, u)$. Denote by

$$t_0 := \inf\{t > 0 : \phi_\varepsilon(x, tu) < \phi(x, tu)\}.$$

Then $t_0 < 1$ and also $t_0 > 0$ because when u' is close to 0, one has

$$\phi(x, u') = \sup\{\langle x^*, u' \rangle - \phi^*(x, u^*) : u^* \in \partial_\varepsilon \phi(x, 0)\}$$
$$= \phi_\varepsilon(x, u').$$

First we wish to prove that there is $u^* \in \partial \phi(x, t_0 u)$ such that

$$\phi(x, 0) + \phi^*(x, u^*) = \varepsilon. \tag{4.11}$$

Indeed, because $\phi_\varepsilon \leq \phi$ and equality holds at $t_0 u$, we have the inclusion

$$\partial \phi_\varepsilon(x, t_0 u) \subseteq \partial \phi(x, t_0 u).$$

Furthermore, because $\phi(x, \cdot)$ is convex and continuous, the set $\partial \phi_\varepsilon(x, t_0 u)$ is nonempty and by definition,

$$\partial \phi_\varepsilon(x, t_0 u) \subseteq \partial_\varepsilon \phi(x, 0).$$

Hence there exists some element u_1^* from $\partial \phi(x, t_0 u) \cap \partial_\varepsilon \phi(x, 0)$. This yields

$$\phi(x, 0) + \phi * (x, u_1^*) \leq \varepsilon.$$

On the other hand, for $t > t_0$ if it is true that

$$\phi(x, tu) > \phi_\varepsilon(x, tu)$$

and $u^* \in \partial \phi(x, tu)$, then this u^* does not belong to the set $\partial_\varepsilon \phi(x, 0)$ (otherwise one would have $\phi(x, tu) = \phi_\varepsilon(x, tu)$), which implies

$$\phi(x, 0) + \phi^*(x, u^*) \geq \varepsilon.$$

By taking a sequence $\{t_k\}$ such that $t_k > t_0$ and $t_k \to t_0$, one may find then an element $u_2^* \in \partial \phi(x, t_0 u)$ such that

$$\phi(x,0) + \phi^*(x, u_2^*) \geq \varepsilon.$$

A convex combination u^* of u_1^* and u_2^* will satisfy (4.11). Now from (4.11) we deduce

$$\varepsilon - \langle u^*, t_0 u \rangle = \phi(x,0) - \phi(x, t_0 u)$$

and by (4.10) one has

$$\langle u*, t_0 u \rangle \geq h(x + t_0 u) - h(x) + \varepsilon - \eta t_0 \|u\|$$
$$\geq \varepsilon - (k + \eta) t_0 \|u\|.$$

Because $0 < t_0 < 1$ and $\|u\| \leq \delta \leq \varepsilon/(2k)$, the above inequality gives that

$$\frac{\langle u^*, t_0 u \rangle}{\|u\|} \geq \frac{\varepsilon}{t_0 \|u\|} - (k + \eta)$$
$$\geq \frac{\varepsilon}{\|u\|} - (k + \eta) \geq k - \eta.$$

Clearly, u^* belongs to the set $\partial\phi(x,0)$, as well as to the sets $\partial_\varepsilon(x, t_0 u)$ and $\partial\phi(x, t_0 u)$, therefore

$$\phi_\varepsilon(x,u) + \eta\|u\| \geq \phi_\varepsilon(x, t_0 u) + \eta\|u\| + (1 - t_0)\langle u^*, u \rangle$$
$$\geq \phi(x, t_0 u) + \eta\|u\| + (1 - t_0)(k - \eta)\|u\|$$
$$\geq \phi(x, t_0 u) + \eta\|u\| \geq f(x + t_0 u).$$

By this the proof is complete.   □

Using LMO-approximations, we have the following characterizations of a local minimum of a locally Lipschitz function.

Lemma 4.3.9 *Assume that h is locally Lipschitz on $\mathbb{R}^n$ and $z \in \mathbb{R}^n$ and that $\phi(x,u)$ is an LMO-approximation of h at z. Let $\beta_\xi(x) = -\min\{\phi^*(x, u^*) : \|u^*\| \leq \xi\}$ for any fixed $\xi > 0$. Then the following conditions are equivalent*

(i) h attains a local minimum at z.
(ii) $0 \in \partial\phi(z,0)$ and β_ξ attains a local minimum at z for any $\xi > 0$.
(iii) $0 \in \partial\phi(z,0)$ and β_ξ attains a local minimum at z for some $\xi > 0$.

Proof. First note that by the definition of conjugate functions one has

$$\phi^*(x, u^*) + \phi(x,0) \geq 0.$$

Therefore,

$$h(x) = \phi(x,0) \geq -\phi^*(x, u^*) \geq \beta_\xi(x). \tag{4.12}$$

To obtain (i) from (iii), we notice that $-\phi^*(z,0) = \phi(z,0)$ whenever $0 \in \partial\phi(z,0)$. Consequently,

$$\beta_\xi(z) \geq -\phi^*(z,0) = \phi(z,0) = h(z).$$

This shows that if z is a local minimizer of β_ξ, then by (4.12),

$$h(x) \geq \beta_\xi(x)\beta_\xi(z)h(z)$$

as soon as x is in a small neighborhood of z.

The implication (ii)→(iii) is evident. Now we show that (i) is obtained from (ii). In view of (i), for each $u \in \mathbb{R}^n$ with $\|u\| = 1$, one has $h(z + tu) - h(z) \geq 0$ for $t > 0$ sufficiently small. According to the definition of LMO-approximations, one deduces

$$\liminf_{t\downarrow 0} \frac{\phi(z,tu) - \phi(z,0)}{t} = \liminf_{t\downarrow 0} \frac{\phi(z,tu) - h(z+tu) + h(z+tu) - hz)}{t}$$

$$\geq \liminf_{t\downarrow 0} \frac{\phi(z,tu) - h(z+tu)}{t}$$

$$+ \liminf_{t\downarrow 0} \frac{h(z+tu) - h(z)}{t}$$

$$\geq 0.$$

Thus the directional derivative $\phi'((z,0);u) \geq 0$ for every direction $u \in \mathbb{R}^n$ and hence $0 \in \partial\phi(z,0)$. Furthermore, let $\xi > 0$ be fixed. It follows from the definition of LMO-approximations that there exists some $\delta_0 > 0$ such that

$$\phi(x,u) + (\xi/2)\|u\| \geq h(x+u) \geq h(z)$$

for $\|x - z\| \leq \delta_0$ and $\|u\| \leq \delta_0$. Then

$$p(x,u) := \phi(x,u) + \xi\|u\| \geq h(z) + (\xi/2)\|u\|. \tag{4.13}$$

Choose $0 < \delta \leq \delta_0$ so small that

$$h(x) \leq h(z) + (\xi/2)\delta_0 \quad \text{whenever } x \in z + \delta B - n.$$

For x as above,

$$p(x,0) = h(x) \leq h(z) + (\xi/2)\delta_0.$$

This inequality together with (4.13) applied to u with $\|u\| = \delta_0$ and the convexity of $p(x,\cdot)$ produces

$$\inf_{u\in\mathbb{R}^n} p(x,u) = \inf_{u\in\delta_0 B_n} p(x,u).$$

Because $\beta_\xi(x) = \inf_{u\in\mathbb{R}^n} p(x,u)$, combining the above equality with (4.12) and (4.13) gives

$$\beta_\xi(x) \geq h(z) \geq \beta_\xi(z)$$

as requested.   □

If g is convex and F is continuous Gâteaux differentiable, then the composite function $f := g \circ F$ is directionally differentiable. Its directional derivative at x is given by

$$f'(x; d) = g'(F(x); \nabla F(x)(d)).$$

Let

$$K(x) := \{u \in \mathbb{R}^n : g(F(x) + t\nabla F(x)(u)) \leq g(F(x)) \text{ for some } t > 0\}$$

and let

$$D(x) := \{u \in \mathbb{R}^n : g'(F(x); \nabla F(x)(u)) \leq 0\}.$$

For $z \in \mathbb{R}^n$, define

$$M_0(z) = \{y^* \in \mathbb{R}^m : y^* \in \partial^C g(F(z)), \ y^* \circ \nabla F(z) = 0\}.$$

Then $M_0(z) \neq \emptyset$ provided $0 \in \partial^C g(F(z)) \circ \nabla F(z)$. Now we state the second-order optimality conditions for the function $g \circ F$.

Theorem 4.3.10 *Let* $a \in \mathbb{R}^n$. *Assume that* g *is a convex function and* F *is Gâteaux differentiable at* a. *Suppose that for each* $y^* \in \mathbb{R}^m$, $\partial^2 L(a, y^*)$ *is a Gâteaux pseudo-Hessian of* $L(\cdot, y^*)$ *at* a *and that* $\partial^2 L(a, \cdot)$ *is upper semicontinuous on* $\mathbb{R}^m$. *If* a *is a local minimizer of* $g \circ F$, *then*

$$\sup\{\langle u, M(u) \rangle : M \in \partial^2 L(a, y^*), \ y^* \in M_0(a)\} \geq 0, \ \forall u \in K(a).$$

Proof. Let $u \in K(a)$. First, observe from Theorem 4.3.1 that

$$0 \in \partial^C g(F(a)) \circ \nabla F(a)$$

as $g \circ F$ attains a local minimum at a. This yields $M_0(a) \neq \emptyset$. Now let $\varepsilon > 0$. Then it follows from Lemma 4.3.9 that the function

$$\rho_\varepsilon(x; u) = g_\varepsilon(\nabla F(a)u + F(x))$$

is an LMO-approximation of f at a, where

$$g_\varepsilon(y) = \sup\{y^{*tr}y - g^*(y^*) : y^* \in \partial_\varepsilon g(F(x))\}.$$

Let $\eta > 0$, and define the function $\phi_{\eta\varepsilon}$ by

$$\phi_{\eta\varepsilon}(x) = \max\{L(x, y^*) : y^* \in M_{\eta\varepsilon}(x)\},$$

where

$$M_{\eta\varepsilon}(x) = \{y^* \in \mathbb{R}^m : y^* \in \partial_\varepsilon g(F(x)), \ \|y^* \circ \nabla F(z)\| \leq \eta\}.$$

By applying the conjugate duality theory, we can get

$$\phi_{\eta\varepsilon}(x) = -\min\{\rho_\varepsilon^*(x, u^*) : \|u^*\| \leq \eta\},$$

where $\rho_\varepsilon^*(x, u^*) = \sup\{\langle u^*, u\rangle - \rho_\varepsilon(x, u) : u \in \mathbb{R}^n\}$ is the Fenchel conjugate of $\rho_\varepsilon(x, \cdot)$. Because f is locally Lipschitz and a is a local minimizer of f, we deduce from Lemma 4.3.7 that $\phi_{\eta\varepsilon}$ attains a local minimum at a, and hence $\phi_{\eta\varepsilon}(x) \geq \phi_{\eta\varepsilon}(a) = g(F(a))$ for any x in a neighborhood of a. Then, from the classical mean value theorem and the definition of the Gâteaux pseudo-Hessian, we get that for t sufficiently small and positive,

$$\begin{aligned}
g(F(a)) \leq \phi_{\eta\varepsilon}(a + tu) &= \sup\{L(a + tu, y^*) : y^* \in M_{\eta\varepsilon}(a)\} \\
&= \sup\{y^{*T} F(a + tu) - g^*(y^*) : y^* \in M_{\eta\varepsilon}(a)\}.
\end{aligned}$$

Let us express

$$\begin{aligned}
\langle y^*, F(a + tu)\rangle - g^*(y^*) &= \langle y^*, F(a)\rangle + \langle y^*, \nabla F(a + su)(tu)\rangle - g^*(y^*) \\
&= \langle y^* F(a)\rangle + \langle y^*, \nabla F(a)(tu)\rangle \\
&\quad + \langle su, A(tu)\rangle + o(s)(tu) - g^*(y^*)
\end{aligned}$$

for some $s \in (0, t)$ and some $A \in \partial^2 L(a, y^*)$. Because $u \in K(a)$ and g is convex, there exists $t_0 > 0$ such that

$$g(F(a) + t\nabla F(a)u) \leq g(F(a)) \ \forall t \in [0, t_0].$$

The basic properties of the Fenchel conjugate function of g give us

$$\langle y^*, (F(a) + t\nabla F(a)u)\rangle - g^*(y^*) \leq g(F(a) + t\nabla F(a)u) \leq g(F(a)), \forall t \in [0, t_0].$$

So, for sufficiently small $t > 0$,

$$\sup\ \{(st)\langle u, A(u)\rangle + o(s)(tu) : y^* \in M_{\eta\varepsilon}(a), \ A \in \partial^2 L(a, \ y^*)\} \geq 0.$$

Thus

$$\sup\ \left\{\langle u, A(u)\rangle + \frac{o(s)u}{s} : y^* \in M_{\eta\varepsilon}(a), \ A \in \partial^2 L(a, \ y^*)\right\} \geq 0.$$

As $t \downarrow 0$, $(o(s)/s) \to 0$ and so, we obtain

$$\sup\ \{\langle u, A(u)\rangle : y^* \in M_{\eta\varepsilon}(a), \ A \in \partial^2 L(a, \ y^*)\} \geq 0.$$

Because also

$$\bigcap_{\eta>0, \varepsilon>0} M_{\eta\varepsilon}(a) = M_0(a)$$

and $\partial^2 L(a, \cdot)$ is upper semicontinuous, the conclusion follows. $\qquad\square$

Corollary 4.3.11 *Let $a \in \mathbb{R}^n$. Assume that g is a convex function and F is Gâteaux differentiable at a. Suppose that for each $y^* \in \mathbb{R}^m$, $\partial^2 L(a, y^*)$ is a bounded Gâteaux pseudo-Hessian of $L(., y^*)$ at a and that $\partial^2 L(a, .)$ is upper semicontinuous on $\mathbb{R}^m$. If a is a local minimizer of $g \circ F$, then*

$$\sup\{\langle u, M(u)\rangle : M \in \partial^2 L(a, y^*), \ y^* \in M_0(a)\} \geq 0, \ \forall u \in \mathrm{cl}(K(a)).$$

Proof. We need only to notice that the conditions of the previous theorem are now true for any $u \in \mathrm{cl}(K(a))$ because $\partial^2 L(a, y^*)$ is bounded for each $y^* \in M_0(a)$. $\qquad\square$

Recall that the point a is a *strict local minimum of order 2* for the function $g \circ F$ if there exists $\varepsilon > 0$ and $r > 0$ such that for each $x \in B_r(a)\backslash\{0\}$,

$$f(x) \geq f(a) + \varepsilon\|x - a\|^2.$$

Theorem 4.3.12 *Let $a \in \mathbb{R}^n$. Assume that g is a convex function and F is continuously Gâteaux differentiable. Suppose that for each $y^* \in \mathbb{R}^m$, $\partial^2 L(\cdot, y^*)$ is a pseudo-Hessian of $L(\cdot, y^*)$. If $M_0(a) \neq \emptyset$ and if for each $u \in D(a)\backslash\{0\}$, there exist $\varepsilon > 0$ and $\delta > 0$ satisfying*

$$\inf_{v \in u + \delta B_n} \sup_{y^* \in M_0(a)} \inf_{M \in \mathrm{co}(\partial^2 L(a + \varepsilon B_n, y^*))} \langle v, M(v)\rangle > 0,$$

then a is a strict local minimum of order 2 for the function $g \circ F$.

Proof. Suppose to the contrary that a is not a strict local minimum of order 2 for $g \circ F$. Then there exist $\{x_k\} \subseteq \mathbb{R}^n$, $x_k \to a$, and $\varepsilon_k \downarrow 0$ as $k \to +\infty$ such that for each k,

$$f(x_k) \leq f(a) + \varepsilon_k\|x_k - a\|^2.$$

We may assume that $u_k := ((x_k - a)/\|x_k - a\|) \to u \in D(a)\backslash\{0\}$ as $k \to +\infty$. It now follows from the definition of the conjugate function that

$$g(F(x_k)) = \sup\{\langle y^*, F(x_k)\rangle - g^*(y^*) : y^* \in \mathbb{R}^n\}$$
$$\geq \sup\{\langle y^*, F(a + t_k u_k)\rangle - g^*(y^*) : y^* \in M_0(a)\},$$

where $t_k = \|x_k - a\| \to 0$ as $k \to \infty$. Now, by the Taylor expansion (see Theorem 2.2.20), there exist $s_k > 0$ with $t_k > s_k$ and $A_k \in \mathrm{co}\partial^2 L(a + s_k u_k, y^*)$ such that

$$\langle y^*, F(a + t_k u_k)\rangle - g^*(y^*) = \langle y^*, F(a)\rangle - g^*(y^*) + \langle y^*, \nabla F(a)(t_k u_k)\rangle$$
$$+ \frac{1}{2}\langle t_k u_k, A_k(t_k u_k)\rangle + \mathrm{o}(t_k^2 u_k),$$

where $\mathrm{o}(t_k^2 u_k)/t_k^2 \to 0$ as $k \to \infty$. Using the fact that $g(F(a)) = \langle y^*, F(a)\rangle - g^*(y^*)$ and $\langle y^*, \nabla F(a)\rangle = 0$, for $y^* \in M_0(a)$, we obtain that

$$\varepsilon_k \geq \sup_{y^* \in M_0(a)} \left\{ \frac{1}{2} \langle u_k, A_k(u_k) \rangle + \frac{o(t_k^2 u_k)}{t_k^2} \right\},$$

where $A_k \in \mathrm{co}(\partial^2 L(a + s_k u_k, y^*))$. Let $\alpha > 0$ be a constant such that

$$\sup_{y^* \in M_0(a)} \inf_{M \in \mathrm{co}(\partial^2 L(a+\varepsilon B_n, y^*))} \langle v, M(v) \rangle \geq \alpha > 0, \ \forall v \in u + \delta B_n.$$

Let k_0 be a sufficiently large integer such that $u_k \in u + \delta B_n$ and $A_k \in \mathrm{co}(\partial^2 L(a + \varepsilon B_n, y^*))$, for $k \geq k_0$. Let k_1 be another integer such that

$$\varepsilon_k - \frac{o(t_k^2 u_k)}{t_k^2} \leq \frac{\alpha}{4} \quad \text{for } k \geq k_1.$$

Hence we get

$$\frac{\alpha}{4} \geq \sup_{y^* \in M_0(a)} \frac{1}{2} \langle u_k, A_k(u_k) \rangle \geq \frac{\alpha}{2},$$

which contradicts the hypothesis and so the conclusion follows.   □

4.4 Multiobjective Programming

Partial Orders and Efficient Points

Let $\mathcal{B}$ be a binary relation in $\mathbb{R}^m$ that can be identified with a subset B of the product space $\mathbb{R}^m \times \mathbb{R}^m$ in the sense that for two points y_1 and $y_2 \in \mathbb{R}^m$, $y_1 \mathcal{B} y_2$ if and only if $(y_1, y_2) \in B$. A binary relation that satisfies the following properties is called a *partial order*.

(i)   Transitivity: $y_1 \mathcal{B} y_2$ and $y_2 \mathcal{B} y_3$ imply $y_1 \mathcal{B} y_3$.
(ii) Reflexivity: $y \mathcal{B} y$ for $y \in \mathbb{R}^m$.
(iii) Antisymmetry: $y_1 \mathcal{B} y_2$ and $y_2 \mathcal{B} y_1$ imply $y_1 = y_2$.

A partial order $\mathcal{B}$ is said to be *linear* if in addition it satisfies

(iv) $y_1 \mathcal{B} y_2$ and $t \geq 0$ imply $t y_1 \mathcal{B} t y_2$.
(v) $y_1 \mathcal{B} y_2$ and $y_3 \mathcal{B} y_4$ imply $(y_1 + y_3) \mathcal{B} (y_2 + y_4)$.

Linear partial orders have quite simple geometric structure. The next result shows that they can be characterized by convex cones.

Proposition 4.4.1 *Suppose that $\mathcal{B}$ is a linear partial order in $\mathbb{R}^m$. Then the set*

$$C_0 := \{ y \in \mathbb{R}^m : y \mathcal{B} 0 \}$$

is a convex and pointed cone. Conversely, if $C \subseteq \mathbb{R}^m$ is a convex and pointed cone, then the relation C defined by

$$y_1 \mathcal{C} y_2 \quad \textit{if and only if } y_1 - y_2 \in C,$$

is a linear partial order in $\mathbb{R}^m$.

Proof. For the first part of the proposition, let y_1 and y_2 be two points of C_0 and let $t \geq 0$. In view of (iv) and (v), one has $ty_1 \in C_0$ and $y_1 + y_2 \in C_0$. Hence C_0 is a convex cone. Furthermore, if $y \in C_0 \cap (-C_0)$, then one has $y\mathcal{B}$ and $0\mathcal{B}y$. The antisymmetry property gives that $y = 0$, by which the cone C_0 is pointed.

The proof of the converse is straightforward by using (i)–(v).   □

¿From now on we consider partial orders generated by convex and pointed cones only. Given such a cone $C \subseteq \mathbb{R}^m$, we use the notation $y_1 \geq_C y_2$ instead of $y_1 - y_2 \in C$. When $y_1 \geq_C y_2$ and $y_1 \neq y_2$, we write $y_1 >_C y_2$, or equivalently $y_1 - y_2 \in C \setminus \{0\}$.

Let $A \subseteq \mathbb{R}^m$ be a nonempty set. A point $a \in A$ is said to be an *efficient (minimal) point* of A with respect to the ordering cone C if there is no $y \in A$ such that $a >_C y$ or equivalently

$$(a - C) \cap A = \{a\}.$$

The set of all efficient points of A with respect to C is denoted by $\mathrm{Min}(A|C)$. When the interior of C is nonempty, efficient points of A with respect to the cone $\mathrm{int}(A) \cup \{0\}$ are traditionally called *weakly efficient points* of A with respect to C, and the set of all weakly efficient points of A is denoted $\mathrm{WMin}(A|C)$. Thus

$$a \in \mathrm{WMin}(A|C) \text{ if and only if } (a - \mathrm{int}(C)) \cap A = \emptyset.$$

First-Order Conditions

Let $f \colon \mathbb{R}^n \to \mathbb{R}^m$, $g \colon \mathbb{R}^n \to \mathbb{R}^p$, and $h \colon \mathbb{R}^n \to \mathbb{R}^q$ be continuous functions. Let the spaces $\mathbb{R}^m$ and $\mathbb{R}^k$ be partially ordered, respectively, by convex, closed and pointed cones C and K with nonempty interiors. We consider the following constrained multiobjective programming problem,

$$(VP) \qquad \mathrm{WMin} \quad f(x)$$
$$\text{subject to } g(x) \leq_K 0$$
$$h(x) = 0.$$

If we denote the feasible solution set by X, then our problem means finding a point $x_0 \in X$ such that the value $f(x_0)$ is a weakly efficient point of the set $f(X)$ with respect to the cone C. A point x_0 is a local weakly efficient solution of (VP) if there is a neighborhood U of x_0 such that $f(x_0)$ is a weakly efficient point of the set $f(X \cap U)$.

Let us equip the product space $\mathbb{R}^m \times \mathbb{R}^p \times \mathbb{R}^q$ with the Euclidean norm: for $\xi \in \mathbb{R}^m, \theta \in \mathbb{R}^p$ and $\gamma \in \mathbb{R}^q$, $\|(\xi, \theta, \gamma)\| = \sqrt{\|\xi\|^2 + \|\theta\|^2 + \|\gamma\|^2}$. And define $H := (f, g, h)$. It is a continuous function from $\mathbb{R}^n$ to $\mathbb{R}^m \times \mathbb{R}^p \times \mathbb{R}^q$. We also denote by T the set of all vectors $\lambda \in (C, K, \{0\})^*$ with $\|\lambda\| = 1$. Here $(C, K, \{0\})^*$ is the positive polar cone of the cone $(C, K, \{0\})$ which consists of vectors λ such that $\langle \lambda, w \rangle \geq 0$ for all vectors w of the cone $(C, K, \{0\})$.

Lemma 4.4.2 *Let $\omega_0 \in \mathbb{R}^m \times \mathbb{R}^k \times \mathbb{R}^l$ be a nonzero vector with $\max_{\lambda \in T} \langle \lambda, \omega_0 \rangle > 0$. Then there exists a unique point $\lambda_0 \in T$ such that*

$$\langle \lambda_0, \omega_0 \rangle = \max_{\lambda \in T} \langle \lambda, \omega_0 \rangle.$$

Moreover, for every $\varepsilon > 0$, there is some $\delta > 0$ such that

$$\max_{\lambda \in T} \langle \lambda, \omega \rangle = \max_{\lambda \in T, \|\lambda - \lambda_0\| \leq \varepsilon} \langle \lambda, \omega \rangle$$

for all ω with $\|\omega - \omega_0\| \leq \delta$.

Proof. That the function $\langle \lambda, \omega_0 \rangle$ attains its maximum on T is obvious because T is compact. Suppose to the contrary that there are two distinct points λ_0 and λ_1 which maximize this function on T. It follows from the hypothesis that $\lambda_1 \neq -\lambda_0$. Let $\lambda_2 := (\lambda_0 + \lambda_1)/\|\lambda_0 + \lambda_1\|$. Then $\lambda_2 \in T$ and

$$\langle \lambda_2, \omega_0 \rangle = \frac{2}{\|\lambda_0 + \lambda_1\|} \langle \lambda_0, \omega_0 \rangle.$$

Because the Euclidean norm is strictly convex, we have

$$\|\lambda_0 + \lambda_1\| < \|\lambda_0\| + \|\lambda_1\| = 2,$$

which yields a contradiction

$$\langle \lambda_2, \omega_0 \rangle > \langle \lambda_0, \omega_0 \rangle.$$

To prove the second part, suppose to the contrary that there is some $\varepsilon_0 > 0$ such that for each $\delta = 1/i, i \geq 1$, one can find a vector ω_i with $\|\omega_i - \omega_0\| \leq 1/i$ satisfying

$$\max_{\lambda \in T} \langle \lambda, \omega_i \rangle \neq \max_{\lambda \in T, \|\lambda - \lambda_0\| \leq \varepsilon_0} \langle \lambda, \omega_i \rangle.$$

Let $\lambda_i \in T$ be a maximizing point of the function $\langle \lambda, \omega_i \rangle$ on T. Then $\|\lambda_i - \lambda_0\| > \varepsilon_0$. We may assume that the sequence $\{\lambda_i\}$ converges to some $\lambda_* \in T$. It follows that $\|\lambda_* - \lambda_0\| \geq \varepsilon_0$. On the other hand, as T is compact, one has

$$\langle \lambda_*, \omega_0 \rangle = \lim_{i \to 0} \langle \lambda_i, \omega_i \rangle = \max_{\lambda \in T} \langle \lambda, \omega_0 \rangle,$$

which shows that λ_* is a maximizing point of the function $\langle \lambda, \omega_0 \rangle$ on T. This contradicts the uniqueness of λ_0 by the first part. The proof is complete.   □

Now we are able to prove a multiplier rule for local solutions of the problem (VP).

Theorem 4.4.3 *Assume that ∂H is a pseudo-Jacobian map of H which is upper semicontinuous at x_0. If x_0 is a local weakly efficient solution of (VP), then there is a vector $\lambda_0 = (\xi_0, \theta_0, \gamma_0) \in T$ such that*

$$0 \in \lambda_0(\overline{co}(\partial H(x_0)) \cup co[(\partial H(x_0))_\infty \setminus \{0\}]),$$

$$\theta_0 g(x_0) = 0.$$

Proof. Let us choose a vector $e \in int(C)$ so that

$$\max_{\xi \in C', \|\xi\| \leq 1} \langle \xi, e \rangle = 1.$$

For each $\varepsilon > 0$, define functions $H_\varepsilon \colon \mathbb{R}^n \to \mathbb{R}^m \times \mathbb{R}^p \times \mathbb{R}^q$ and $P_\varepsilon \colon \mathbb{R}^n \to \mathbb{R}$ as follows.

$$H_\varepsilon(x) := (f(x) - f(x_0) + \varepsilon e, g(x), h(x)),$$
$$P_\varepsilon(x) := \max_{\lambda \in T} \langle \lambda, H_\varepsilon(x) \rangle.$$

It is clear that these functions are continuous. Let $U \subset \mathbb{R}^n$ be a neighborhood that exists by the definition of the local weakly efficient solution x_0. We claim that

$$P_\varepsilon(x) > 0 \text{ for all } x \in U.$$

Indeed, suppose that there is $x \in U$ such that $P_\varepsilon(x) \leq 0$. Setting $\lambda = (0, 0, \beta) \neq 0$, we obtain $\beta h(x) \leq 0$ for all $\beta \in \mathbb{R}^l \setminus \{0\}$ and hence $h(x) = 0$. Taking $\lambda = (0, \gamma, 0), \gamma \in K' \setminus \{0\}$, we obtain $\gamma(g(x)) \leq 0$ for all $\gamma \in K' \setminus \{0\}$, which implies $g(x) \in -K$. By a similar argument, choosing $\lambda = (\xi, 0, 0)$, we have $\xi(f(x) - f(x_0) + \varepsilon e) \leq 0$ for all $\xi \in C' \setminus \{0\}$. Because $e \in int(C)$, we derive $f(x) - f(x_0) \in int(C)$. This contradicts the fact that x_0 is a local weakly efficient solution of (VP).

Furthermore, because $P_\varepsilon(x_0) = \varepsilon < \inf P_\varepsilon + \varepsilon$, by Ekeland's variational principle (Lemma 3.5.5), there is x_ε such that $\|x_0 - x_\varepsilon\| < \sqrt{\varepsilon}$, and

$$P_\varepsilon(x_\varepsilon) < P_\varepsilon(x) + \sqrt{\varepsilon}\|x - x_\varepsilon\| \text{ for all } x \neq x_\varepsilon.$$

In particular, the net $\{x_\varepsilon\}$ converges to x_0 as ε tends to 0, and x_ε provides a minimum of the function

$$Q_\varepsilon(x) := P_\varepsilon(x) + \sqrt{\varepsilon}\|x - x_\varepsilon\|.$$

According to the optimality condition (Theorem 2.1.13), if $\partial Q_\varepsilon(x_\varepsilon)$ is a pseudo-Jacobian of Q_ε at x_ε, then

$$0 \in \overline{\text{co}}(\partial Q_\varepsilon(x_\varepsilon)). \tag{4.14}$$

Our aim is to find a suitable pseudo-Jacobian of Q_ε. This can be done if we are able to find a suitable pseudo-Jacobian $\partial P_\varepsilon(x_\varepsilon)$ of P_ε because the set $\sqrt{\varepsilon}B_n$ is a pseudo-Jacobian of the function $x \mapsto \sqrt{\varepsilon}\|x - x_\varepsilon\|$ at x_ε. By the sum rule (Theorem 2.1.1), the set $\partial P_\varepsilon(x_\varepsilon) + \sqrt{\varepsilon}B_n$ is a pseudo-Jacobian of Q_ε at x_ε. Because the function H_ε is the sum of H and the constant function $x \mapsto (-f(x_0) + \varepsilon e, 0, 0)$, $\partial H(x_\varepsilon)$ is a pseudo-Jacobian of H_ε at x_ε. Moreover, for $\varepsilon > 0$, let λ_ε be the unique vector that maximizes the function $\langle \lambda, H_\varepsilon(x_\varepsilon) \rangle$ on T (by Lemma 4.4.2). We claim that for each integer $r \geq 1$, there is some $\varepsilon(r) > 0$ such that for every $\varepsilon \in (0, \varepsilon(r)]$ the set

$$L_\varepsilon := \{\lambda(M + \tfrac{1}{r}N) : \lambda \in T, \|\lambda - \lambda_\varepsilon\| \leq \varepsilon, M \in \partial H(x_0), N \in B\},$$

where we abbreviate $B_{(m+k+l) \times n}$ by B (we keep this shortened notation during this proof), is a pseudo-Jacobian of P_ε at x_ε. Indeed, let $\delta > 0$ be a positive number that exists by virtue of Lemma 4.4.2. Because H_ε is continuous, there is some $t_0 > 0$ such that

$$\|H_\varepsilon(x_\varepsilon) - H_\varepsilon(x)\| < \delta \quad \text{for all } x \in U \quad \text{with} \quad \|x - x_\varepsilon\| \leq t_0.$$

For every $u \in \mathbb{R}^n$, we deduce from Lemma 4.4.2 that

$$
\begin{aligned}
& P_\varepsilon(x_\varepsilon + tu) - P_\varepsilon(x_\varepsilon) \\
&= \max_{\lambda \in T}\langle \lambda, H_\varepsilon(x_\varepsilon + tu) \rangle - \max_{\lambda \in T}\langle \lambda, H_\varepsilon(x_\varepsilon) \rangle \\
&= \max_{\lambda \in T, \|\lambda - \lambda_\varepsilon\| \leq \varepsilon}\langle \lambda, H_\varepsilon(x_\varepsilon + tu) \rangle - \max_{\lambda \in T, \|\lambda - \lambda_\varepsilon\| \leq \varepsilon}\langle \lambda, H_\varepsilon(x_\varepsilon) \rangle \\
&\leq \max_{\lambda \in T, \|\lambda - \lambda_\varepsilon\| \leq \varepsilon}\langle \lambda, H_\varepsilon(x_\varepsilon + tu) - H_\varepsilon(x_\varepsilon) \rangle
\end{aligned}
$$

for every $t \geq 0$ with $\|tu\| \leq t_0$. Applying the mean value theorem (Theorem 2.2.2), we find for each such t, a matrix $M_t \in \overline{\text{co}}(\partial H[x_\varepsilon, x_\varepsilon + tu]) + (1/2r)B$ such that

$$H_\varepsilon(x_\varepsilon + tu) - H_\varepsilon(x_\varepsilon) = M_t(tu).$$

Because ∂H is upper semicontinuous at x_0 and $\lim_{\varepsilon \to 0} x_\varepsilon = x_0$, for each $r \geq 1$, there is some $\varepsilon(r) > 0$ such that for every $\varepsilon \in (0, \varepsilon(r)]$ one has

$$\overline{\text{co}}(\partial H[x_\varepsilon, x_\varepsilon + tu]) \subset \overline{\text{co}}(\partial H(x_0)) + \frac{1}{2r}B$$

for t sufficiently small. It follows that

$$P_\varepsilon^+(x_\varepsilon, u) \leq \limsup_{t\downarrow 0} \max_{\lambda \in T, \|\lambda - \lambda_\varepsilon\| \leq \varepsilon} \langle \lambda, M_t(u) \rangle$$
$$\leq \sup_{M \in \overline{\text{co}}(\partial H(x_0)), N \in B, \lambda \in T, \|\lambda - \lambda_\varepsilon\| \leq \varepsilon} \langle \lambda, (M + \tfrac{1}{r}N)(u) \rangle$$
$$\leq \sup_{\xi \in L_\varepsilon} \langle \xi, u \rangle.$$

Similarly,

$$(-P_\varepsilon)^+(x_\varepsilon, u) \leq \sup_{\xi \in L_\varepsilon} (-\langle \xi, u \rangle).$$

Consequently, L_ε is a pseudo-Jacobian of P_ε at x_ε. Summing up the above, we conclude that for each $r \geq 1$, there is $\varepsilon(r) > 0$ such that for $0 < \varepsilon \leq \varepsilon(r)$, the set

$$\partial Q_\varepsilon(x_\varepsilon) := L_\varepsilon + \sqrt{\varepsilon} B_n$$

is a pseudo-Jacobian of Q_ε at x_ε. We may choose $\varepsilon(r) \downarrow 0$ as $r \to \infty$. Relation (4.14) becomes

$$0 \in \overline{\text{co}}(\partial Q_\varepsilon(x_\varepsilon)) \subset \overline{\text{co}}(L_\varepsilon) + \sqrt{\varepsilon} B_n$$
$$\subset \text{co}\{\lambda M : \lambda \in T, \|\lambda - \lambda_\varepsilon\| \leq \varepsilon, M \in \partial H(x_0)\}$$
$$+ \text{co}\{\tfrac{1}{r}\lambda N : \lambda \in T, \|\lambda - \lambda_\varepsilon\| \leq \varepsilon, N \in B\} + 2\sqrt{\varepsilon} B_n.$$

Taking into account the fact that B, B_n, and T are all compacts, there exist vectors

$$\xi_r \in \text{co}\{\lambda M : \lambda \in T, \|\lambda - \lambda_\varepsilon(r)\| \leq \varepsilon(r), M \in \partial H(x_0)\}$$

such that

$$\lim_{r \to \infty} \xi_r = 0.$$

We apply Caratheodory's theorem to express the vectors ξ_r as

$$\xi_r = \sum_{j=1}^{n+1} a_{rj} \lambda_{rj} M_{rj},$$

where $\sum_{j=1}^{n+1} a_{rj} = 1$, $a_{rj} \geq 0$, $\lambda_{rj} \in T$ with $\|\lambda_{rj} - \lambda_{\varepsilon(r)}\| \leq \varepsilon(r)$, and $M_{rj} \in \partial H(x_0)$, $j = 1, \ldots, n+1$.

Because T is compact, without loss of generality, we may assume that the sequence $\{\lambda_{\varepsilon(r)}\}$ converges to some $\lambda_0 \in T$. Then

$$\lim_{r \to \infty} \lambda_{rj} = \lambda_0 \text{ for all } j = 1, \ldots, n+1.$$

Moreover, by taking a subsequence if necessary, we may also assume that the sequences $\{a_{rj}\}_r$ converge to $a_{0j}, j = 1, \ldots, n+1$, and that

$$\xi_r = \sum_{j \in I_1} a_{rj} \lambda_{rj} M_{rj} + \sum_{j \in I_2} a_{rj} \lambda_{rj} M_{rj} + \sum_{j \in I_3} a_{rj} \lambda_{rj} M_{rj},$$

where the above sums have the following properties.

1. For each $j \in I_1$, the sequence $\{M_{rj}\}_r$ is bounded and converges to some $M_{0j} \in \partial H(x_0)$.
2. For each $j \in I_2$, the sequence $\{M_{rj}\}_r$ is unbounded, but the sequence $\{a_{rj} M_{rj}\}_r$ is bounded and converges to some M_{*j}.
3. For each $j \in I_3$, the sequence $\{a_{rj} M_{rj}\}_r$ is unbounded and there is some $j_0 \in I_3$ such that the sequences $\{a_{rj} M_{rj} / \|a_{rj_0} M_{rj_0}\|\}_r$ converge to some $M_{\infty j}, j \in I_3$.

Let us first consider the case where I_3 is nonempty. By dividing ξ_r by $\|a_{rj_0} M_{rj_0}\|$ and passing to the limit when r tends to ∞, we obtain

$$0 = \lim_{r \to \infty} \frac{\xi_r}{\|a_{rj_0} M_{rj_0}\|} = \lim_{r \to \infty} \sum_{j \in I_3} \lambda_{rj} \frac{a_{rj} M_{rj}}{\|a_{rj_0} M_{rj_0}\|} = \lambda_0 \sum_{j \in I_3} M_{\infty j}.$$

In the latter sum, we have $M_{\infty j} \in [\partial H(x_0)]_\infty$ and $M_{\infty j_0} \neq 0$. Hence

$$0 \in \lambda_0 \mathrm{co}([\partial H(x_0)]_\infty \setminus \{0\}). \tag{4.15}$$

It remains to consider the case where I_3 is empty. For $j \in I_2$, one has $a_{0j} = 0$, which implies that $\sum_{j \in I_1} a_{0j} = 1$ and $M_{*j} \in [\partial H(x_0)]_\infty$. Thus

$$0 = \lim_{r \to \infty} \xi_r = \lambda_0 \Big(\sum_{i \in I_1} a_{0j} M_{0j} + \sum_{j \in I_2} M_{*j} \Big)$$
$$\in \lambda_0 (\mathrm{co}[\partial H(x_0)] + \mathrm{co}[(\partial H(x_0))_\infty]) \subset \lambda_0 \overline{\mathrm{co}}(\partial H(x_0)).$$

This and (4.15) establish the multiplier rule. As to the complementary slackness $\theta_0 g(x_0) = 0$, we observe that if $g_i(x_0) < 0$, then the vector λ_ε must have the corresponding component $\theta_{\varepsilon i} = 0$, and when passing to the limit, we obtain $\theta_{0i} = 0$ as requested. $\square$

Next we present another proof of Theorem 4.4.3 which is based on the open mapping theorem (Corollary 3.5.7).

Second proof of Theorem 4.4.3. Consider the continuous function $\phi :$ $\mathbb{R}^n \to \mathbb{R}^k \times \mathbb{R}^m \times \mathbb{R}^l$ defined by $\phi(x) = (f(x) - f(x_0), g(x), h(x))$ for $x \in \mathbb{R}^n$. Because x_0 is a local weakly efficient solution, the origin of the product space $\mathbb{R}^m \times \mathbb{R}^p \times \mathbb{R}^q$ cannot be an interior point of the set $\phi(x_0 + \varepsilon B_n) + C \times K \times \{0_l\}$ for sufficiently small $\varepsilon > 0$. Moreover, as ∂H is also a pseudo-Jacobian map of ϕ, in view of Corollary 3.5.7, there is at least one element M of the set $\overline{\mathrm{co}}(\partial H(x_0)) \cup \mathrm{co}((\partial H(x_0))_\infty \setminus \{0\})$ that is not $(\phi(0) + K \times \{0_q\})$-surjective on $x_0 + \varepsilon B_n$ at x_0. Because the set $M(C - x_0) + \phi(x_0) + C \times K \times \{0_q\}$ is convex, one can find a nonzero vector $(\alpha, \xi, \gamma) \in \mathbb{R}^m \times \mathbb{R}^p \times \mathbb{R}^q$ such that

$$0 \leq \langle (\alpha, \xi, \gamma), M(x - x_0) + \phi(0) + (y, z, 0) \rangle$$

for all $x \in \mathbb{R}^n$, $y \in C$ and $z \in K$. By setting $x = x_0$ and $z = 0$ in the above inequality, we deduce $\alpha \in C^*$. Similarly, we obtain $\xi \in (g(x_0) + K)^*$ by setting $x = x_0$ and $y = 0$, and $0 = M^{tr}(\alpha, \xi, \gamma)$ by setting $y = 0$ and $z = 0$. $\qquad \square$

The following modified version of Theorem 4.4.3 is useful in the situations when some of the components of the data admit bounded pseudo-Jacobians.

Corollary 4.4.4 *Assume that $H = (H_1, H_2)$ and $\partial H_i, i = 1, 2$ are pseudo-Jacobian maps of H which are upper semicontinuous at x_0. If x_0 is a local weakly efficient solution of (VP), then there is a vector $\lambda_0 = (\xi_0, \theta_0, \gamma_0) \in T$ such that $\theta_0 g(x_0) = 0$ and*

$$0 \in \lambda_0 \left(\overline{\text{co}}(\partial H_1(x_0)) \cup \text{co}[(\partial H_1(x_0))_\infty \setminus \{0\}], \right.$$
$$\left. \overline{\text{co}}(\partial H_2(x_0)) \cup \text{co}[(\partial H_2(x_0))_\infty \setminus \{0\}] \right).$$

Proof. Use the product rule (Theorem 2.1.5) and the proof of Theorem 4.4.3. $\qquad \square$

Example 4.4.5 Let us now apply Theorem 4.4.3 to a particular problem in which the data are Gâteaux differentiable but not necessarily locally Lipschitz. For this purpose, let us define for a Gâteaux differentiable function $\phi : R^n \to R^m$ the following sets,

$$\hat{\nabla}\phi(x) = \{\lim \nabla\phi(x_i) : x_i \to x\}$$
$$\nabla^\infty \phi(x) = \{\lim t_i \nabla\phi(x_i) : x_i \to x, t_i \downarrow 0\}.$$

Actually $\hat{\nabla}\phi(x)$ is the upper limit of the set $\{\nabla\phi(x')\}$ when $x' \to x$ in the sense of Kuratowski–Painleve, and $\nabla^\infty \phi(x)$ is the outer horizon limit of $\{\nabla\phi(x')\}$ when $x' \to x$ as we have defined in Section 1.4. When ϕ has a locally bounded derivative around x, one has $\nabla^\infty \phi(x) = \{0\}$, and $\hat{\nabla}\phi(x)$ is a compact set. This is the case when ϕ is locally Lipschitz. When $m = 1$ and ϕ is locally Lipschitz, the set $\hat{\nabla}\phi(x)$ is exactly the B-subdifferential of ϕ at x, and $\text{co}(\hat{\nabla}\phi(x))$ is the Clarke generalized subdifferential.

Corollary 4.4.6 *Assume that x_0 is a local weakly efficient solution of (VP) and the functions $f, g,$ and h are Gâteaux differentiable in a neighborhood of x_0. Then there exists a vector $\lambda_0 = (\xi_0, \theta_0, \gamma_0) \in T$ such that $\theta_0 g(x_0) = 0$ and*

$$0 \in \lambda_0 \{\overline{\mathrm{co}}(\tilde{\nabla} H(x_0)) \cup \mathrm{co}[\nabla^\infty H(x_0) \setminus \{0\}]\}.$$

Proof. We may assume without loss of generality that $H = (f, g, h)$ is differentiable at every $x \in R^n$ with $\|x - x_0\| \leq 1$. For every $k \geq 1$, let us construct a pseudo-Jacobian of H as follows

$$\partial H(x) = \begin{cases} L(\mathbb{R}^n, \mathbb{R}^m) & \text{if } \|x - x_0\| \geq \frac{1}{k}, \\ \{\nabla H(x)\} & \text{if } 0 < \|x - x_0\| < \frac{1}{k}, \\ \mathrm{cl}\{\nabla H(x') : \|x' - x_0\| < \frac{1}{k}\} & \text{if } x = x_0. \end{cases}$$

It is clear that the set-valued map $x \mapsto \partial H(x)$ is a pseudo-Jacobian map of H which is upper semicontinuous at x_0. According to Theorem 4.4.3, there is a vector $\lambda_k = (\xi_k, \theta_k, \gamma_k) \in T$ such that

$$0 \in \lambda_k \{\overline{\mathrm{co}}(\partial H(x_0)) \cup \mathrm{co}[(\partial H(x_0))_\infty \setminus \{0\}]\},$$

$$\theta_k g(x_0) = 0.$$

By taking a subsequence if necessary, we need only consider cases

(a) There exist $\alpha_{kj} \geq 0, x_{kj} \in \mathbb{R}^n, j = 1, \ldots, mn + 1$, and $m \times n$-matrices b_k with

$$\sum_{j=1}^{mn+1} \alpha_{kj} = 1, \|x_{kj} - x_0\| < \frac{1}{k}, j = 1, \ldots, mn + 1, \|b_k\| \leq 1$$

such that

$$0 = \lambda_k \Big\{ \sum_{j=1}^{mn+1} \alpha_{kj} \nabla H(x_{kj}) + \frac{1}{k} b_k \Big\}.$$

(b) There exist $\alpha_{kj} \geq 0, \beta_{kj} \geq 0, x_{kj} \in \mathbb{R}^n, j = 1, \ldots, mn + 1$ and $m \times n$-matrices b_k with

$$\sum_{j=1}^{mn+1} \alpha_{kj} = 1, \|x_{kj} - x_0\| < \frac{1}{k}, \|\nabla H(x_{kj})\| \geq k, j = 1, \ldots, mn + 1, \|b_k\| \leq 1$$

such that

$$0 = \lambda_k \Big\{ \sum \alpha_{kj} \beta_{kj} \nabla H(x_{kj}) + \frac{1}{k} b_k \Big\}.$$

We may assume that $\{\lambda_k\}$ converges to some $\lambda_0 \in T$ because T is compact. By using an argument similar to that in the proof of Theorem 4.4.3, we derive from (a) that either

$$0 \in \lambda_0 \overline{\text{co}}(\tilde{\nabla} H(x_0)) \quad \text{or} \quad 0 \in \lambda_0 \text{co}[\nabla^\infty H(x) \setminus \{0\}].$$

and from (b) that

$$0 \in \lambda_0 \text{co}[\nabla^\infty H(x) \setminus \{0\}].$$

This completes the proof. □

Example 4.4.7 Consider the following biobjective problem in $\mathbb{R}^5$:

$$\text{WMin} \quad (-x_2 + x_3 + (x_5)^2, x_2 + (x_4)^2)$$

$$\text{subject to} \quad x_5 \geq 0$$

$$(x_1)^{2/3}\text{sign}(x_1) + (x_2)^4 - x_3 = 0$$

$$(x_1)^{1/3} + (x_2)^2 - x_4 = 0$$

and the ordering cone of $\mathbb{R}^2$ is the positive octant $\mathbb{R}^2_+$. The function $H = (f, g, h)$, where

$$f(x) := (-x_2 + x_3 + (x_5)^2, x_2 + (x_4)^2),$$
$$g(x) := x_5,$$
$$h(x) := ((x_1)^{2/3}\text{sign}(x_1) + (x_2)^4 - x_3, (x_1)^{1/3} + (x_2)^2 - x_4),$$

is not Lipschitz at $x = (x_1, \ldots, x_5)$ with $x_1 = 0$. It is not hard to see that the set

$$\partial H(x) := \left\{ \begin{pmatrix} 0 & -1 & 1 & 0 & 2x_5 \\ 0 & 1 & 0 & 2x_4 & 0 \\ 0 & 0 & 0 & 0 & 1 \\ \frac{2}{3}(x_1)^{-1/3}\text{sign}(x_1) & 4(x_2)^3 & -1 & 0 & 0 \\ \frac{1}{3}(x_1)^{-2/3} & 2x_2 & 0 & -1 & 0 \end{pmatrix} \right\}$$

is a pseudo-Jacobian of H at $x = (x_1, \ldots, x_5)$ with $x_1 \neq 0$, and the set

$$\partial H(x) := \left\{ \begin{pmatrix} 0 & -1 & 1 & 0 & 2x_5 \\ 0 & 1 & 0 & 2x_4 & 0 \\ 0 & 0 & 0 & 0 & 1 \\ \alpha & 4(x_2)^3 & -1 & 0 & 0 \\ \alpha^2 & 2x_2 & 0 & -1 & 0 \end{pmatrix} : \alpha \geq 0 \right\}$$

is a pseudo-Jacobian of H at x with $x_1 = 0$. Moreover, the set-valued map $x \mapsto \partial H(x)$ is upper semicontinuous.

Let us first consider $x \in \mathbb{R}^5$ with $x_1 \neq 0$. Observe that H is continuously differentiable at x with $\partial H(x) = \{\nabla H(x)\}$ and the multiplier rule is written as

$$0 = \lambda_0 \nabla H(x).$$

In particular, we derive the following equation that a local weakly efficient solution must satisfy,

$$2(x_1)^{-1/3}\text{sign}(x_1)(1 - 4x_2x_4) + (x_1)^{-2/3}(1 - 4(x_2)^3) = 0.$$

Because the problem is continuously differentiable in a small neighborhood of x, this result can easily be obtained by the classical necessary optimality condition.

Now we consider the case where $x \in \mathbb{R}^5$ has $x_1 = 0$. Set $H_1 = (f, g)$ and $H_2 = h$. The function H_1 is continuously differentiable and the map $x' \mapsto \{\nabla H_1(x')\}$ is an upper semicontinuous pseudo-Jacobian map of H_1. The function H_2 is neither differentiable nor locally Lipschitz at x. Defining

$$\partial H_2(x) := \left\{ \begin{pmatrix} \alpha & 4(x_2)^3 & -1 & 0 & 0 \\ \alpha^2 & 2x_2 & 0 & -1 & 0 \end{pmatrix} : \alpha \geq 1 \right\},$$

we see that the set-valued map $x' \mapsto \nabla H_2(x')$ for x' having the first component nonzero and $x' \mapsto \partial H_2(x')$ for the other x', is an upper semicontinuous pseudo-Jacobian map of H_2. The recession cone of $\partial H_2(x)$ is given by

$$(\partial H_2(x))_\infty = \left\{ \begin{pmatrix} 0 & 0 & 0 & 0 & 0 \\ \alpha & 0 & 0 & 0 & 0 \end{pmatrix} : \alpha \geq 0 \right\}.$$

According to Theorem 4.4.3, a local weakly efficient solution must satisfy either of the following conditions

(i) $0 = (\xi_0, \theta_0)\nabla H_1(x)$ and $0 \in \gamma_0 \partial H_2(x)$.
(ii) $0 = (\xi_0, \theta_0)\nabla H_1(x)$ and $0 \in \gamma_0[(\partial H_2(x))_\infty \setminus \{0\}]$.

Let us look for instance at $x = 0$. Condition (i) implies $\xi_0 = (0,0), \theta_0 = 0$, and $\gamma_0 = (0,0)$. In other words at $x = 0$ there is no multiplier $\lambda_0 \in T$ that satisfies (i). However, the multiplier λ_0 with $\xi_0 = (0,0), \theta_0 = 0$, and $\gamma_0 = (1,0)$ satisfies (ii), which means that $x = 0$ is susceptible to be a local weakly efficient solution. Using a scalarization method, we now show that the point $x = 0$ is in fact a local solution of the biobjective problem. Let $\lambda = (\lambda_1, \lambda_2)$ be a nonzero vector of the positive octant $\mathbb{R}^2_+$. Consider the following mathematical programming problem (P),

$$\begin{aligned} \text{min} \quad & \lambda \circ f(x) \\ \text{subject to} \quad & g(x) \geq 0 \\ & h(x) = 0. \end{aligned}$$

This problem is called a scalarized problem of the biobjective problem. It is plain that every local optimal solution of the problem (P) is a local weakly efficient solution of the bi-objective problem. By taking $\lambda_1 = \lambda_2 = 1$, the problem (P) is equivalent to the problem (P'):

$$\text{min} \qquad (x_1)^{2/3}(\text{sign}(x_1) + 1) + (x_2)^2(1 + 2(x_1)^{1/3}) + (x_2)^4 + (x_5)^2$$
$$\text{subject to } x_5 \geq 0.$$

When $\|x\| < 1/8$, one has

$$1 + 2(x_1)^{1/3} \geq 0.$$

Therefore, the local minimum of (P') is attained at $x = 0$. In other words $x = 0$ is a local optimal solution of (P'), hence it is a local weakly efficient solution of the bi-objective problem.

Second-Order Conditions

We study the following multiobjective problem,

$$(VP) \qquad\qquad \text{VMin} \qquad f(x)$$
$$\text{subject to } x \in S,$$

where $f: \mathbb{R}^n \to \mathbb{R}^m$ is of class C^1; that is, it is continuously differentiable, and S is a nonempty subset of $\mathbb{R}^n$.

Some notations are in order. For $x_0 \in S$, the first-order and the second-order tangent cone to S at x_0 are defined, respectively, by

$$T_1(S, x_0) := \{u \in \mathbb{R}^n : \exists t_i > 0, x_i = x_0 + t_i u + o(t_i) \in S\},$$

$$T_2(S, x_0) := \left\{(u, v) \in \mathbb{R}^n \times \mathbb{R}^n : \exists t_i > 0, x_i = x_0 + t_i u + \frac{1}{2}t_i^2 v + o(t_i^2) \in S\right\}.$$

We also set

$$\Lambda := \{\xi \in C^* : \|\xi\| = 1\},$$

and for $\delta > 0$,

$$S_\delta(x_0) = \{t(x - x_0) : t \geq 0, x \in S \text{ and } \|x - x_0\| \leq \delta\}.$$

Theorem 4.4.8 *Assume that f is a continuously differentiable function, $x_0 \in S$ is a local weakly efficient solution of the problem (VP), and $\partial^2 f$ is a pseudo-Hessian map of f which is upper semicontinuous at x_0. Then for each $(u, v) \in T_2(S, x_0)$, one has*

(i)   *There is $\lambda \in \Lambda$ such that $\langle \lambda, \nabla f(x_0)(u)\rangle \geq 0$.*
(ii) *When $\nabla f(x_0)(u) = 0$, There is $\lambda' \in \Lambda$ such that*

either $\langle \lambda', \nabla f(x_0)(v) + M(u, u)\rangle \geq 0$ *for some* $M \in \overline{\text{co}}(\partial^2 f(x_0))$

or $\langle \lambda', M_*(u, u)\rangle \geq 0$ *for some* $M_* \in (\text{co}(\partial^2 f(x_0)))_\infty \setminus \{0\}$.

If, in addition, the cone C is polyhedral, then (i) holds and when $\langle \lambda, \nabla f(x_0)(u)\rangle = 0$, the inequalities of (ii) are true for $\lambda' = \lambda$.

Proof. Let $(u, v) \in T_2(S, x_0)$, say

$$x_i = x_0 = t_i u + \frac{1}{2} t_i^2 v + o(t_i^2) \in S \qquad (4.16)$$

for some sequence $\{t_i\}$ of positive numbers converging to 0. Because x_0 is a local weakly efficient solution, there is some $i_0 \geq 1$ such that

$$f(x_i) - f(x_0) \in (-\mathrm{int}(C))^c \quad \text{for } i \geq i_0. \qquad (4.17)$$

Because f is continuously differentiable, we derive

$$f(x_i) - f(x_0) = \nabla f(x_0)(x_i - x_0) + o(x_i - x_0).$$

This and (4.17) imply that

$$\nabla f(x_0)(u) \in (-\mathrm{int}(C))^c$$

which is equivalent to (i).

Now let $\nabla f(x_0)(u) = 0$. First observe that by the upper semicontinuity of $\partial^2 f$ at x_0, for every $\varepsilon > 0$, there is $\delta > 0$ such that

$$\partial^2 f(x) \subseteq \partial^2 f(x_0) + \varepsilon B \quad \text{for each } x \text{ with } \|x - x_0\| < \delta,$$

where B is the closed unit ball in the space of matrices in which $\partial^2 f$ takes its values. Consequently, there is $i_1 \geq i_0$ such that

$$\overline{\mathrm{co}}(\partial^2 f[x_0, x_i]) \subseteq \mathrm{co}(\partial^2 f(x_0)) + 2\varepsilon B \quad \text{for every } i \geq i_1.$$

We apply the Taylor expansion to find $M_i \in \mathrm{co}(\partial^2 f(x_0)) + 2\varepsilon B$ such that

$$f(x_i) - f(x_0) = \nabla f(x_0)(x_i - x_0) + \frac{1}{2} M_i(x_i - x_0, x_i - x_0), i \geq i_1.$$

Substituting (4.16) into this equality, we derive

$$f(x_i) - f(x_0) = \frac{1}{2} t_i^2 (\nabla f(x_0)(v) + M_i(u, v)) + \alpha_i,$$

where $\alpha_i = \frac{1}{2} M_i \left(\frac{1}{2} t_i^2 v + o(t_i^2), t_i u + \frac{1}{2} t_i^2 v + o(t_i^2) \right) + \nabla f(x_0)(o(t_i^2))$. This and (4.17) show

$$\nabla f(x_0)(v) + M_i(u, v) + \alpha_i / t_i^2 \in (-\mathrm{int}(C))^c, \ i \geq i_1. \qquad (4.18)$$

Consider the sequence $\{M_i\}$. If it is bounded, we may assume that it converges to some $M_0 \in \overline{\mathrm{co}}(\partial^2 f(x_0)) + 2\varepsilon B$. Then $\alpha_i / t_i^2 \to 0$ as $i \to \infty$ and (4.18) gives

$$\nabla f(x_0)(v) + M_0(u, u) \in (-\mathrm{int}(C))^c.$$

Because ε is arbitrary, the latter inclusion yields the existence of $M \in \overline{\text{co}}(\partial^2 f(x_0))$ such that

$$\nabla f(x_0)(v) + M(u, u) \in (-\text{int}(C))^c,$$

which is equivalent to the first inequality in (ii).
If $\{M_i\}$ is unbounded, say $\lim_{i \to \infty} \|M_i\| = \infty$, we may assume that

$$\lim_{i \to \infty} \frac{M_i}{\|M_i\|} = M_* \in (\text{co}(\partial^2 f(x_0)))_\infty \setminus \{0\}.$$

By dividing (4.18) by $\|M_i\|$ and passing to the limit when $i \to \infty$, we deduce

$$M_*(u, u) \in (-\text{int}(C))^c,$$

which is equivalent to the second inequality in (ii).

Now assume that C is polyhedral. It follows from (4.17) that there is some $\lambda \in \Lambda$ such that

$$\langle \lambda, f(x_i) - f(x_0) \rangle \geq 0$$

for infinitely many i. By taking a subsequence instead if necessary, we may assume this for all $i = 1, 2, \ldots$ Because f is continuously differentiable, we deduce

$$\langle \lambda, \nabla f(x_0)(u) \rangle \geq 0.$$

Assume that $\langle \lambda, \nabla f(x_0)(u) \rangle = 0$. Then using the same argument as in the first part, we can find $M_i \in \text{co}(\partial^2 f(x_0)) + 2\varepsilon B$ such that

$$0 \leq \langle \lambda, f(x_i) - f(x_0) \rangle = \langle \lambda, \frac{1}{2} t_i^2 (\nabla f(x_0)(v) + M_i(u, u)) + \alpha_i \rangle,$$

from which the two last inequalities of the theorem follow.   □

Now let us study the problem where S is explicitly given by the following system,

$$g(x) \leq 0$$
$$h(x) = 0,$$

where $g \colon \mathbb{R}^n \to \mathbb{R}^p$ and $h \colon \mathbb{R}^n \to \mathbb{R}^q$ are given. In other words, we consider the constrained problem

(CP) WMin $f(x)$
        subject to $g(x) \leq 0$
                $h(x) = 0.$

Let $\xi \in C', \beta \in \mathbb{R}^p$, and $\gamma \in \mathbb{R}^q$. Define the Lagrangian function L by

$$L(x, \xi, \beta, \gamma) := \langle \lambda, f(x) \rangle + \langle \beta, g(x) \rangle + \langle \gamma, h(x) \rangle$$

and set

$$S_0 := \{x \in \mathbb{R}^n : g_i(x) = 0 \text{ if } \beta_i > 0, g_i(x) \leq 0 \text{ if } \beta_i = 0, \text{ and } h(x) = 0\}.$$

In the sequel, when (ξ, β, γ) is fixed, we write $L(x)$ instead of $L(x, \xi, \beta, \gamma)$ and ∇L means the gradient of $L(x, \xi, \beta, \gamma)$ with respect to the variable x.

Theorem 4.4.9 *Assume that f, g, and h are continuously differentiable functions and C is a polyhedral convex cone. If $x_0 \in S$ is a local weakly efficient solution of the problem (CP), then there is a nonzero vector $(\xi_0, \beta, \gamma) \in C' \times \mathbb{R}_+^p \times \mathbb{R}^q$ such that*

$$\nabla L(x_0, \xi_0, \beta, \gamma) = 0$$

and for each $(u, v) \in T_2(S_0, x_0)$, there is some $\xi \in \Lambda$ such that either

$$\nabla L(x_0, \xi, \beta, \gamma)(u) > 0$$

or

$$\nabla L(x_0, \xi, \beta, \gamma)(u) = 0,$$

in which case either

$$\nabla L(x_0, \xi, \beta, \gamma)(v) + M(u, u) \geq 0 \quad \text{for some } M \in \overline{\text{co}}(\partial^2 L(x_0, \xi, \beta, \gamma))$$

or

$$M_*(u, u) \geq 0 \quad \text{for some } M_* \in (\text{co}(\partial^2 L(x_0, \xi, \beta, \gamma)))_\infty \setminus \{0\},$$

provided $\partial^2 L$ is a pseudo-Hessian map of L that is upper semicontinuous at x_0.

Proof. The first condition about the existence of (ξ_0, β, γ) is already known from Theorem 4.4.3 and is true for any convex closed cone C with a nonempty interior. Let now $(u, v) \in T_2(S_0, x_0)$. Let $x_i = x_0 + t_i u + \frac{1}{2} t_i^2 v + o(t_i^2) \in S_0$ for some $t_i > 0, t_i \to 0$ as $i \to \infty$. Because x_0 is a local weakly efficient solution of (CP), there is some $i_0 \geq 1$ such that

$$f(x_i) - f(x_0) \in (-\text{int}(C))^c, \quad \text{for } i \geq i_0.$$

Moreover, as C is polyhedral, there exists $\xi \in \Lambda$ such that

$$\langle \xi, f(x_i) - f(x_0) \rangle \geq 0 \tag{4.19}$$

for infinitely many i. We may assume this for all $i \geq i_0$. Since $\partial^2 L$ is upper semicontinuous at x_0, by applying the Taylor expansion to L we can find

$$M_i \in \mathrm{co}(\partial^2 L(x_0)) + 2\varepsilon B,$$

where ε is an arbitrarily fixed positive number, such that

$$L(x_i) - L(x_0) = \nabla L(x_0)(x_i - x_0) + \frac{1}{2} M_i(x_i - x_0, x_i - x_0)$$

for i sufficiently large. Substituting the expression $x_i - x_0 = t_i u + \frac{1}{2} t_i^2 v + o(t_i^2)$ into the above equality and taking (4.19) into account, we derive

$$0 \leq t_i \nabla L(x_0)(u) + \frac{t_i^2}{2} (\nabla L(x_0)(v) + M_i(u, u)) + \alpha_i,$$

where $\alpha_i = \frac{1}{2} M_i(\frac{1}{2} t_i^2 v + o(t_i^2), t_i u + \frac{1}{2} t_i^2 v + o(t_i^2)) + \nabla L(x_0)(o(t_i^2))$. This, in particular, implies $\nabla L(x_0)(u) \geq 0$.

When $\nabla L(x_0)(u) = 0$, we also derive

$$0 \leq \nabla L(x_0)(v) + M_i(u, u) + \alpha_i / t_i^2,$$

which by the same reason as discussed in the proof of Theorem 4.4.8, yields the requested inequalities. $\square$

We notice that the second conclusion of Theorem 4.4.8 and the conclusion of Theorem 4.4.9 are no longer true if C is not polyhedral. Here is a counterexample when the data are smooth.

Example 4.4.10 Define $f \colon \mathbb{R} \to \mathbb{R}^3$ by

$$f(t) := -(t + t^2 \cos t, t + t \cos t, t \sin t).$$

We consider $\mathbb{R}^3$ partially ordered by the cone C,

$$C := \{(x, y, z) \in \mathbb{R}^3 : x^2 \geq y^2 + z^2, x \geq 0\}.$$

We consider the following three-objective problem,

$$\text{WMin} \quad f(t)$$
$$\text{subject to } t \in [0, \infty).$$

It is clear that $t = 0$ is a local efficient solution of the problem. At this point, $\nabla f(0) = -(1, 2, 0)$ and $\nabla^2 f(0) = -(2, 0, 2)$. A simple calculation confirms that equation

$$\langle \lambda, \nabla f(0) \rangle = 0, \ \lambda \in \Lambda$$

holds for either $\lambda = (2, -1, 3^{1/2})/8^{1/2}$ or $\lambda = (2, -1, -3^{1/2})/8^{1/2}$. For these values of λ and for the vector $(u, v) = (1, 0) \in T_2(S, 0)$, we have

$$\langle \lambda, \nabla f(0)(v) + \nabla^2 f(0)(u, u) \rangle < 0,$$

which shows that the conclusion of Theorem 4.4.8 (Theorem 4.4.9) does not hold.

In the following we provide some sufficient optimality conditions. First we consider the problem (VP) in which no explicit constraints are given.

Theorem 4.4.11 *Assume that f is a continuously differentiable function, and $\partial^2 f$ is a pseudo-Jacobian map of f which is upper semicontinuous at $x_0 \in S$. Then each of the following conditions is sufficient for x_0 to be a locally unique efficient solution of the problem (VP)*

(i) For each $u \in T_1(S, x_0) \setminus \{0\}$, there is some $\xi \in \Lambda$ such that

$$\langle \xi, \nabla f(x_0)(u) \rangle > 0.$$

(ii) There is $\delta > 0$ such that for each $v \in S_\delta(x_0)$ and $u \in T_1(S, x_0)$, one has

$$\langle \xi_0, \nabla f(x_0)(v) \rangle \geq 0 \quad \text{for some } \xi_0 \in \Lambda$$

and

$$\langle \xi, M(u, u) \rangle > 0$$

for every $\xi \in \Lambda$ and for every $M \in \overline{\text{co}}(\partial^2 f(x_0)) \cup [(\text{co}(\partial^2 f(x_0)))_\infty \setminus \{0\}]$.

Proof. Suppose to the contrary that x_0 is not a locally unique efficient solution of (VP). There exists a sequence $\{x_i\}$, $x_i \in S$ such that $x_i \to x_0$ and

$$f(x_i) - f(x_0) \in -C. \tag{4.20}$$

We may assume that $(x_i - x_0)/\|x_i - x_0\| \to u \in T_1(S, x_0)$ as $i \to \infty$. By dividing (4.20) by $\|x_i - x_0\|$ and passing to the limit, we deduce

$$\nabla f(x_0)(u) \in -C.$$

This contradicts condition (i) and shows the sufficiency of this condition.

For the second condition, let us apply the Taylor expansion to find $M_i \in \text{co}(\partial^2 f(x_0)) + 2\varepsilon B$ for an arbitrarily fixed $\varepsilon > 0$ such that

$$f(x_i) - f(x_0) = \nabla f(x_0)(x_i - x_0) + \frac{1}{2} M_i(x_i - x_0, x_i - x_0). \tag{4.21}$$

Observe that the first inequality of (ii) implies

$$\nabla f(x_0)(x_i - x_0) \in (-\text{int}(C))^c$$

for i sufficiently large. For such i, there is $\xi_i \in \Lambda$ such that

$$\langle \xi_i, \nabla f(x_0)(x_i - x_0) \rangle \geq 0.$$

On the other hand, (4.20) shows that

$$\langle \xi_i, f(x_i) - f(x_0) \rangle \leq 0.$$

This and (4.21) imply

$$\langle \xi_i, M_i(x_i - x_0, x_i - x_0) \rangle \leq 0 \quad \text{for } i \text{ sufficiently large.}$$

Furthermore, because Λ is compact, we may assume $\xi_i \to \xi \in \Lambda$. By considering separately the case when $\{M_i\}$ is bounded and the case when $\{M_i\}$ is unbounded (as in the proof of Theorem 4.4.8), we deduce

$$\langle \xi, M(u, u) \rangle \leq 0 \quad \text{for some } M \in \overline{\text{co}}(\partial^2 f(x_0)) \cup [(\text{co}(\partial^2 f(x_0)))_\infty \setminus \{0\}],$$

which contradicts (ii). The proof is complete. $\qquad \square$

Theorem 4.4.12 *Assume that f is a continuously differentiable function and $\partial^2 f$ is a pseudo-Hessian map of f. If there is some $\delta > 0$ such that for every $v \in S_\delta(x_0)$ one has*

$$\langle \xi_0, \nabla f(x_0)(v) \rangle \geq 0 \quad \text{for some } \xi_0 \in \Lambda$$

and

$$\langle \xi, M(u, v) \rangle \geq 0 \quad \text{for all } \xi \in \Lambda, M \in \partial^2 f(x) \quad \text{with } \|x - x_0\| \leq \delta,$$

then x_0 is a local weakly efficient solution of the problem (VP).

Proof. Suppose to the contrary that x_0 is not a local weakly efficient solution of (VP). There is $\bar{x} \in S$ with $\|\bar{x} - x_0\| \leq \delta$ such that

$$f(\bar{x}) - f(x_0) \in -\text{int}(C). \tag{4.22}$$

Set $v = \bar{x} - x_0$. Then $v \in S_\delta(x_0)$. The first inequality of the hypothesis implies

$$\nabla f(x_0)(v) \in (-\text{int}(C))^c$$

and the second one implies

$$M(v, v) \in C \text{ for every } M \in \partial^2 f(x), \|x - x_0\| \leq \delta.$$

Because C is convex and closed, the latter inclusion gives, in particular, that

$$\overline{\mathrm{co}}(\partial^2 f(x)) \subseteq C.$$

Using the Taylor expansion, we derive

$$f(\overline{x}) - f(x_0) \in \nabla f(x_0)(v) + \frac{1}{2}\overline{\mathrm{co}}\{\partial^2 f[x_0, \overline{x}](v, v)\}$$
$$\subseteq (-\mathrm{int}(C))^c + C \subseteq (-\mathrm{int}(C))^c,$$

which contradicts (4.22). The proof is complete. $\square$

Now we proceed to sufficient conditions for the problem (CP) in which explicit constraints are given in form of equality and inequality systems.

Theorem 4.4.13 *Assume that f, g, and h are continuously differentiable functions and for every $u \in T_1(S, x_0) \setminus \{0\}$ there is some $(\xi, \beta, \gamma) \in \Lambda \times \mathbb{R}_+^p \times \mathbb{R}^q$ such that*

$$\nabla L(x_0, \xi, \beta, \gamma) = 0, \quad \beta g(x_0) = 0,$$

and

$$M(u, u) > 0 \quad \text{for each } M \in \overline{\mathrm{co}}(\partial^2 L(x_0)) \cup ((\mathrm{co}(\partial^2 L(x_0)))_\infty \setminus \{0\}),$$

where $\partial^2 L$ is a pseudo-Jacobian map of L which is upper semicontinuous at x_0. Then x_0 is a locally unique efficient solution of the problem (CP).

Proof. Suppose to the contrary that x_0 is not a locally unique solution of (CP). Then there exists a sequence $\{x_i\}$, $x_i \in S$ such that $x_i \to x_0$ and $f(x_i) - f(x_0) \in -C$. We may assume $(x_i - x_0)/\|x_i - x_0\| \to u \in T_1(S, x_0)$. It follows that

$$L(x_i) - L(x_0) \leq 0 \quad \text{for all } i \geq 1.$$

Applying the Taylor expansion to L and by the upper semicontinuity of $\partial^2 L$, we obtain

$$L(x_i) - L(x_0) - \nabla L(x_0)(x_i - x_0) \in \frac{1}{2}\overline{\mathrm{co}}\{\partial^2 L[x_0, x_i](x_i - x_0, x_i - x_0)\}$$
$$\subseteq \frac{1}{2}(\mathrm{co}(\partial^2 L(x_0)) + \|x_i - x_0\|B)(x_i - x_0, x_i - x_0),$$

for i sufficiently large. Here and later on we use the notation B for $B_{(m+k+l) \times n}$. These relations yield

$$M_i(x_i - x_0, x_i - x_0) \leq 0$$

for some $M_i \in \mathrm{co}(\partial^2 L(x_0)) + \|x_i - x_0\| B$ with i sufficiently large. By the same argument as in the proof of Theorem 4.4.3, we derive the existence of some matrix $M \in \overline{\mathrm{co}}(\partial^2 L(x_0)) \cup ((\mathrm{co}(\partial^2 L(x_0)))_\infty \setminus \{0\})$ such that

$$M(u, u) \leq 0,$$

which contradicts the hypothesis. $\square$

Theorem 4.4.14 *Assume that f, g, and h are continuously differentiable functions and that there is $\delta > 0$ such that for each $v \in S_\delta(x_0)$, one can find a vector $(\xi, \beta, \gamma) \in \Lambda \times \mathrm{I\!R}^p_+ \times \mathrm{I\!R}^q$ and a pseudo-Hessian map $\partial^2 L(x, \xi, \beta, \gamma)$ of L such that*

$$\nabla L(x_0, \xi, \beta, \gamma) = 0, \quad \beta g(x_0) = 0$$

and

$$M(u, u) \geq 0 \quad \text{for every } M \in \partial^2 L(x, \xi, \beta, \gamma) \quad \text{with } \|x - x_0\| \leq \delta.$$

Then x_0 is a local weakly efficient solution of the problem (CP).

Proof. The proof is similar to the proof of Theorem 4.1. $\square$

We now give an example which shows that the recession Hessian matrices in Theorem 4.4.9 cannot be removed when the data of the problem are of class C^1. Examples that show the importance of the recession Hessian matrices in the theorems of Section 4.4 on sufficient conditions can be constructed in a similar way.

Example 4.4.15 Let us consider the following two-objective problem,

$$\text{WMin } (x, x^{4/3} - y^4)$$
$$\text{subject to } -x^2 + y^4 \leq 0.$$

The partial order of $\mathrm{I\!R}^2$ is given by the positive octant $\mathrm{I\!R}^2_+$. It is easy to see that $(0, 0)$ is a local efficient solution of this problem. By taking $\xi_0 = (0, 1)$ and $\beta = 1$, the Lagrangian function of the problem is

$$L((x, y), \xi_0, \beta) = x^{4/3} - y^4 - x^2 + y^4 = x^{4/3} - x^2$$

and satisfies the necessary condition

$$\nabla L((0, 0), \xi_0, \beta) = (0, 0).$$

The set S_0 is given by

$$S_0 = \{(x, y) \in \mathrm{I\!R}^2 : x^2 = y^4\}.$$

Let us take $u = (0, 1)$ and $v = (-2, 0)$. It is clear that $(u, v) \in T_2(S_0, (0, 0))$. According to Theorem 4.4.9, there is some $\xi = (\xi_1, \xi_2) \in \mathbb{R}_+^2$ with $\|\xi\| = 1$ such that $\nabla L((0, 0), \xi, \beta)(u) \geq 0$. Actually we have

$$\nabla L((0, 0), \xi, \beta) = (\xi_1, 0).$$

Hence $\nabla L((0, 0), \xi_o, \beta)(u) = 0$, and the second-order conditions of that theorem must hold. First observe that if $\xi_2 = 0$, then

$$\partial^2 L(x, y) := \left\{ \begin{pmatrix} -2 & 0 \\ 0 & 12y^2 \end{pmatrix} \right\}$$

is a pseudo-Hessian map of L, which is upper semicontinuous at $(0, 0)$. It is not hard to verify that the second-order condition of Theorem 4.4.9 does not hold for this ξ. Consequently, $\xi_2 > 0$. Let us define

$$\partial^2 L(x, y) := \left\{ \begin{pmatrix} \frac{4}{9}\xi_2 x^{-2/3} - 2 & 0 \\ 0 & 12(1 - \xi_2)y^2 \end{pmatrix} \right\}, \quad \text{for } x \neq 0,$$

and

$$\partial^2 L(0, y) := \left\{ \begin{pmatrix} \frac{4}{9}\xi_2 \alpha - 2 & 0 \\ 0 & 12(1 - \xi_2)y^2 - 1/\alpha \end{pmatrix} : \alpha \geq \frac{9}{\xi_2} \right\}.$$

A direct calculation confirms that the set-valued map $(x, y) \to \partial^2 L(x, y)$ is a pseudo-Hessian map of L which is upper semicontinuous at $(0, 0)$. Moreover, for each $M \in \overline{\text{co}}(\partial^2 L(0, 0))$, one has

$$\nabla L(0, 0)(v) + M(u, u) = -2\xi_1 - \frac{1}{\alpha} < 0,$$

which shows that the first inequality of the second-order condition of Theorem 4.4.9 is not true. The recession cone of $\partial^2 L(0, 0)$ is given by

$$(\partial^2 L(0, 0))_\infty = \left\{ \begin{pmatrix} \alpha & 0 \\ 0 & 0 \end{pmatrix} : \alpha \geq 0 \right\}.$$

By choosing

$$M_* = \begin{pmatrix} 1 & 0 \\ 0 & 0 \end{pmatrix} \in (\text{co}(\partial^2 L(0, 0)))_\infty \setminus \{0\}$$

we have $M_*(u, u) \geq 0$.

5

Monotone Operators and Nonsmooth Variational Inequalities

In this chapter we present various characterizations of monotone and generalized monotone operators in terms of pseudo-Jacobians. We obtain conditions for the uniqueness of solutions of nonsmooth continuous variational inequalities problems. We provide finally a solution method for nonlinear nonsmooth complementarity problems.

5.1 Generalized Monotone Operators

The monotonicity of vector-valued maps plays a crucial role in the study of complementarity problems, variational inequality problems, and equilibrium problems just as the convexity of real-valued maps does in mathematical programming. In this section, we characterize the monotonicity of continuous maps in terms of pseudo-Jacobian matrices.

Monotone Operators

Let S be a nonempty, open and convex subset of $\mathbb{R}^n$ and let $F : S \rightrightarrows \mathbb{R}^n$ be a set-valued map. We say that F is a *monotone operator* on S if for every two points x and y in S, and for every element $\xi \in F(x)$ and $\zeta \in F(y)$ one has

$$\langle \xi, \, y - x \rangle + \langle \zeta, \, x - y \rangle \leq 0,$$

or equivalently

$$\sup_{\xi \in F(x), \zeta \in F(y)} \langle \xi - \zeta, x - y \rangle \geq 0.$$

If these inequalities are strict whenever x and y are distinct, the map F is called *strictly monotone*.

A special case is when $n = 1$ and F is single-valued. Let $S = (a, b) \subseteq \mathbb{R}$ be an interval and f a real-valued function on S. Then f is a monotone operator on S if and only if for each $x, y \in S$ with $x < y$ one has

$$f(x)(y - x) + f(y)(x - y) \leq 0,$$

or equivalently

$$f(x) \leq f(y).$$

Thus, f is monotone if and only if it is nondecreasing. Similarly, f is strictly monotone if and only if it is increasing.

Here are some elementary properties of monotone operators. We make use of the notations $\overline{co}F$ for the map whose value at every point $x \in S$ is the closed convex hull of $F(x)$. A set-valued map $F_1 : S \rightrightarrows \mathbb{R}^n$ is said to be a *submap* (or *suboperator*) of F if $F_1(x) \subseteq F(x)$ for every $x \in S$.

Proposition 5.1.1 *Assume that F and G are monotone operators on a nonempty, open, and convex subset S of $\mathbb{R}^n$. Then the following assertions are true.*

(i)   *The operators λF with $\lambda \geq 0$, $\overline{co}F$, $F \cup G$, and $F + G$ are monotone.*
(ii)   *Every suboperator of F is monotone.*

Proof. These assertions are immediate from the definition. We take up, for instance, the sum $F + G$. Let x and y be two points of S and $\xi \in (F+G)(x), \zeta \in (F+G)(y)$. Then there are $\xi_1 \in F(x), \xi_2 \in G(x), \zeta_1 \in F(y)$ and $\zeta_2 \in G(y)$ such that $\xi = \xi_1 + \xi_2$ and $\zeta = \zeta_1 + \zeta_2$. Then, by the monotonicity of F and G, one derives

$$\langle \xi, y - x \rangle + \langle \zeta, x - y \rangle = \langle \xi_1, y - x \rangle + \langle \xi_2, x - y \rangle + \langle \zeta_1, x - y \rangle + \langle \zeta_2, x - y \rangle \leq 0.$$

Hence $F + G$ is monotone.   □

Similar assertions are available for strictly monotone operators. Now we characterize single-valued monotone operators by means of pseudo-Jacobians. We say that a pseudo-Jacobian ∂f of a vector function $f : S \to \mathbb{R}^m$ is *densely regular* on S if there exists a dense subset $S_0 \subseteq S$ such that

(a) $\partial f(x)$ is regular at every $x \in S_0$,
(b)   The pseudo-Jacobian $\partial f(x)$ of f at every $x \notin S_0$ is contained in the set consisting of all limits $\lim_{k \to \infty} M_k$, where $M_k \in \partial f(x_k)$ and $\{x_k\}$ is a sequence in S_0 converging to x.

An $n \times n$-matrix M is said to be *positive semidefinite* (respectively, *positive definite*) if for all vector $v \in \mathbb{R}^n \setminus \{0\}$ one has

$$\langle v, M(v) \rangle \geq 0 \quad (\text{respectively}, \quad \langle v, M(v) \rangle > 0).$$

A necessary and sufficient condition for a matrix to be positive definite is that its principal minors be positive. When a matrix is not positive definite, it is positive semidefinite if and only if its determinant is zero and all the minors formed by deleting rows and columns of the same indices are nonnegative.

Theorem 5.1.2 *Let $F : S \to \mathbb{R}^n$ be a continuous map that admits a pseudo-Jacobian $\partial F(x)$ for each $x \in S$. If for each $x \in S$, the matrices of $\partial F(x)$ are positive semidefinite, then F is monotone.*

Conversely, if F is monotone and if the pseudo-Jacobian ∂F is densely regular on S, then for each $x \in S$ the matrices of $\partial F(x)$ are positive semidefinite.

Proof. Let x, $y \in S$ be arbitrary; set $u = y - x$. By the mean value theorem (Theorem 2.2.2),

$$F(x + u) - F(x) \in \overline{\text{co}} \left(\partial F([x, \ x + u])u \right),$$

and so

$$\langle F(x + u) - F(x), \ u \rangle \in \langle \overline{\text{co}} \left(\partial F([x, \ x + u])u \right), \ u \rangle.$$

Thus there exists $z \in [x, \ x + u]$ and $N \in \overline{\text{co}} \left(\partial F(z) \right)$ such that

$$
\begin{aligned}
\langle F(x + u) - F(x), \ u \rangle &= \langle N(u), u \rangle \\
&\geq \inf_{M \in \overline{\text{co}} \left(\partial F(z) \right)} \langle M(u), u \rangle \\
&= \inf_{M \in \partial F(z)} \langle M(u), u \rangle \\
&\geq 0.
\end{aligned}
$$

This shows that F is monotone.

For the converse, suppose to the contrary that

$$\langle M_0(u_0), u_0 \rangle < 0,$$

for some $x_0, u_0 \in S$ and $M_0 \in \partial F(x_0)$. If $x_0 \in S_0$, then by regularity,

$$(u_0 F)^-(x_0; u_0) = \inf_{M \in \partial F(x_0)} \langle M(u_0), u_0 \rangle < 0.$$

So, there exists t sufficiently small and positive such that

$$\langle u_0, F(x_0 + t u_0) \rangle - \langle u_0, F(x_0) \rangle < 0.$$

This contradicts the monotonicity of F.

If, on the other hand, $x_0 \notin K$, then by hypothesis we can find a sequence $\{x_n\} \subset K$, $x_n \to x_0$ and $M_n \in \partial F(x_n)$ such that

$$\lim_{n\to\infty} M_n = M_0.$$

So for n_0 sufficiently large, $M_{n_0} \in \partial F(x_{n_0})$ and $\langle M_{n_0}(u_0), u_0 \rangle < 0$. Hence

$$(u_0 F)^-(x_{n_0}, u_0) = \inf_{M \in \partial F(x_{n_0})} \langle M(u_0), u_0 \rangle < 0.$$

Then, for sufficiently small $t > 0$,

$$\langle u_0, F(x_{n_0} + tu_0) \rangle - \langle u_0, F(x_{n_0}) \rangle < 0.$$

This again contradicts the monotonicity of F, and so the proof is complete.
□

It is worth noting that the conclusion of the above theorem is no longer true without the regularity condition. This can be seen by choosing $L(\mathbb{R}^n, \mathbb{R}^n)$ as a pseudo-Jacobian at each point. A similar result for strictly monotone operators can be developed.

Theorem 5.1.3 *Assume that $F: S \to \mathbb{R}^n$ is a continuous map and ∂F is a pseudo-Jacobian map of F such that for every $x \in S$, the set $\overline{\text{co}}(\partial F(x)) \cup ((\text{co}(\partial F(x)))_\infty \setminus \{0\})$ consists of positive definite matrices only. Then F is strictly monotone on S.*

Proof. Suppose to the contrary that F is not strictly monotone, that is, there are x_0 and $y_0 \in S$ such that

$$\langle F(x_0) - F(y_0), x_0 - y_0 \rangle \leq 0. \tag{5.1}$$

We consider the scalar function $x \longmapsto \langle F(x), x_0 - y_0 \rangle$. It follows that the closure of the set

$$Q(x) := \{M(x_0 - y_0) : M \in \partial F(x)\}$$

is a pseudo-Jacobian of $\langle F(\cdot), x_0 - y_0 \rangle$ at x. We apply the mean value theorem to this function on the interval $[x_0, y_0]$. There exists $c \in (x_0, y_0)$ and $\xi_i \in \text{co}(Q(c))$ such that

$$\langle F(x_0) - F(y_0), x_0 - y_0 \rangle = \lim_{i\to\infty} \langle \xi_i, x_0 - y_0 \rangle. \tag{5.2}$$

Because $\text{co}(Q(c)) = [\text{co}(\partial F(c))](x_0 - y_0)$, there is $M_i \in \text{co}(\partial F(c))$ such that

$$\xi_i = M_i(x_0 - y_0).$$

If the sequence $\{M_i\}$ is bounded, we may assume that it converges to some $M_0 \in \overline{\text{co}}(\partial F(c))$. Then by (5.2), inequality (5.1) becomes

$$\langle F(x_0) - F(y_0), x_0 - y_0 \rangle = \langle M_0(x_0 - y_0), x_0 - y_0 \rangle \le 0 .$$

This contradicts the hypothesis that M_0 is positive definite.

Now suppose that $\{M_i\}$ is unbounded. We may assume that

$$\lim_{i \to \infty} \|M_i\| = \infty \text{ and } \lim_{i \to \infty} M_i \, / \, \|M_i\| = M_* \in \left(\overline{\mathrm{co}}(\partial F(c)) \right)_\infty \setminus \{0\}.$$

It follows from (5.2) that

$$\langle M_*(x_0 - y_0), x_0 - y_0 \rangle = \lim_{i \to \infty} \left\langle \frac{M_i}{\|M_i\|} (x_0 - y_0), x_0 - y_0 \right\rangle = 0,$$

which contradicts the hypothesis. The proof is complete. $\qquad\square$

The converse of Theorem 5.1.3 is no longer true. For instance, let $F : \mathbb{R} \to \mathbb{R}$ be defined by $F(x) = x^3$. Then F is strictly monotone on $\mathbb{R}$. Nevertheless, the gradient ∇F, which is a regular pseudo-Jacobian of F, has no positive definite elements at $x = 0$. As a special case of Theorem 5.1.3 we see that if F is locally Lipschitz, then monotonicity of F is characterized by positive semidefiniteness of the Jacobian matrices.

Corollary 5.1.4 *Let $F : S \to \mathbb{R}^n$ be a locally Lipschitz map. Then F is monotone if and only if for each $x \in S$ the matrices $M \in \partial^C F(x)$ are positive semidefinite. Moreover, if for every $x \in S$, the Clarke generalized Jacobian $\partial^C F(x)$ consists of positive definite matrices only, then F is strictly monotone on S.*

Proof. Let $x \in S$ be arbitrary. Because F is locally Lipschitz by Rademacher's Theorem there exists a dense subset K of S on which ∇F exists. Define

$$\partial F(x) = \begin{cases} \{\nabla F(x)\} & x \in K, \\ \{\lim_{k \to \infty} \nabla F(x_k) : x_n \to x, \{x_k\} \subset K\} & x \notin K. \end{cases}$$

Then $\partial F(x)$ is a pseudo-Jacobian of F at x. If F is monotone, then the hypotheses of Theorem 5.1.2 are satisfied, and so the matrices $M \in \partial F(x)$ are positive semidefinite. Hence, the matrices $M \in \mathrm{co}(\partial F(x)) = \partial^C F(x)$ are positive semidefinite too.

Conversely, if for each $x \in S$ the matrices $M \in \partial^C F(x)$ are positive semidefinite, then the monotonicity of F follows from Theorem 5.1.2 Because $\partial^C F(x)$ is a pseudo-Jacobian of F at x. The last assertion is immediate from Theorem 5.1.3. $\qquad\square$

Comonotonicity

In order to develop methods for solving complementarity problems, we need some more notions related to the monotonicity behavior of maps. We say that a set-valued map $F : S \rightrightarrows \mathbb{R}^n$ is *strongly monotone* with modulus $\alpha > 0$ on S if for each $x, y \in S$,

$$\langle \xi - \zeta, \ y - x \rangle \geq \alpha \|y - x\|^2 \text{ for all } \xi \in F(x), \zeta \in F(y).$$

It is clear that strongly monotone maps are strictly monotone and that the converse is not true in general (see Example 5.1.6 below). Similarly to the case of monotone operators, one can easily prove that if F is strongly monotone, then the operators λF with $\lambda > 0$, $\overline{\mathrm{co}}F$ and every suboperator of F are strongly monotone. Moreover, if F is strongly monotone and G is monotone, then their sum $F + G$ is strongly monotone. Let us now characterize strongly monotone single-valued operators.

Proposition 5.1.5 *Assume that* $F : S \rightarrow \mathbb{R}^n$ *is a continuous operator, where S is a nonempty open and convex subset of $\mathbb{R}^n$. If F admits a pseudo-Jacobian ∂F such that*

$$\alpha := \inf_{\|u\|=1, M \in \{\partial F(x):x\in S\}} \langle M(u), u \rangle > 0,$$

then F is strongly monotone with modulus α on S.

Conversely, if F is strongly monotone with modulus β on S, then every pseudo-Jacobian ∂F of F satisfies

$$\inf_{\|u\|=1} \sup_{M \in \{\partial F(x):x\in S\}} \langle M(u), u \rangle \geq \beta.$$

In particular, when F is Gâteaux differentiable, it is strongly monotone on S if and only if its Jacobian is uniformly positive definite in the sense that $\inf_{\|u\|=1, x \in S} \langle \nabla F(x)(u), u \rangle > 0$.

Proof. We wish to prove that F is strongly monotone with modulus α. Suppose to the contrary that there exist two points x and y of S such that

$$\langle F(x) - F(y), x - y \rangle < \alpha \|x - y\|^2.$$

According to the mean value theorem, one can find some positive numbers $\lambda_1, \ldots, \lambda_k$ whose sum equals 1 and matrices $M_1, \ldots, M_k \in \partial F([x, y])$ such that

$$\left\langle \sum_{i=1}^{k} \lambda_i M_i(x - y), x - y \right\rangle < \alpha \|x - y\|^2.$$

There exists at least one index i such that

$$\langle M_i(x - y), x - y \rangle < \alpha \|x - y\|^2.$$

This contradicts the assumptions.

Conversely, let $u \in \mathbb{R}^n$ with $\|u\| = 1$ and let $x \in S$. By strong monotonicity, one has that

$$\langle F(x + tu) - F(x), tu \rangle \geq \beta t^2 \text{ for every } t \in (0, 1).$$

We deduce that

$$\sup_{M \in \partial F(x)} \langle M(u), u \rangle \geq (u \circ F)^+(x, u) = \limsup_{t \downarrow 0} \frac{\langle F(x + tu) - F(x), u \rangle}{t} \geq \beta$$

and the proof is complete. $\square$

Example 5.1.6 Let $f : \mathbb{R} \to \mathbb{R}$ be a monotone function. This means that, for any $(x, u) \in \mathbb{R} \times \mathbb{R}$, and for all $t \geq 0$,

$$(f(x + tu) - f(x))u \geq 0. \tag{5.3}$$

If $u \in \mathbb{R}$ and

$$\liminf_{t \downarrow 0} \frac{|f(x + tu) - f(x)|}{t} > 0, \tag{5.4}$$

then the monotonicity of f yields the existence of some $\alpha > 0$ such that

$$(f(x + tu) - f(x))u = |u||f(x + tu) - f(x)| \geq \alpha |u| t \tag{5.5}$$

for all $t \geq 0$ sufficiently small. Obviously, this is a much stronger property than (5.3). For example, consider the function f, defined by

$$f(x) := \begin{cases} x^{1/k} & \text{if } x \geq 0, \\ 0 & \text{otherwise} \end{cases}$$

for some $k > 1$. This function is not locally Lipschitz at $x = 0$ and (5.4) is satisfied for $(x, u) := (0, 1)$, where the left-hand side of (5.4) attains $+\infty$. Moreover, f is monotone but not strongly monotone on $\mathbb{R}$. On the other hand, on $[0, 1]$ we have for $u := 1$,

$$f(0 + t) - f(0) = t^{1/k} \geq t \qquad \forall t \in [0, 1].$$

Thus the function f is strongly monotone on $[0, 1]$ and, in addition, has the property (5.5) for $(x, u) := (0, 1)$ for all $t \in [0, 1]$.

Our observation in this one-dimensional example leads us to the following notion that characterizes a corresponding behavior of directional monotonicity in the multi-dimensional case.

A map $F : \mathbb{R}^n \to \mathbb{R}^n$ is called *comonotone* at $x \in \mathbb{R}^n$ in the direction $u \in \mathbb{R}^n$ if there exists some $\gamma_{(x,u)} > 0$ so that

$$\langle F(x + tu) - F(x), u\rangle \geq \gamma_{(x,u)} \|F(x + tu) - F(x)\|$$

holds for all $t \geq 0$ sufficiently small.

Later we show that the comonotonicity of the monotone map F is particularly important in those directions in which

$$\limsup_{t\downarrow 0} \frac{\|F(x + tu) - F(x)\|}{t} = +\infty. \tag{5.6}$$

We now investigate how the notion of comonotonicity of F relates to the known monotonicity properties of F and how it can be characterized by means of pseudo-Jacobians of F. For this purpose, let us introduce the concept of cocoercivity. A map $F : \mathbb{R}^n \to \mathbb{R}^n$ is called *cocoercive* on $\mathbb{R}^n$ if there exists $\alpha > 0$ such that

$$\langle F(y) - F(x), y - x\rangle \geq \alpha\|F(y) - F(x)\|^2 \qquad \forall x, y \in \mathbb{R}^n.$$

The map $F : \mathbb{R}^n \to \mathbb{R}^n$ is called *cocoercive at* $x \in \mathbb{R}^n$ *in the direction* $u \in \mathbb{R}^n$ if there exist some $\alpha_{(x,u)} > 0$ so that

$$\langle F(x + tu) - F(x), tu\rangle \geq \alpha_{(x,u)}\|F(x + tu) - F(x)\|^2$$

for all $t \geq 0$ sufficiently small.

Given a point $x \in \mathbb{R}^n$ and a direction $u \in \mathbb{R}^n$, the following theorem illustrates the general relationship between comonotonicity and cocoercivity.

Theorem 5.1.7 *If $F : \mathbb{R}^n \to \mathbb{R}^n$ is cocoercive at $x \in \mathbb{R}^n$ in the direction $u \in \mathbb{R}^n$ and if*

$$\liminf_{t\downarrow 0} \frac{\|F(x + tu) - F(x)\|}{t} > 0, \tag{5.7}$$

then F is comonotone at x in the direction u.

If $F : \mathbb{R}^n \to \mathbb{R}^n$ is comonotone at $x \in \mathbb{R}^n$ in the direction $u \in \mathbb{R}^n$ and if

$$\limsup_{t\downarrow 0} \frac{\|F(x + tu) - F(x)\|}{t} < +\infty, \tag{5.8}$$

then F is cocoercive at x in the direction u.

Proof. The cocoercivity of F at x in the direction u implies that there is some $\alpha_{(x,u)} > 0$ so that

$$\langle F(x + tu) - F(x), u\rangle \geq \alpha_{(x,u)} \frac{\|F(x + tu) - F(x)\|}{t}\|F(x + tu) - F(x)\|$$

for all $t > 0$ sufficiently small. Using (5.7) we see that F must be comonotone at x in the direction u. Conversely, let us consider the case where F is comonotone at x in the direction u. Set

$$h_* := \limsup_{t \downarrow 0} \frac{\|F(x+tu) - F(x)\|}{t}.$$

Then, by (5.8), $0 \leq h_* < \infty$. If $h_* = 0$, we easily get, for some $\gamma_{(x,u)} > 0$,

$$\langle F(x+tu) - F(x), tu \rangle \geq \gamma_{(x,u)} t \|F(x+tu) - F(x)\| \geq \gamma_{(x,u)} \|F(x+tu) - F(x)\|^2$$

for all $t > 0$ sufficiently small. If $0 < h_* < \infty$, then it follows that, for some $\gamma_{(x,u)} > 0$,

$$\langle F(x+tu) - F(x), tu \rangle \geq \gamma_{(x,u)} h_*^{-1} h_* t \|F(x+tu) - F(x)\|$$
$$\geq \tfrac{1}{2} \gamma_{(x,u)} h_*^{-1} \|F(x+tu) - F(x)\|^2$$

for all $t > 0$ sufficiently small. Thus F is cocoercive at x in the direction u. $\square$

Note that the left-hand side in (5.7) may be equal to $+\infty$. It can be seen from (5.4) and (5.6) that this case is of particular importance for the analysis in Section 5.4.

Theorem 5.1.8 *Let $F: \mathbb{R}^n \to \mathbb{R}^n$ be a continuous map. Assume that F admits a pseudo-Jacobian map ∂F. Let $(x, u) \in \mathbb{R}^n \times \mathbb{R}^n$ with $u \neq 0$. If there exist numbers $\alpha_{(x,u)} > 0$ and $t_{(x,u)} > 0$ such that*

$$\langle u, M(u) \rangle \geq \alpha_{(x,u)} \|u\| \|M(u)\| \text{ for all } M \in \overline{co}(\partial F[x, x + t_{(x,u)} u]), \quad (5.9)$$

then F is comonotone at x in the direction u.

Proof. Let $t \in [0, t_{(x,u)}]$ be arbitrary but fixed. Then it follows from the mean value theorem (Theorem 2.2.2) that $N \in \overline{co}(\partial F[x, x+tu])$ exists with

$$F(x + tu) - F(x) = tN(u). \quad (5.10)$$

This together with (5.9) yields

$$\langle F(x + tu) - F(x), u \rangle = \langle u, tN(u) \rangle \geq \alpha_{(x,u)} \|u\| \|tN(u)\|.$$

Now, using (5.10) again, we get

$$\langle F(x + tu) - F(x), u \rangle \geq \gamma_{(x,u)} \|F(x + tu) - F(x)\|$$

with $\gamma_{(x,u)} := \alpha_{(x,u)} \|u\|$. $\square$

Quasimonotone Operators

Let S be a nonempty, open, and convex subset of $\mathbb{R}^n$. We say that a set-valued map $F : S \rightrightarrows \mathbb{R}^n$ is quasimonotone on S if for each $x, y \in S$ and for each $\xi \in F(x), \zeta \in F(y)$, one has

$$\min\{\langle \xi, y - x \rangle, \ \langle \zeta, \ x - y \rangle\} \leq 0,$$

or equivalently

$$\sup_{\xi \in F(x), \zeta \in F(y)} \min\{\langle \xi, y - x \rangle, \langle \zeta, x - y \rangle\} \leq 0.$$

Because the variables ξ and ζ are independent in the expressions under min and sup, we may interchange sup and min to obtain another equivalent form of quasimonotonicity

$$\min\{\sup_{\xi \in F(x)} \langle \xi, y - x \rangle, \ \sup_{\zeta \in F(y)} \langle \zeta, x - y \rangle\} \leq 0.$$

When $n = 1$ quasimonotone single-valued operators have quite simple structure. Indeed, let $S = (a, b) \subseteq \mathbb{R}$ with $a < b$, and let $f : S \to \mathbb{R}$ be continuous. Set $c := \inf\{t \in (a, b) : f(t) > 0\}$. Then it is easy to verify that f is quasimonotone on S if and only if it takes nonpositive values on (a, c) and nonnegative values on (c, b). Note that a can be $-\infty$, b can be $+\infty$, and c can be $\pm\infty$.

Some elementary properties of quasimonotone operators are given next.

Proposition 5.1.9 *Assume that F is an operator on a nonempty, open, and convex subset S of $\mathbb{R}^n$. Then the following assertions are true.*

(i)   If F is monotone, then it is quasimonotone.
(ii)   If F is quasimonotone, then the operators λF with $\lambda \geq 0$, $\overline{co}F$, and every suboperator of F is quasimonotone.

Proof. This is immediate from the definitions of monotone and quasimonotone operators. $\square$

We notice that a quasimonotone operator is not necessarily monotone; the sum and the union of two quasimonotone operators are not necessarily quasimonotone either.

Example 5.1.10 Define two single-valued operators F and G on $\mathbb{R}$ by

$$F(x) = \begin{cases} -2x & \text{if } x \leq 0 \\ x & \text{else,} \end{cases} \quad \text{and} \quad G(x) = \begin{cases} x & \text{if } x \leq 0 \\ -2x & \text{else.} \end{cases}$$

Direct verification shows that these operators are quasimonotone, but not monotone on $\mathbb{R}$. Their sum and union are given by

$$(F+G)(x) = -x$$
$$(F \cup G)(x) = \{x, -2x\}.$$

By taking $x = -1$ and $y = 1$, we have

$$\min\{\langle (F+G)(x), y-x \rangle, \langle (F+G)(y), x-y \rangle\} = 2 > 0$$
$$\min\{\sup_{\xi \in (F \cup G)(x)} \langle \xi, y-x \rangle, \sup_{\zeta \in (F \cup G)(y)} \langle \xi, x-y \rangle\} = 4 > 0,$$

and therefore these operators are not quasimonotone.

Here are some characterizations of single-valued quasimonotone operators.

Theorem 5.1.11 *Assume that $F: S \to \mathbb{R}^n$ is continuous and admits a pseudo-Jacobian $\partial F(x)$ at each $x \in S$. If F is quasimonotone, then*

(i) $\langle F(x), u \rangle = 0$ implies $\sup_{M \in \partial F(x)} \langle M(u), u \rangle \geq 0$,
(ii) $\langle F(x), u \rangle = 0$ and $\langle F(x + t'u), u \rangle > 0$ for some $t_1 < 0$ imply the existence of $t_2 > 0$ such that $\langle F(x+tu), u \rangle \geq 0$ for all $t \in [0, t_2]$.

Proof. Suppose (i) does not hold. Then there exist $x, u \in S$ such that

$$\langle F(x), u \rangle = 0 \quad \text{and} \quad \sup_{M \in \partial F(x)} \langle M(u), u \rangle < 0.$$

Thus from the definition of pseudo-Jacobian we get

$$(uF)^+(x, u) \leq \sup_{M \in \partial F(x)} \langle M(u), u \rangle < 0$$

and

$$(-uF)^+(x, -u) \leq \sup_{M \in \partial F(x)} \langle M(u), u \rangle < 0.$$

Hence, for sufficiently small $t > 0$,

$$\langle u, F(x+tu) - F(x) \rangle < 0$$

and

$$\langle -u, F(x + t(-u)) - F(x) \rangle < 0.$$

These give us that

$$\langle u, F(x+tu) \rangle < 0 \quad \text{and} \quad \langle u, F(x-tu) \rangle > 0.$$

Thus

$$\langle F(x+tu), (x-tu) - (x+tu) \rangle > 0$$

and

$$\langle F(x - tu),\ (x + tu) - (x - tu) \rangle > 0.$$

This contradicts the quasi-monotonicity of F, and so (i) holds.

Furthermore, if (ii) does not hold, then there exists $t_0 > 0$ such that $\langle F(x), u \rangle = 0$, $\langle F(x + t'u), u \rangle > 0$ for some $t' < 0$ and $\langle F(x + t_0 u), u \rangle < 0$. Let $x_0 = x + t'u$ and let $y_0 = x + t_0 u$. Then we have

$$\langle F(y_0), x_0 - y_0 \rangle = \langle F(x + t_0 u),\ (t' - t_0)u \rangle > 0,$$

$$\langle F(x_0), y_0 - x_0 \rangle = \langle F(x + t'u),\ (t_0 - t')u \rangle > 0.$$

These inequalities contradict the quasimonotonicity of F. $\square$

In general, it is not true that quasimonotonicity of F implies

$$\inf_{M \in \partial F(x)} \langle M(u), u \rangle \geq 0$$

for each $x, u \in S$ as in the differentiable case. Moreover, the conditions (i) and (ii) may not be sufficient without certain restrictions on the pseudo-Jacobian. This can be seen by taking $\partial F(x) = L(\mathbb{R}^n, \mathbb{R}^n)$ for each $x \in S$. We now obtain sufficient conditions under the additional hypotheses that pseudo-Jacobians are bounded and densely regular.

Theorem 5.1.12 *Let $F : S \to \mathbb{R}^n$ be a continuous map that admits a bounded and densely regular pseudo-Jacobian ∂F on S. Assume that the following conditions hold for every $x, u \in \mathbb{R}^n$.*

(i) $\langle F(x), u \rangle = 0$ implies $\max_{M \in \partial F(x)} \langle M(u), u \rangle \geq 0$.
(ii) $\langle F(x), u \rangle = 0$, $0 \in \{\langle u, M(u) \rangle : M \in \partial F(x)\}$ and $\langle F(x + t'u), u \rangle > 0$ for some $t' < 0$ imply the existence of $t_0 > 0$ such that $\langle F(x+tu), u \rangle \geq 0$ for all $t \in [0, t_0]$.

Then F is quasimonotone.

Proof. Suppose there exist $x, y \in S$ such that

$$\langle F(x), y - x \rangle > 0 \text{ and } \langle F(y), x - y \rangle > 0.$$

Let $u = y - x$ and let $g(t) = \langle F(x + tu), u \rangle$. Then g is continuous, $g(0) > 0$ and $g(1) < 0$. So, there exists $t_1 \in (0, 1)$ such that

$$g(t_1) = 0 \text{ and } g(t) < 0 \text{ for all } t \in (t_1, 1).$$

Define $x_1 = x + t_1 u$. Then, $g(t_1) = \langle F(x_1), u \rangle = 0$ and $(uF)^-(x_1, u) \leq 0$. Now we claim that

$$0 \in \{\langle u, M(u) \rangle : M \in \partial F(x_1)\}.$$

To see this, first consider the case where $x_1 \in S_0$. If $\langle u, M(u) \rangle > 0$ for each $M \in \partial F(x_1)$, then by regularity of $\partial F(x_1)$ we get a contradiction because

$$0 < \min_{M \in \partial F(x_1)} \langle M(u), u \rangle = \inf_{M \in \partial F(x_1)} \langle M(u), u \rangle = (uF)^-(x_1, u) \leq 0.$$

If $\langle u, M(u) \rangle < 0$ for each $M \in \partial F(x_1)$, then by (i) we get a contradiction because

$$0 > \max_{M \in \partial F(x_1)} \langle M(u), u \rangle \geq 0.$$

Now consider the case where $x_1 \notin S_0$. Then for each $M \in \partial F(x_1)$ we can find a sequence $\{y_k\} \subset S_0, y_k \to x_1$, $M_k \in \partial F(y_k)$ such that $\lim_{k \to \infty} M_k = M$. As in the above case, the claim holds by applying the arguments in the two subcases to $M_{k_0} \in \partial F(y_{k_0}), y_{k_0} \in S_0$, for sufficiently large k_0. By continuity of g, there exists $t' < 0$ such that

$$g(t_1 + t') = \langle F(x_1 + t'u), u \rangle > 0.$$

Condition (ii) gives us that there exists $t_0 > 0$ such that

$$g(t_1 + t) = \langle F(x_1 + tu), u \rangle \geq 0 \text{ for all } t \in [0, t_0].$$

This contradicts the condition that $g(t) < 0$ for all $t \in (t_1, 1)$. Hence F is quasimonotone. □

As a special case, we obtain a characterization of quasimonotone locally Lipschitz maps.

Corollary 5.1.13 *Assume $F \colon S \to \mathbb{R}^n$ is locally Lipschitz on S. Then F is quasimonotone if and only if the following conditions hold for each $x, u \in S$.*

(i) $\langle F(x), u \rangle = 0$ implies $\max_{M \in \partial^C F(x)} \langle M(u), u \rangle \geq 0$.

(ii) $\langle F(x), u \rangle = 0, 0 \in \{\langle u, Au \rangle : A \in \partial^C F(x)\}$ and $\langle F(x+t'u), u \rangle > 0$ for some $t' < 0$ imply the existence of $t_0 > 0$ such that $\langle F(x + tu), u \rangle \geq 0$ for all $t \in [0, t_0]$.

Proof. The conclusion follows from Theorem 5.1.11 and Theorem 5.1.12 by noting that

$$\partial F(x) = \begin{cases} \{\nabla F(x)\} & x \in K, \\ \{\lim_{n \to \infty} \nabla F(x_n) : x_n \to x, \ \{x_n\} \subset K\} & x \notin K, \end{cases}$$

where K is a dense subset of S on which F is differentiable, is a pseudo-Jacobian of F at x that satisfies the hypotheses of the previous theorem and observing that $\partial^C F(x) = \mathrm{co}(\partial F(x))$. □

Corollary 5.1.14 *Assume* $F: S \to \mathbb{R}^n$ *is differentiable on* S. *Then* F *is quasimonotone if and only if the following conditions hold for each* $x, u \in \mathbb{R}^n$.

(i) $\langle F(x), u \rangle = 0$ *implies* $\langle u, \nabla F(x)u \rangle \geq 0$.
(ii) $\langle F(x), u \rangle = \langle u, \nabla F(x)u \rangle = 0$ *and* $\langle F(x + t'u), u \rangle > 0$ *for some* $t' < 0$ *imply the existence of* $t_0 > 0$ *such that* $\langle F(x + tu), u \rangle \geq 0$ *for all* $t \in [0, t_0]$.

Proof. Because F is differentiable, $\{\nabla F(x)\}$ is a regular and bounded pseudo-Jacobian for each $x \in S$. So, the conclusion follows from Theorems 5.1.11 and 5.1.12. □

Pseudomonotone Operators

Let $F : S \to \mathbb{R}^n$ be a set-valued map, where as before S is a nonempty, open and convex subset of $\mathbb{R}^n$. It is said to be *pseudomonotone* on S if for each $x, y \in S$ and $\xi \in F(x), \zeta \in F(y)$, one has

$$\langle \xi, y - x \rangle > 0 \text{ implies } \langle \zeta, y - x \rangle > 0, \tag{5.11}$$

or equivalently

$$\min\{\langle \xi, y - x \rangle, \langle \zeta, x - y \rangle\} < 0$$

whenever one of the terms under min is nonzero.

It can be seen that in the definition above, the strict inequalities of (5.11) can be replaced by inequalities

$$\langle \xi, y - x \rangle \geq 0 \text{ implies } \langle \zeta, y - x \rangle \geq 0.$$

Here are some elementary properties of pseudomonotone operators.

Proposition 5.1.15 *Assume that* F *is an operator on a nonempty, open, and convex subset* S *of* $\mathbb{R}^n$. *Then the following assertions are true.*

(i)   *If* F *is monotone, then it is pseudomonotone.*
(ii)   *If* F *is pseudomonotone, then it is quasimonotone.*
(ii)   *If* F *is pseudomonotone, then the operators* λF *with* $\lambda \geq 0$, $\overline{co}F$, *and every suboperator of* F *are pseudomonotone.*

Proof. This follows from the definitions of pseudomonotone and quasi-monotone operators. □

The operator F given in Example 5.1.10 is quasimonotone, but not pseudomonotone. Indeed, with $x = -1, y = 0$ one has $\langle F(x), y - x \rangle =$

$2 > 0$ and $\langle F(y), x - y \rangle = 0$. The operator G of the same example is pseudomonotone, but it is not nondecreasing (hence not monotone).

For the case $n = 1$ and F is single-valued, one can easily prove that F is pseudomonotone on an open interval (a, b) if and only if there is a point $c \in [a, b]$ such that F is nonpositive on (a, c) and strictly positive on (c, b). Here we understand that $(a, c) = \emptyset$ if $a = c$. When n is arbitrary and F is single-valued, some characterizations of pseudomonotonicity can be obtained by using pseudo-Jacobians.

Theorem 5.1.16 *Assume $F : S \to \mathbb{R}^n$ is a continuous map and admits a pseudo-Jacobian $\partial F(x)$ at each $x \in S$. If F is pseudomonotone, then $\langle F(x), u \rangle = 0$ implies that*

(i) $\sup_{M \in \partial F(x)} \langle M(u), u \rangle \geq 0$.
(ii)   *There exists $t_0 > 0$, such that $\langle F(x + tu), u \rangle \geq 0$, for all $t \in [0, t_0]$.*

Proof. Pseudomonotonicity implies quasimonotonicity therefore (i) follows from Theorem 5.1.11. If (ii) does not hold, then there exist $x \in S$, and $t' > 0$ such that $\langle F(x), u \rangle = 0$ and $\langle F(x + t'u), u \rangle < 0$. Define $y = x + t'u$. Then

$$\langle F(x), y - x \rangle = \langle F(x), t'u \rangle = 0. \tag{5.12}$$

On the other hand,

$$\langle F(y), x - y \rangle = \langle F(x + t'u), -t'u \rangle > 0.$$

Now it follows from pseudomonotonicity that $\langle F(x), y - x \rangle > 0$. This contradicts (5.12). $\qquad\square$

Theorem 5.1.17 *Let $F : S \to \mathbb{R}^n$ be a continuous map that admits a bounded and densely regular pseudo-Jacobian ∂F on S. Assume that the following conditions hold for every $x, u \in \mathbb{R}^n$.*

(i) $\langle F(x), u \rangle = 0$ *implies* $\max_{M \in \partial F(x)} \langle M(u), u \rangle \geq 0$.
(ii) $\langle F(x), u \rangle = 0$ *and* $0 \in \{ \langle u, M(u) \rangle : M \in \partial F(x) \}$ *imply the existence of $t_0 > 0$ such that $\langle F(x + tu), u \rangle \geq 0$ for all $t \in [0, t_0]$.*

Then F is pseudomonotone.

Proof. Suppose F is not pseudomonotone. Then there exist $x, y \in S$ such that

$$\langle F(x), y - x \rangle \geq 0 \text{ and } \langle F(y), x - y \rangle > 0.$$

Let $u = y - x$ and $g(t) = \langle F(x + tu), u \rangle$. Then g is continuous, $g(0) \geq 0$ and $g(1) < 0$. So, there exists $t_1 \in [0, 1]$ such that

$$g(t_1) = 0 \text{ and } g(t) < 0 \text{ for all } t \in (t_1, 1]. \tag{5.13}$$

Define $x_1 = x + t_1 u$. As in the proof of Theorem 5.1.12, $\langle F(x_1), u \rangle = 0$, $(uF)^-(x_1, u) \leq 0$ and

$$0 \in \{\langle u, M(u) \rangle : M \in \partial F(x_1)\}.$$

Now it follows from (ii) that there exists $t_0 > 0$ such that

$$\langle F(x_1 + tu), u \rangle \geq 0, \ \forall t \in [0, t_0].$$

Thus $g(t_1 + t) = \langle F(x_1 + tu), u \rangle \geq 0$ for sufficiently small t close to t_0. This is a contradiction to (5.13), and hence F is pseudomonotone.   □

Corollary 5.1.18 *Assume $F: S \to \mathbb{R}^n$ is locally Lipschitz on S. Then F is pseudomonotone if and only if the following conditions hold for each $x, u \in S$.*

(i) $\langle F(x), u \rangle = 0$ implies $\max_{M \in \partial_C F(x)} \langle M(u), u \rangle \geq 0$.

(ii) $\langle F(x), u \rangle = 0$ and $0 \in \{\langle u, M(u) \rangle : M \in \partial_C F(x)\}$ imply the existence of $t_0 > 0$ such that $\langle F(x + tu), u \rangle \geq 0$ for all $t \in [0, t_0]$.

Proof. The proof follows along the same line of arguments as in Corollary 5.1.13, and so the details are left to the reader.   □

Corollary 5.1.19 *Assume $F: S \to \mathbb{R}^n$ is differentiable on S. Then F is pseudomonotone if and only if the following conditions hold for each $x, u \in \mathbb{R}^n$.*

(i) $\langle F(x), u \rangle = 0$ implies $\langle u, \nabla F(x)u \rangle \geq 0$.

(ii) $\langle F(x), u \rangle = \langle u, \nabla F(x)u \rangle = 0$ implies the existence of $t_0 > 0$ such that $\langle F(x + tu), u \rangle \geq 0$ for all $t \in [0, t_0]$.

Proof. Because F is differentiable, $\{\nabla F(x)\}$ is a bounded regular pseudo-Jacobian for each $x \in S$. So the conclusion follows from Theorem 5.1.16 and Theorem 5.1.17.   □

5.2 Generalized Convex Functions

Let $S \subseteq \mathbb{R}^n$ be a nonempty, open, and convex set and let $\phi: S \to \mathbb{R}$ be a continuous function. Recall that ϕ is convex on S if for each pair of distinct points x and y in S and for every number $t \in (0, 1)$, one has

$$\phi(tx + (1 - t)y) \leq t\phi(x) + (1 - t)\phi(y).$$

If this inequality is strict, one says that ϕ is strictly convex. As we have seen in the first chapter, convex functions are locally Lipschitz around and directionally differentiable at any interior point of the effective domain. Another important feature of convex functions is that for them any local minimum point is also global. When a function is strictly convex, it attains its minimum at most at one point. Now we wish to characterize convexity of ϕ by means of pseudo-differentials and pseudo-Hessians of ϕ.

Proposition 5.2.1 *Assume that $\partial\phi\colon S \rightrightarrows L(\mathbb{R}^n, \mathbb{R})$ is a pseudo-differential of ϕ on S. If $\partial\phi$ is monotone, then the function ϕ is convex.*

Conversely, if ϕ is convex and $\partial\phi$ is a densely regular pseudo-differential of ϕ on S, then $\partial\phi$ is monotone.

Proof. Assume that $\partial\phi$ is a monotone pseudo-differential of ϕ on S. Suppose to the contrary that ϕ is not convex; that is, there are some points $a, b \in S$ and $c = (1 - \lambda)a + \lambda b$ for some $\lambda \in (a, b)$ such that

$$\phi(c) > (1 - \lambda)\phi(a) + \lambda(b).$$

Choose a number α such that

$$\phi(c) - \phi(a) > \alpha > \lambda(\phi(b) - \phi(a)).$$

In view of Corollary 2.2.6, there exist some $x \in (a, c), y \in (c, b)$, and $\xi \in \mathrm{co}(\partial\phi(x)), \zeta \in \mathrm{co}(\partial\phi(y))$ such that

$$\langle \xi, c - a \rangle > \alpha,$$
$$\langle \zeta, b - a \rangle < \frac{\alpha}{\lambda}.$$

Expressing $c - a = \lambda(b - a)$ and summing up the latter inequalities give

$$\langle \xi, c - a \rangle + \langle \zeta, a - c \rangle > 0.$$

Because $c - a = t(y - x)$ for some positive t, this inequality implies

$$\langle \xi, y - x \rangle + \langle \zeta, x - y \rangle > 0,$$

which contradicts the monotonicity of $\partial\phi$.

Conversely, assume that ϕ is convex and $\partial\phi$ is a densely regular pseudo-differential of ϕ on S. Let $x, y \in S$ and $\xi \in \partial\phi(x), \zeta \in \partial\phi(y)$. Then there exist two sequences $\{x_k\}, \{y_k\}$ (both in S_0) converging to x and y, and sequences $\xi_k \in \partial\phi(x_k), \zeta_k \in \partial\phi(y_k)$ converging to ξ and ζ respectively. (Here, if x is a point at which $\partial\phi$ is regular, one takes $x_k = x$ and $\xi_k = \xi$; and similarly for y and ζ.) Because at x_k and y_k the pseudo-Jacobian of ϕ is regular, one has that

$$\langle \xi_k, y_k - x_k \rangle \le \phi^+(x_k, y_k - x_k),$$
$$\langle \zeta_k, x_k - y_k \rangle \le \phi^+(y_k, x_k - y_k).$$

Because ϕ is convex, in view of Lemma 1.4.2 we have

$$\phi'(x_k; y_k - x_k) \le \phi(y_k) - \phi(x_k)$$
$$\phi'(y_k, x_k - y_k) \le \phi(x_k) - \phi(y_k).$$

We deduce that

$$\langle \xi_k, y_k - x_k \rangle + \langle \zeta_k, x_k - y_k \rangle \le 0.$$

When k tends to ∞, this inequality gives

$$\langle \xi, y - x \rangle + \langle \zeta, x - y \rangle \le 0$$

by which $\partial\phi$ is monotone. $\square$

Corollary 5.2.2 *A continuous function ϕ on a nonempty open and convex set S is convex if and only if it is locally Lipschitz and its Clarke subdifferential is a monotone operator on S.*

Proof. If ϕ is convex, then by Lemma 1.4.2 it is locally Lipschitz on S. Moreover, in view of Proposition 1.4.7, its Clarke subdifferential $\partial^C f$ coincides with the convex subdifferential $\partial^{ca} f$, which is a regular pseudo-differential. Hence, by Proposition 5.1.1, $\partial^C f$ is monotone on S. The converse is immediate from the said proposition because the Clarke subdifferential is a pseudo-Jacobian. $\square$

A second-order characterization of convex functions can be obtained from the first-order characterization of monotone operators given in the previous section.

Corollary 5.2.3 *Let $\phi: S \to \mathbb{R}$ be a C^1-function that admits a pseudo-Hessian $\partial^2\phi(x)$ at each $x \in \mathbb{R}^n$. If the matrices of $\partial^2\phi(x)$ are positive semidefinite, then ϕ is convex on S.*

Conversely, if ϕ is convex and the pseudo-Hessian $\partial^2\phi$ is a densely pseudo-Jacobian of ∇f on S, then for each $x \in S$ the matrices $M \in \partial^2\phi(x)$ are positive semidefinite.

Proof. Apply Theorem 5.1.2 and Proposition 5.2.1. $\square$

Corollary 5.2.4 *Let $\phi: S \to \mathbb{R}$ be $C^{1,1}$. Then ϕ is convex if and only if for each $x \in S$ the matrices $M \in \partial_H^2 f(x)$ are positive semidefinite.*

Proof. The conclusion follows from Corollaries 5.1.4 and 5.2.2.   □

For strictly convex functions we have the following characterizations.

Proposition 5.2.5 *Let $\phi\colon S \to \mathbb{R}$ be a continuous function. Then each of the conditions below is sufficient for ϕ to be strictly convex.*

(i) ϕ admits a bounded pseudo-differential that is strictly monotone on S.

(ii) ϕ is of class $C^{1,1}$ and admits a pseudo-Hessian $\partial^2\phi$ for which all elements of the sets $\overline{\mathrm{co}}(\partial^2\phi(x)) \cup ((\mathrm{co}(\partial^2\phi(x)))_\infty \setminus \{0\})$, $x \in S$ are positive definite matrices.

Conversely, if ϕ is strictly convex and if $\partial\phi$ is a regular pseudo-differential of ϕ on S, then $\partial\phi$ is strictly monotone.

Proof. We need only to prove the strict convexity of ϕ under the first condition because, in view of Theorem 5.1.3, the second condition implies the first one. Let x and y be two distinct points in S and let $t \in [0,1]$. In view of the mean value theorem (Corollary 2.2.6) and as the pseudo-differential is bounded, one can find two points $a \in [x, tx + (1-t)y], b \in [tx + (1-t)y, y]$ and two elements $\xi \in \partial\phi(a), \zeta \in \partial\phi(b)$ such that

$$\phi(x) - \phi(tx + (1-t)y) = \langle \xi, (1-t)(y-x) \rangle$$
$$\phi(y) - \phi(tx + (1-t)y) = \langle \zeta, t(x-y) \rangle.$$

Multiplying the first inequality by t and the second by $(1-t)$ and summing them up gives

$$t\phi(x) + (1-t)\phi(y) - \phi(tx + (1-t)y) = t(1-t)\langle \xi - \zeta, x - y \rangle.$$

Because $\partial\phi$ is strictly monotone, the expression on the right-hand side of the above equality is strictly positive. This shows that ϕ is strictly convex.

For the second part of the proposition, let x and y be two distinct points in S. It follows from the strict convexity of ϕ that

$$\phi^+(x; y - x) < \phi(y) - \phi(x)$$
$$\phi^+(y; x - y) < \phi(x) - \phi(y).$$

Because $\partial\phi$ is regular, by summing up the latter inequalities, we obtain

$$\sup_{\xi\in\partial\phi(x)} \langle \xi, y - x \rangle + \sup_{\zeta\in\partial\phi(y)} \langle \zeta, x - y \rangle = \phi^+(x; y - x) + \phi^+(y; x - y) < 0.$$

By this, $\partial\phi$ is strictly monotone.   □

Note that the second condition stated in the previous proposition is not necessary for ϕ to be strictly convex even when the pseudo-Hessian is regular. The function $\phi(x) = x^4$ is strictly convex on $\mathbb{R}$, its second derivative is a regular pseudo-Hessian that takes the value zero at $x = 0$.

Quasiconvex Functions

Let S be a nonempty, open, and convex subset of $\mathbb{R}^n$. Let $\phi\colon S \to \mathbb{R}$ be a continuous function. We say that ϕ is *quasiconvex* on S if for every points x and y of S and for every $\lambda \in [0,1]$ one has

$$\phi(\lambda x + (1 - \lambda)y) \leq \max\{\phi(x); \phi(y)\}.$$

It is plain that convex functions are quasiconvex and that the converse is not true. Quasiconvex functions can be characterized by convexity of lower level sets. Namely, ϕ is quasiconvex if and only if its lower level sets

$$\{x \in S : \phi(x) \leq t\}, \quad t \in \mathbb{R},$$

are convex sets. Other characterizations of quasiconvexity are expressed in terms of pseudo-Jacobians.

Proposition 5.2.6 *Assume that $\partial\phi\colon S \rightrightarrows L(\mathbb{R}^n, \mathbb{R})$ is a pseudo-differential of ϕ on S. If $\partial\phi$ is quasimonotone, then the function ϕ is quasiconvex.*

Conversely, if ϕ is quasiconvex and $\partial\phi$ is a densely regular pseudo-differential of ϕ on S, then $\partial\phi$ is quasimonotone.

Proof. Suppose that $\partial\phi$ is a quasimonotone pseudo-differential of ϕ on S and that ϕ is not quasiconvex. There exist three points a, b, and c in S with $c = (1 - \lambda)a + \lambda b$ for some $\lambda \in (0, 1)$ such that

$$\phi(c) > \max\{\phi(a), \phi(b)\}.$$

By using the mean value theorem (Corollary 2.2.6), one can find points $x \in (a, c), y \in (c, b)$, and $\xi \in \partial\phi(x), \zeta \in \partial\phi(y)$ such that

$$\langle \xi, c - a \rangle > \frac{1}{2}(\phi(c) - \phi(a)) > 0,$$

$$\langle \zeta, c - b \rangle > \frac{1}{2}(\phi(c) - \phi(b)) > 0.$$

There exist two positive numbers t_1 and t_2 satisfying $c - a = t_1(y - x)$ and $c - b = t_2(x - y)$. Substituting these expressions into the two latter inequalities gives

$$\langle \xi, y - x \rangle > 0,$$
$$\langle \zeta, x - y \rangle > 0.$$

This contradicts the quasimonotonicity of $\partial\phi$.

Conversely, let $\partial\phi$ be a densely regular pseudo-differential of the quasiconvex function ϕ on S. Let x and y be two arbitrary distinct points of S and let $\xi \in \partial\phi(x)$ and $\zeta \in \partial\phi(y)$. First consider the case when $\partial\phi$ is regular at x and y. We may assume $\phi(x) \geq \phi(y)$. Then for every $t \in (0,1)$, one has

$$\phi(x + t(y - x)) \leq \phi(x).$$

This and the regularity of $\partial\phi$ imply

$$
\begin{aligned}
\langle \xi, y - x \rangle &= \phi^+(x; y - x) \\
&= \limsup_{t\downarrow 0} \frac{\phi(x + t(y - x)) - \phi(x)}{t} \\
&\leq 0.
\end{aligned}
$$

Hence

$$\min\{\langle \xi, y - x \rangle, \langle \zeta, x - y \rangle\} \leq 0. \tag{5.14}$$

Now we take up the case where $\partial\phi$ is not regular at x and at y. Then there exist sequences $\{x_k\}$, $\{y_k\}$ in S_0 converging to x and y, and sequences $\xi_k \in \partial\phi(y_k), \zeta_k \in \partial\phi(x_k)$ converging to ξ and ζ. According to the proof above, we obtain

$$\min\{\langle \xi_k, y_k - x_k \rangle; \langle \zeta_k, x_k - y_k \rangle\} \leq 0.$$

Passing to the limit when k tends to ∞ in this inequality gives us (5.14). Hence $\partial\phi$ is quasimonotone. □

Corollary 5.2.7 *Let $f: S \to \mathbb{R}$ be a C^1-function that admits a pseudo-Hessian $\partial^2 f(x)$ at each $x \in \mathbb{R}^n$. If f is quasiconvex, then for each $x, u \in S$ with $\langle \nabla f(x),\ u \rangle = 0$,*

$$\sup_{M \in \partial^2 f(x)} \langle M(u), u \rangle \geq 0.$$

Proof. The conclusion follows from Theorem 5.1.11 by replacing F by ∇f and noting that f is quasiconvex if and only if ∇f is quasimonotone. □

Pseudoconvex Functions

Let $\phi : S \to \mathbb{R}$ be a continuous function, where S is a nonempty open and convex subset of $\mathbb{R}^n$. We say that ϕ is *pseudoconvex* on S if for any two points x and y of S with $\phi(y) > \phi(x)$, there exist two positive numbers β and $\delta \in (0, 1]$ such that

$$\phi(y) \geq \phi(\lambda x + (1 - \lambda)y) + \lambda\beta, \text{ for all } \lambda \in (0, \delta).$$

Notice that convex functions are pseudoconvex and pseudoconvex functions are quasiconvex. The converse is not true in general. For instance, the function $\phi : \mathbb{R} \to \mathbb{R}$ defined by

$$\phi(x) = \begin{cases} 2x & \text{if } x \leq 0, \\ x & \text{else} \end{cases}$$

is pseudoconvex, but not convex, whereas the function $\psi(x) = x^3$ is quasiconvex, but not pseudoconvex.

Proposition 5.2.8 *Assume that $\partial\phi : S \rightrightarrows L(\mathbb{R}^n, \mathbb{R})$ is a pseudo differential of ϕ on S. If $\partial\phi$ is bounded and pseudomonotone, then the function ϕ is pseudoconvex.*

Conversely, if ϕ is pseudoconvex and $\partial\phi$ is a regular pseudo-differential of ϕ on S, then $\partial\phi$ is pseudomonotone.

Proof. Let $\partial\phi$ be a pseudomonotone differential of ϕ on S. Suppose to the contrary that ϕ is not pseudoconvex. Then there exist two points x and y of S with $\phi(y) > \phi(x)$ such that for each $k = 1, 2, \ldots$, one can find some $\lambda_k \in (0, 1/k)$ satisfying

$$\phi(y) < \phi(y + \lambda_k(x - y)) + \frac{\lambda_k}{k}.$$

This implies that

$$\phi^+(y; x - y) \geq \limsup_{k \to \infty} \frac{\phi(y + \lambda_k(x - y)) - \phi(y)}{\lambda_k} \geq 0.$$

By the definition of pseudo-differential, we deduce that

$$\sup_{\xi \in \partial\phi(y)} \langle \xi, x - y \rangle \geq 0.$$

Because $\partial\phi(y)$ is bounded, there exists some $\xi \in \partial\phi(y)$ such that

$$\langle \xi, x - y \rangle \geq 0. \tag{5.15}$$

On the other hand, as $\phi(y) > \phi(x)$, in virtue of the mean value theorem, there are some $z \in (x, y)$ and $\zeta \in \partial\phi(z)$ such that

$$\langle \zeta, y - x \rangle > 0.$$

This and (5.15) contradict the pseudomonotonicity hypothesis.

Conversely, assume $\partial\phi$ is a regular pseudo-differential of the pseudoconvex function ϕ. Let x and y be arbitrary points of S. If $\langle \xi, y - x \rangle \leq 0$ for all $\xi \in \partial\phi(x)$, there is nothing to prove. So, assume that

$$\langle \xi_0, y - x \rangle > 0 \quad \text{for some} \quad \xi_0 \in \partial \phi(x).$$

Then

$$\phi^+(x; y - x) = \sup_{\xi \in \partial \phi(x)} \langle \xi, y - x \rangle > 0.$$

Thus there is some $t \in (0, 1)$ such that

$$\phi(x + t(y - x)) > \phi(x).$$

As pseudoconvex functions are quasiconvex, one derives

$$\phi(y) \geq \phi(x + t(y - x)) > \phi(x).$$

Then there are some positive numbers β and $\delta \in (0, 1)$ such that

$$\phi(y + t(x - y)) - \phi(y) \leq -t\beta \text{for} t \in (0, \delta),$$

which implies that

$$\phi^+(y; x - y) \leq \beta < 0.$$

The regularity hypothesis shows that

$$\langle \xi, x - y \rangle \leq -\beta < 0 \text{for all } \xi \in \partial \phi(y).$$

Thus $\partial \phi$ is pseudomonotone and the proof is complete. $\qquad \square$

It is interesting to notice that in contrast to the case of convex and quasiconvex functions, the conclusion of Theorem 5.1.17 is not true when regularity is substituted by dense regularity. To see this, let us define a function $\phi : \mathbb{R} \to \mathbb{R}$ by

$$\phi(x) = \begin{cases} -x & \text{if} \quad x \leq 0, \\ x & \text{if} \quad 0 < x \leq 1, \\ (x - 1)^2 + 1 & \text{else.} \end{cases}$$

This function is pseudoconvex and locally Lipschitz on $\mathbb{R}$. Its Clarke subdifferential $\partial^C \phi$ is a regular pseudo-differential at any point $x \in \mathbb{R} \setminus \{1\}$, hence densely regular on $\mathbb{R}$. Despite this, $\partial^C \phi$ is not pseudomonotone because for $x = 0$ and $y = 1$, by taking $\xi = 1 \in \partial^C \phi(x)$ and $\zeta = 0 \in \partial^C \phi(y)$, one has $\langle \xi, y - x \rangle > 0$, but $\langle \zeta, x - y \rangle = 0$.

5.3 Variational Inequalities

Let K be a nonempty closed convex set in the n-dimensional Euclidean space $\mathbb{R}^n$ and let f and $g : \mathbb{R}^n \to \mathbb{R}^n$ be nonlinear continuous operators. The general variational inequality problem that is associated with f, g, and K, denoted $V(f, g, K)$, consists of finding $x_0 \in R^n$ with $g(x_0) \in K$ such that

$$\langle f(x_0), g(x) - g(x_0) \rangle \geq 0 \text{ for every } x \in \mathbb{R}^n \text{ with } g(x) \in K .$$

A particular case of $V(f, g, K)$ is when g is the identity operator that is known as the Hartman–Stampacchia variational inequality. It is, in fact, an extension of an optimality condition in nonlinear programming. Let us consider the following constrained minimization problem.

(P) $\qquad$ minimize $\phi(x)$
$\qquad\qquad$ subject to $x \in K$,

where ϕ is a real-valued differentiable function on $\mathbb{R}^n$. According to Theorem 2.1.16, if $x_0 \in K$ is a local minimizer of ϕ on K, then

$$\langle \nabla\phi(x_0), x - x_0 \rangle \geq 0 \text{ for all } x \in K .$$

This is the Hartman–Stampacchia variational inequality in which the gradient $\nabla\phi$ is used in the role of f. Of course, not every vector function f can be expressed as a gradient map, so it is not always possible to express the Hartman-Stampacchia problem in the form of optimality conditions.

A counterpart of the Hartman–Stampacchia inequality is the so-called Minty variational inequality which consists of finding a point x_0 of K such that

$$\langle f(x), x - x_0 \rangle \geq 0 \text{ for all } x \in K .$$

In general, the solution set of the Hartman-Stampacchia problem and the one of the Minty problem are distinct. However, they coincide under a certain monotonicity assumption.

Proposition 5.3.1 *Let $K \subseteq \mathbb{R}^n$ be a nonempty closed and convex set, and let $f : \mathbb{R}^n \to \mathbb{R}^n$ be a continuous map that is pseudomonotone on K. Then every solution to the Hartman–Stampacchia variational inequality is a solution to the Minty variational inequality and vice versa.*

Proof. If $x_0 \in K$ is not a solution to the Minty variational inequality, then one can find a point x of K such that

$$\langle f(x), x_0 - x \rangle > 0.$$

Because f is pseudomonotone, we deduce

$$\langle f(x_0), x - x_0 \rangle < 0,$$

which shows that x_0 is not a solution to the Hartman–Stampacchia variational inequality. Conversely, if $x_0 \in K$ is not a solution to the Hartman–Stampacchia problem, then the latter inequality holds for some $x \in K$. The continuity of f implies the existence of a positive ϵ such that

$$\langle f(x'), x - x_0 \rangle < 0 \text{ for all } x' \in K \cap (x_0 + \epsilon B_n).$$

Choose a positive t less than $\min\{1, \epsilon/\|x - x_0\|\}$ and $x' = x_0 + t(x - x_0)$. Then x' belongs to $K \cap (x_0 + \epsilon B_n)$. Consequently,

$$\langle f(x'), x' - x_0 \rangle = t \langle f(x'), x - x_0 \rangle < 0.$$

By this, x_0 cannot be a solution to the Minty variational inequality problem. $\qquad\square$

By defining a set-valued map $G : K \rightrightarrows K$ by

$$G(x) := \{y \in K : \langle f(x), x - y \rangle \geq 0\},$$

we easily prove that the Minty variational inequality is equivalent to the following intersection problem.

Find $x_0 \in K$ such that $x_0 \in \bigcap_{x \in K} G(x)$.

Likewise the variational inequality problem $V(f, g, K)$ is equivalent to the intersection problem.

Find $x_0 \in K$ such that $x_0 \in \bigcap_{x \in K} F(x)$,

where $F \colon K \rightrightarrows K$ is given by

$$F(x) := \{y \in K : \langle f(y), g(x) - g(y) \rangle \geq 0\}.$$

Thus the existence of solutions to variational inequalities is exactly the existence of intersection points for a suitably defined set-valued map from K to itself. It is clear that if g is the identity map and if f is pseudomonotone, then F is a submap of G. Hence any solution of the Hartman–Stampacchia problem is also a solution of the Minty problem. Conversely, with g being the identity map, if $-f$ is pseudomonotone, then G is a submap of F, and hence any solution of the Minty problem is a solution of the Hartman–Stampacchia problem. Without pseudomonotonicity the two problems have distinct solution sets. Now we focus our efforts on the question of the uniqueness of solutions to the problem $V(f, g, K)$ by using pseudo-Jacobian matrices.

Critical Cones

Given a point $x \in \mathbb{R}^n$ with $g(x) \in K$, one defines the critical cone of (f, g) at x as the set

$$C_{(f,g)}(K, x) := \{v \in T(K, g(x)) : \langle f(x), v \rangle = 0\}.$$

In other words, the critical cone is the intersection of the tangent cone to K at $g(x)$ and the orthogonal subspace of the vector $f(x)$. We write $C_f(K, x)$ for the critical cone when g is the identity map. The positive polar cone of the critical cone is the set

$$[C_{(f,g)}(K, x)]^* := \{\xi \in \mathbb{R}^n : \langle \xi, v \rangle \geq 0 \text{ for all } v \in C_{(f,g)}(K, x)\}.$$

Under certain assumptions, the critical cone and its positive polar cone can be computed by solving a system of linear equations and inequalities. Let us consider the case where K is explicitly represented by constraints

$$g_i(x) \leq 0, \ i = 1, \ldots, p$$
$$h_j(x) = 0, \ j = 1, \ldots, q.$$

The active index set at a point x is denoted by $I(x)$. It consists of the indices $i \in \{1, \ldots, p\}$ satisfying $g_i(x) = 0$. We know that if g_i and h_j are differentiable and if the gradient vectors $\nabla g_i(x), i \in I(x)$ and $\nabla h_j(x), j = 1, \ldots, q$ are linearly independent, then the tangent cone to K at $x \in K$ is the solution set to the system

$$\langle \nabla g_i(x), v \rangle \leq 0, \ i \in I(x)$$
$$\langle \nabla h_j(x), v \rangle = 0, \ j = 1, \ldots, q.$$

Then the critical cone of (f, g) at x_0 with $y_0 := g(x_0) \in K$ is given by the system

$$\langle f(x_0), v \rangle = 0$$
$$\langle \nabla g_i(y_0), v \rangle \leq 0, \ i \in I(x)$$
$$\langle \nabla h_j(y_0), v \rangle = 0, \ j = 1, \ldots, q.$$

It is now easy to compute the positive polar cone of the critical cone. Namely, a vector ξ belongs to the cone $[C_{(f,g)}(K, x)]^*$ if and only if there exist some numbers $\lambda_i \geq 0, i \in I(y_0)$, and $\mu_1, \ldots, \mu_q, \mu$ such that

$$-\xi = \sum_{i \in I(y_0)} \lambda_i \nabla g_i(y_0) + \sum_{j=1}^{q} \mu_j \nabla h_j(y_0) + \mu f(x_0).$$

The "if" part is clear. The "only if" part easily follows from the separation theorem. Further observe that if x_0 is a solution to the problem $V(f, g, K)$

and if K is contained in the image of g, then the first equality $\langle f(x_0), v \rangle = 0$ in the system determining the critical cone can be relaxed to the inequality

$$\langle f(x_0), v \rangle \leq 0.$$

This is because when x_0 solves the problem $V(f, g, K)$, one has

$$\langle f(x_0), x - y_0 \rangle \geq 0 \text{ for all } x \in K,$$

which, in view of convexity of K, implies the converse inequality

$$\langle f(x_0), v \rangle \geq 0 \text{ for all } v \in T(K, g(x_0)).$$

In this case, the coefficient μ corresponding to $f(x_0)$ in the expression of the vector ξ may take nonnegative values only.

Local Uniqueness of Solutions

We say that a solution x_0 of $V(f, g, K)$ is *locally unique* if there is a neighborhood of x_0 such that no other solutions of the problem are inside this neighborhood. A nonempty subset A of $\mathbb{R}^n$ is said to be *polyhedral* if it is the intersection of a finite number of closed half-spaces. In other words, A is polyhedral when there exist a finite number of vectors $a_1, \ldots, a_k$ of $\mathbb{R}^n$ and numbers $\alpha_1, cdots, \alpha_k$ such that A is the solution set of the system of inequalities

$$\langle a_i, x \rangle \geq \alpha_i, \ i = 1, \ldots, k.$$

The following properties of a polyhedral set $A \subseteq \mathbb{R}^n$ are of use.

(a) $T(A, x) = \text{cone}(A - x)$ for every $x \in A$.
(b)   For each $x_0 \in A$, there is a neighborhood U of x_0 such that $\text{cone}(A - x_0) \subseteq \text{cone}(A - x)$ for all $x \in U \cap A$.

We keep the notation $\tilde{\partial} f(x_0) = \partial f(x_0) \cup ((\partial f(x_0))_\infty \setminus \{0\})$ where $\partial f(x_0)$ is a subset of $L(\mathbb{R}^n, \mathbb{R}^m)$.

Theorem 5.3.2 *Let $K \subseteq \mathbb{R}^n$ be a nonempty closed convex set, let $f, g : \mathbb{R}^n \to \mathbb{R}^n$ be continuous with g being onto K, and let $\partial f(x_0)$ and $\partial g(x_0)$ be Fréchet pseudo-Jacobians of f and g at x_0, respectively. If x_0 is a solution of $V(f, g, K)$, then each of the following conditions is sufficient for x_0 to be locally unique.*

(i)   *K is polyhedral and for every $M \in \tilde{\partial} f(x_0)$ and $N \in \tilde{\partial} g(x_0)$, one has*

$$\langle M(v), N(v) \rangle > 0$$

for all $v \in \mathbb{R}^n \setminus \{0\}$ with $N(v) \in C_{(f,g)}(K, x_0)$, $M(v) \in [C_{(f,g)}(K, x_0)]^*$.

(ii) K is polyhedral and for every $M \in \tilde{\partial} f(x_0)$ and $N \in \tilde{\partial} g(x_0)$, one has

$$\langle M(v), N(v) \rangle > 0$$

for all $v \in \mathbb{R}^n \backslash \{0\}$ with $N(v) \in C_{(f,g)}(K, x_0)$ and $f(x_0) + M(v) \in [T(K, g(x_0))]^*$.

(iii) For every $M \in \tilde{\partial} f(x_0)$ and $N \in \tilde{\partial} g(x_0)$, one has

$$\langle M(v), N(v) \rangle > 0$$

for all $v \in \mathbb{R}^n \backslash \{0\}$ with $N(v) \in C_{(f,g)}(K, x_0)$.

Proof. We first show that (i) implies (ii). Indeed, let $v \in \mathbb{R}^n \backslash \{0\}$, $M \in \tilde{\partial} f(x_0)$, and $N \in \tilde{\partial} g(x_0)$ satisfy $N(v) \in C_{(f,g)}(K, x_0)$ and $f(x_0) + M(v) \in [T(K, g(x_0))]^*$. It suffices to prove that $M(v) \in [C_{(f,g)}(K, x_0)]^*$. For, let $u \in C_{(f,g)}(K, x_0)$, which means that $u \in T(K, g(x_0))$ and $\langle f(x_0), u \rangle = 0$. Then

$$0 \le \langle f(x_0) + M(v), u \rangle = \langle M(v), u \rangle \ ,$$

by which $M(v) \in [C_{(f,g)}(K, x_0)]^*$.

Now assume (ii). Suppose to the contrary that x_0 is not a locally unique solution. One can find a sequence $\{x_i\}$ of solutions of $V(f, g, K)$ that converges to x_0. By considering a subsequence if necessary, one may assume that $\{(x_i - x_0)/\|x_i - x_0\|\}$ converges to some $v \ne 0$. Because x_i and x_0 are solutions of $V(f, g, K)$, the following relations hold true.

$$f(x_0) \in [T(K, g(x_0))]^*, \quad f(x_i) \in [T(K, g(x_i))]^*, \tag{5.16}$$

$$\langle f(x_0), g(x_i) - g(x_0) \rangle \ge 0, \quad \langle f(x_i) , \ g(x_i) - g(x_0) \rangle \ge 0 \ . \tag{5.17}$$

By property (b) of polyhedral sets, there is $i_0 \ge 1$ such that

$$[T(K, g(x_i))]^* \subseteq [T(K, g(x_0))]^*$$

and hence

$$f(x_i) - f(x_0) \in [T(K, g(x_0))]^* - f(x_0) \text{ for } i \ge i_0 \ . \tag{5.18}$$

Furthermore, because $\partial f(x_0)$ and $\partial g(x_0)$ are Fréchet pseudo-Jacobians of f and g at x_0, one can find $M_i \in \partial f(x_0)$ and $N_i \in \partial g(x_0)$ such that

$$f(x_i) - f(x_0) = M_i(x_i - x_0) + r_1(x_i - x_0),$$

$$g(x_i) - g(x_0) = N_i(x_i - x_0) + r_2(x_i - x_0),$$

where $r_1(x_i - x_0)/\|x_i - x_0\| \to 0$ and $r_2(x_i - x_0)/\|x_i - x_0\| \to 0$ as $i \to \infty$. Substituting these expressions into (5.2) and (5.18) we obtain

$$M_i(x_i - x_0) + r_1(x_i - x_0) \in [T(K, g(x_0))]^* - f(x_0), \qquad (5.19)$$

$$\langle f(x_0), \ N_i(x_i - x_0) + r_2(x_i - x_0) \rangle \geq 0, \qquad (5.20)$$

$$\langle f(x_i), \ N_i(x_i - x_0) + r_2(x_i - x_0) \rangle \leq 0 \qquad (5.21)$$

for $i \geq i_0$. If $\{M_i\}$ is bounded, then we may assume that it converges to some $M \in \partial f(x_0)$. Dividing (5.19) by $\|x_i - x_0\|$ and passing to the limit when $i \to \infty$, we deduce

$$M(v) \in \text{cone} \left([T(K, g(x_0))]^* - f(x_0)\right).$$

Consequently, there is some $t > 0$ such that

$$f(x_0) + M(tv) \in [T(K, g(x_0))]^* . \qquad (5.22)$$

If $\{M_i\}$ is unbounded, then we may assume that $\lim_{i \to \infty} \|M_i\| = \infty$ and $\{M_i/\|M_i\|\}$ converges to some $M \in (\partial f(x_0))_\infty \backslash \{0\}$. Upon dividing (5.19) by $\|M_i\| \cdot \|x_i - x_0\|$ and letting $i \to \infty$, we deduce relation (5.22) too. Further consider the sequence $\{N_i\}$. If it is bounded, we may assume that it converges to some $N \in \partial g(x_0)$. Dividing (5.20) by $\|x_i - x_0\|$ and taking the limit as $i \to \infty$, we have

$$\langle f(x_0), N(v) \rangle \geq 0.$$

Similarly (5.21) implies the inverse inequality, and hence

$$\langle f(x_0), N(v) \rangle = 0. \qquad (5.23)$$

Moreover, (5.20) and (5.21) give

$$\langle f(x_i) - f(x_0), g(x_i) - g(x_0) \rangle \leq 0 , \qquad (5.24)$$

which yields

$$\langle M(v), N(v) \rangle \leq 0 . \qquad (5.25)$$

Relations (5.22), (5.23), and (5.25) contradict the hypothesis of (ii). Now, if $\{N_i\}$ is unbounded, then we may assume that $\lim_{i \to \infty} \|N_i\| = \infty$ and $\{N_i/\|N_i\|\}$ converges to some $N \in (\partial g(x_0))_\infty \backslash \{0\}$. Dividing (5.20) and (5.21) by $\|N_i\| \cdot \|x_i - x_0\|$ and by either $\|N_i\| \cdot \|x_i - x_0\|^2$ when $\{M_i\}$ is bounded, or $\|M_i\| \|N_i\| \|x_i - x_0\|^2$ when $\{M_i\}$ is unbounded and taking the limit as $i \to \infty$, we can obtain (5.23) and (5.25) as well, which together with (5.22) contradict the hypothesis of (ii).

Finally, let (iii) hold. If x_0 is not a locally unique solution, then there is a sequence $\{x_i\}$ of solutions converging to x_0 such that (5.16) and (5.17) are satisfied. These imply (5.20) and (5.21), which give (5.23) and (5.25) by the same argument as above. Relations (5.23) and (5.24) contradict the hypothesis of (iii). $\qquad \square$

We notice that for the Hartman–Stampacchia variational inequality, the conditions of Theorem 5.3.2 are written in the following form.

(i') K is polyhedral and for each $v \in C_f(K, x_0)$ and $M \in [C_f(K, x_0)]^*$, the relation $\langle M(v), v \rangle = 0$ implies $v = 0$.

(ii') K is polyhedral and for every $M \in \tilde{\partial} f(x_0)$ and $v \in C_f(K, x_0) \setminus \{0\}$ the relation $f(x_0) + M(v) \in [T(K, x_0)]^*$ implies $\langle M(v), v \rangle > 0$.

(iii') Every matrix $M \in \tilde{\partial} f(x_0)$ is strictly positive on $C_f(K, x_0)$, i.e., $\langle M(v), v \rangle > 0$ for all $v \in C_f(K, x_0) \setminus \{0\}$.

When f and g are locally Lipschitz, Clarke's generalized Jacobian can be used as a Fréchet pseudo-Jacobian and in this case the recession cones $(\partial^C f(x_0))_\infty$ and $(\partial^C g(x_0))_\infty$ are trivial, and they do not play any role in the conclusion of the theorem.

Linearized Problems

Let M and N be $n \times n$-matrices and let $x_0 \in \mathbb{R}^n$ with $g(x_0) \in K$. We define f_M and $g_N : R^n \longrightarrow \mathbb{R}^n$ by

$$f_M(x) := f(x_0) + M(x - x_0) \,,$$
$$g_N(x) := g(x_0) + N(x - x_0) \,.$$

The general variational inequality problem $V(f_M, g_N, K)$ is called a *linearized problem* of $V(f, g, K)$ at x_0.

Theorem 5.3.3 *Let K be a polyhedral cone, let f and $g : \mathbb{R}^n \longrightarrow \mathbb{R}^n$ be continuous with g being onto K, and let $\partial f(x_0)$ and $\partial g(x_0)$ be Fréchet pseudo-Jacobians of f and g at x_0 with g_N being onto K for each $N \in \tilde{\partial} g(x_0)$. If x_0 is a locally unique solution of the linearized problem $V(f_M, g_N, K)$ for every $M \in \tilde{\partial} f(x_0)$ and $N \in \tilde{\partial} g(x_0)$, then it is a locally unique solution of $V(f, g, K)$.*

Proof. First we easily notice that because K is a polyhedral cone, a point $x_* \in \mathbb{R}^n$ is a solution of $V(f, g, K)$ if and only if

$$g(x_*) \in K, \quad f(x_*) \in K^*, \quad \text{and} \quad \langle f(x_*), g(x_*) \rangle = 0 \,. \tag{5.26}$$

Suppose to the contrary that x_0 is not a locally unique solution of $V(f, g, K)$. There exists a sequence $\{x_i\}$ of solutions of $V(f, g, K)$ that converges to x_0. We may assume that $\lim_{i \to \infty} (x_i - x_0)/\|x_i - x_0\| = v$. The following relations are immediate.

$$\langle f(x_0), g(x_i) - g(x_0) \rangle, \ \geq 0 \langle f(x_i), g(x_0) - g(x_i) \rangle \geq 0, \tag{5.27}$$
$$\langle f(x_i) - f(x_0), g(x_i) - g(x_0) \rangle \leq 0. \tag{5.28}$$

It follows from the definition that there exist $M_i \in \tilde{\partial} f(x_0)$ and $N_i \in \tilde{\partial} g(x_0)$ such that

$$f(x_i) - f(x_0) = M_i(x_i - x_0) + r_1(x_i - x_0),$$
$$g(x_i) - g(x_0) = N_i(x_i - x_0) + r_2(x_i - x_0).$$

where $r_1(x_i - x_0)/||x_i - x_0|| \to 0$ and $r_2(x_i - x_0)/||x_i - x_0|| \to 0$ as $i \to \infty$.

First consider the case when $\{M_i\}$ and $\{N_i\}$ are bounded. We may assume that they converge to $M \in \partial f(x_0)$ and $N \in \partial g(x_0)$, respectively. We wish to prove that there is some $\delta_0 > 0$ such that

$$g_N(x_0 + \delta v) \in K \tag{5.29}$$

$$f_M(x_0 + \delta v) \in K^* \tag{5.30}$$

$$\langle f_M(x_0 + \delta v) \, , \, g_N(x_0 + \delta v) \rangle = 0 \text{ for } 0 \le \delta < \delta_0 \, . \tag{5.31}$$

According to (5.26), these relations show that for each $\delta \in (0, \delta_0)$, the point $x_0 + \delta v$ is a solution of the linearized problem $V(f_M, g_N, K)$, which contradicts the hypothesis of the theorem. Thus our aim is to establish (5.29), (5.30) and (5.31). For (5.29), observe that $g(x_i) \in K$ and therefore

$$N_i(x_i - x_0) + r_2(x_i - x_0) \in K - g(x_0) \, . \tag{5.32}$$

Dividing both sides of (5.32) by $||x_i - x_0||$ and passing to the limit when $i \to \infty$, we derive

$$N(v) \in \text{cone} \, (K - g(x_0)) \, .$$

As K is a polyhedral set, there is some $\delta_1 > 0$ such that

$$N(\delta v) \in K - g(x_0) \quad \text{for } \delta \in [0, \delta_1],$$

which means that (5.29) holds for all $\delta \in [0, \delta_1)$. For (5.30) we apply (5.26) to x_i to obtain

$$M_i(x_i - x_0) + r_1(x_i - x_0) \in K^* - f(x_0) \, . \tag{5.33}$$

Dividing both sides of (5.33) by $||x_i - x_0||$, and passing to the limit when $i \to \infty$, and using the fact that K^* is polyhedral, we derive

$$M(v) \in T(K^*, f(x_0)) \, .$$

Again, because K^* is polyhedral, there is some $\delta_0 \in (0, \delta_1)$ such that

$$f(x_0) + M(\delta v) \in K^* \quad \text{for all} \quad \delta \in [0, \delta_0) \, ,$$

which means that (5.30) holds for all $\delta \in [0, \delta_0)$. Finally, for (5.31) we deduce from (5.27) and (5.28) that

$$\langle f(x_0), N(x_0) \rangle = 0 \tag{5.34}$$

$$\langle M(v), N(v) \rangle \le 0 \, . \tag{5.35}$$

Applying (5.26) to x_i and x_0 yields

$$
\begin{aligned}
0 &= \langle f(x_i), g(x_i) \rangle \\
&= \langle f(x_0) + M_i(x_i - x_0) + r_1(x_i - x_0), g(x_i) \rangle \\
&= \langle f(x_0), g(x_i) \rangle + \langle M_i(x_i - x_0) + r_1(x_i - x_0), g(x_i) \rangle \\
&= \langle f(x_0), g(x_i) - g(x_0) \rangle + \langle M_i(x_i - x_0) + r_1(x_i - x_0), g(x_i) \rangle \ .
\end{aligned}
$$

Dividing this by $||x_i - x_0||$, passing to the limit as $i \to \infty$, and using (5.34), we obtain

$$
\langle M(v), g(x_0) \rangle = 0 \ . \tag{5.36}
$$

Furthermore, because g is onto K and x_i is a solution of the problem $V(f, g, K)$, one has

$$
\begin{aligned}
0 &\le \langle f(x_i), g(x_0) + N(\delta v) - g(x_i) \rangle \\
&\le \langle f(x_0) + M_i(x_i - x_0) + r_1(x_i - x_0), g(x_0) - g(x_i) \rangle \\
&\quad + \langle f(x_0), N(\delta v) \rangle + \langle M_i(x_i - x_0) + r_1(x_i - x_0), N(\delta v) \rangle \ .
\end{aligned}
$$

This and (5.34) yield

$$
\begin{aligned}
0 &\le \langle f(x_0) + M_i(x_i - x_0) + r_1(x_i - x_0), g(x_0) - g(x_i) \rangle \\
&\quad + \langle M_i(x_i - x_0) + r_1(x_i - x_0), N(\delta v) \rangle.
\end{aligned}
$$

By dividing both sides of the latter inequality by $||x_i - x_0||$, passing to the limit when $i \to \infty$, and using (5.34), we derive

$$
\langle M(v), N(\delta v) \rangle \ge 0.
$$

This together with (5.35) gives

$$
\langle M(v), N(v) \rangle = 0 \ .
$$

Combining (5.26), (5.34), and (5.36) with the above equality, we obtain (5.31). Hence contradiction.

Consider now the case when $\{M_i\}$ is bounded and $\{N_i\}$ is unbounded. We may assume that $\lim_{i \to \infty} ||N_i|| = \infty$ and $\{N_i/||N_i||\}$ converges to some $N \in (\partial g(x_0))_\infty \backslash \{0\}$. By dividing both sides of (5.32) by $||N_i|| \, ||x_i - x_0||$ one derives (5.29) by the same argument. Similarly, (5.34) and (5.35) are obtained for this N, and (5.31) follows. The case when $\{M_i\}$ is unbounded, or both $\{M_i\}$ and $\{N_i\}$ are unbounded, is treated in the same way. $\qquad\square$

We remark that if f and g are H-differentiable with H-differentials $\partial f(x_0)$ and $\partial g(x_0)$ at x_0, respectively, then one may assume that there are two matrices $M \in \partial f(x_0)$ and $N \in \partial g(x_0)$ such that all the terms of the sequences $\{M_i\}$ and $\{N_i\}$ in the proof of Theorem 5.3.3 (and Theorem

5.3.2 too) coincide with M and N, respectively. Consequently, in these theorems, the sets $\partial f(x_0)$ and $\partial g(x_0)$ can be used instead of $\tilde{\partial}(x_0)$ and $\tilde{\partial}g(x_0)$.

Global Uniqueness of Solutions

Let us denote by K_0 the convex hull of the inverse image of K under g; that is,

$$K_0 = \text{co}(\{x \in \mathbb{R}^n : g(x) \in K\}).$$

When g is the identity operator, one has $K_0 = K$.

Theorem 5.3.4 *Assume that f and $g : \mathbb{R}^n \to \mathbb{R}^n$ are continuous with g being onto K, and ∂f and ∂g are pseudo-Jacobian maps of f and g, respectively. Further assume that for each*

$$M \in \bigcup_{x \in K_0} \overline{\text{co}}(\partial f(x)) \cup ((\text{co}(\partial f(x)))_\infty \backslash \{0\}),$$

$$N \in \bigcup_{x \in K_0} \overline{\text{co}}(\partial g(x)) \cup ((\text{co}(\partial g(x)))_\infty \backslash \{0\}),$$

the matrix $N \circ M$ is positive definite. Then problem $V(f, g, K)$ has at most one solution.

Proof. Suppose to the contrary that the problem has two distinct solutions x_0 and y_0. Then $[x_0, y_0] \subseteq K_0$ and

$$\langle f(x_0) - f(y_0), g(x_0) - g(y_0)\rangle \leq 0 . \tag{5.37}$$

We consider the scalar function $x \mapsto \langle f(x), g(x_0) - g(y_0)\rangle$. It is evident that the closure of the set

$$F(x) := \{M(g(x_0) - g(y_0)) : M \in \partial f(x)\}$$

is a pseudo-Jacobian of $\langle f(\cdot), g(x_0) - g(y_0)\rangle$ at x. Let us apply the mean value theorem to this scalar function on $[x_0, y_0]$. There exists $c \in (x_0, y_0)$ and $\xi_i \in \text{co}(F(c))$ such that

$$\langle f(x_0) - f(y_0), g(x_0) - g(y_0)\rangle = \lim_{i \to \infty} \langle \xi_i, g(x_0) - g(x_0)\rangle . \tag{5.38}$$

Because $\text{co}(F(c)) = [\text{co}(\partial f(c))](x_0 - y_0)$, we can find $M_i \in \text{co}(\partial f(c))$ such that

$$\xi_i = M_i(x_0 - y_0) .$$

If $\{M_i\}$ is bounded, we may assume that it converges to some $M_0 \in \overline{\text{co}}(\partial f(c))$. Then (5.37) and (5.38) imply

$$\langle f(x_0) - f(y_0), g(y_0) \rangle = \langle M_0(x_0 - y_0), g(x_0) - g(y_0) \rangle \leq 0. \qquad (5.39)$$

If $\{M_i\}$ is unbounded, then we may assume that

$$\lim_{i \to \infty} \|M_i\| = \infty \quad \text{and} \quad \lim_{i \to \infty} M_i/\|M_i\| = M_0 \in (\mathrm{co}(\partial f(c)))_\infty \backslash \{0\}.$$

Equality (5.38) gives

$$\langle M_0(x_0 - y_0), g(x_0) - g(y_0) \rangle = \lim_{i \to \infty} \Big\langle \frac{M_i}{\|M_i\|}(x_0 - y_0), g(x_0) - g(x_0) \Big\rangle \leq 0.$$
$$(5.40)$$

Let us now consider the scalar function $x \mapsto \langle M_0(x_0 - y_0), g(x) \rangle$ where M_0 is the matrix obtained above from $\overline{\mathrm{co}}(\partial f(c)) \cup (\mathrm{co}(\partial f(c)))_\infty \backslash \{0\}$. Arguing in the same way as in the case for the function $x \longmapsto \langle f(x), g(x_0) - g(y_0) \rangle$, we find $d \in (x_0, y_0)$ and $N_i \in \mathrm{co}(\partial g(d))$ such that

$$\langle M_0(x_0 - y_0), g(x_0) - g(y_0) \rangle = \lim_{i \to \infty} \langle M_0(x_0 - y_0), N_i(x_0 - y_0) \rangle.$$

This together with (5.39) and (5.40) yield the existence of some $N_0 \in \overline{\mathrm{co}}(\partial g(d)) \cup ((\mathrm{co}(\partial g(d)))_\infty \backslash \{0\})$ such that

$$\langle M_0(x_0 - y_0), N_0(x_0 - y_0) \rangle \leq 0.$$

This contradicts the positive definiteness of the matrix $N_0 \circ M_0$ by the hypothesis of the theorem. The proof is complete. $\qquad \square$

The relation (5.37) tells us that for the Hartman–Stampacchia variational inequality the global uniqueness is guaranteed when f is strictly monotone. Under the hypothesis of Theorem 5.3.4 with g being the identity map, the map f is strictly monotone and the uniqueness also follows.

A Particular Case

A particular situation that deserves attention is when g is invertible with the inverse g^{-1}. The general problem can be replaced by the Hartman–Stampacchia problem whose cost operator is $f \circ g^{-1}$. In fact, these two problems are equivalent in the sense that $x_0 \in \mathbb{R}^n$ is a solution (respectively, a locally unique solution) of the problem $V(f, g, K)$ if and only if $g(x_0)$ is a solution (respectively, a locally unique solution) of the Hartman–Stampacchia one. We now show that under a reasonable hypothesis on g, the conditions of Theorem 5.3.2 can be given in a simpler form.

Proposition 5.3.5 *Assume that the following conditions hold.*

(a) g admits an inverse g^{-1} that is locally Lipschitz at $y_0 = g(x_0) \in K$.

(b) $\partial f(x_0)$ and $\partial g(x_0)$ are bounded Fréchet pseudo-Jacobians of f and g at x_0, respectively.

(c) $\partial g(x_0)$ consists of nonsingular matrices only.

Then the set $Q := \{MN^{-1} : M \in \partial f(x_0), N \in \partial g(x_0)\}$ is a Fréchet pseudo-Jacobian of $f \circ g^{-1}$ at y_0 and

(i) Elements of Q are positive definite if and only if the matrices of the form $N^{tr}M$ with $M \in \partial f(x_0)$ and $N \in \partial g(x_0)$ are positive definite.

(ii) Each of the conditions of Theorem 5.3.2 is equivalent to the corresponding condition of (i')-(iii') (described after the proof of that theorem) in which the Fréchet pseudo-Jacobian Q of the function $f \circ g^{-1}$ at y_0 is used.

Proof. The fact that Q is a Fréchet pseudo-Jacobian of $f \circ g^{-1}$ is obtained from Proposition 2.2.15 and Proposition 2.5.6. Furthermore, given two $n \times n$-matrices M and N with N invertible, it is plain that MN^{-1} is positive definite if and only if $N^T M$ is positive definite. It remains only to prove the last assertion of the proposition. We observe first that the cone $C_{(f,g)}(K, x)$ is exactly the cone $C_{f \circ g^{-1}}(K, g(x))$ given by:

$$C_{f \circ g^{-1}}(K, g(x)) = \{v \in T(K, g(x)) : \langle f \circ g^{-1}(g(x)), v \rangle = 0\}.$$

Let us consider the condition (i) of Theorem 5.3.2. Let $v \in \mathbb{R}^n \setminus \{0\}$. Then

$$N(v) \in C_{(f,g)}(K, x_0) \quad \text{and} \quad M(v) \in [C_{(f,g)}(K, x_0)]^*$$

if and only if

$$N(v) \in T(K, y_0), \ \langle f(y_0), N(v) \rangle = 0 \text{ and } M(v) \in [C_{(f,g)}(K, x_0)]^*.$$

By denoting $\bar{v} = N(v)$, the above is equivalent to

$$\bar{v} \in T(K, y_0) \setminus \{0\}, \ \bar{v} \in C_{f \circ g^{-1}}(K, y_0) \text{ and } MN^{-1}(\bar{v}) \in [C_{f \circ g^{-1}}(K, y_0)]^*.$$

This and the equality

$$\langle \bar{v}, MN^{-1}(\bar{v}) \rangle = \langle M(v), N(v) \rangle$$

show the equivalence between the condition (i) of Theorem 5.3.2 and (i'). For the other conditions, the proof is similar. □

Examples

We now provide some examples to illustrate the uniqueness criteria developed in this section. The first example shows that, in general, the problem $V(f, g, K)$ cannot be reduced to the Hartman–Stampacchia model with $g =$

id (the identity map), and the use of the Clarke generalized Jacobian does not permit us to obtain a satisfactory result. The second example shows a typical situation when the operator f is not locally Lipschitz, so that a suitable pseudo-Jacobian must be chosen when applying Theorem 5.3.2. The last example shows that when dealing with non-Lipschitz problems, recession pseudo-Jacobian matrices cannot be neglected.

Example 5.3.6 Let $K = [0,1] \times [0,1] \subseteq \mathbb{R}^2$ and let us define f and $g : \mathbb{R}^2 \to \mathbb{R}^2$ by

$$f(x,y) = (h(x), y) \text{ and } g(x,y) = (x, h(y)) \text{ for } (x,y) \in \mathbb{R}^2,$$

where $h(x)$ is given by

$$h(x) = \begin{cases} 1 & \text{if } x \geq 1, \\ 2x - 1/3^k & \text{if } x \in [2/3^{k+1}, 1/3^k], \quad k = 0, 1, \ldots, \\ 1/3^{k+1} & \text{if } x \in [1/3^{k+1}, 2/3^{k+1}], \quad k = 0, 1, \ldots. \\ 0 & \text{if } x = 0. \end{cases}$$

and $h(x) = -h(-x)$ for $x < 0$.

The point $(0,0)$ is a solution of the general variational inequality problem $V(f, g, K)$. At this solution the critical cone $C_{(f,g)}(K, (0,0))$ coincides with the positive quadrant $\mathbb{R}^2_+$. Define

$$\partial f(0,0) = \left\{ \begin{pmatrix} \alpha & 0 \\ 0 & 1 \end{pmatrix} : \alpha \in [\tfrac{1}{2}, 1] \right\}$$

$$\partial g(0,0) = \left\{ \begin{pmatrix} 1 & 0 \\ 0 & \alpha \end{pmatrix} : \alpha \in [\tfrac{1}{2}, 1] \right\}.$$

A simple calculation confirms that $\partial f(0,0)$ and $\partial g(0,0)$ are Fréchet pseudo-Jacobians of f and g at $(0,0)$, respectively. Clearly, with these Fréchet pseudo-Jacobians, the condition (iii) of Theorem 5.3.2 is verified, by which $(0,0)$ is a locally unique solution as expected. We observe that the function g is not invertible, so the method of converting the general problem to the classical one that we describe above does not work. Moreover, Clarke's generalized Jacobians of f and g at $(0,0)$ are given by

$$\partial_C f(0,0) = \left\{ \begin{pmatrix} \alpha & 0 \\ 0 & 1 \end{pmatrix} : \alpha \in [0,2] \right\}$$

$$\partial_C g(0,0) = \left\{ \begin{pmatrix} 1 & 0 \\ 0 & \alpha \end{pmatrix} : \alpha \in [0,2] \right\}.$$

It is evident that the condition (iii) of Theorem 5.3.2 does not hold when these Jacobians are used as Fréchet pseudo-Jacobians.

Example 5.3.7 In this example we consider the Hartman–Stampacchia problem $V(f, \mathrm{id}, K)$ with $K = [0, 1] \subseteq \mathbb{R}$ and $f(x) = x^{1/3}$. The function f is not locally Lipschitz at $x = 0$. We set

$$\partial f(0) = \{\alpha \in R : \alpha \geq 1\} \, .$$

It is easy to see that $\partial f(0)$ is a Fréchet pseudo-Jacobian of f at $x = 0$. The recession cone of this set is given by

$$(\partial f(0))_\infty = \{\alpha \in \mathbb{R} : \alpha \geq 0\} \, .$$

The critical cone $C_f(K, 0)$ coincides with $\mathbb{R}_+$. Moreover, every element of the set

$$\partial f(0) \cup [(\partial f(0))_\infty \setminus \{0\}]$$

is strictly positive on $C_f(K, 0) \setminus \{0\}$. Therefore, by Theorem 5.3.2, we conclude that $x = 0$ is a locally unique solution of $V(f, K)$.

Example 5.3.8 Let $K = \mathbb{R}_+^2$ and let $f : K \to \mathbb{R}^2$ be defined by

$$f(x, y) = (-x + y^{1/3}, -x^3 + y).$$

Problem $V(f, \mathrm{id}, K)$ has $(0, 0)$ as a solution that is not locally unique. At this solution the critical cone $C_f(K, (0, 0))$ coincides with the positive quadrant $\mathbb{R}_+^2$. Define

$$\partial f(0, 0) = \left\{ \begin{pmatrix} -1 & \alpha \\ 0 & 1 \end{pmatrix} : \alpha \geq 1 \right\} \, .$$

A direct calculation shows that $\partial f(0, 0)$ is a Fréchet pseudo-Jacobian of f at $(0, 0)$ and that condition (ii) of Theorem 5.3.2 is verified for all matrices of $\partial f(0, 0)$. However, that condition is violated on the recession part. In fact, let

$$M = \begin{pmatrix} 0 & 1 \\ 0 & 0 \end{pmatrix} \in (\partial f(0, 0))_\infty \setminus \{0\} \, .$$

For $v = (1, 0) \in C_f(K, (0, 0))$, one has $f(x_0) + M(v) \in [T(K, x_0)]^*$, but $\langle v, M(v) \rangle = 0$.

5.4 Complementarity Problems

Let F be a vector-valued function from $\mathbb{R}^n$ into itself. The nonlinear complementarity problem associated with F is commonly given in the form (CP):

Find $x \in \mathbb{R}^n$ satisfying

$$x \geq 0, \ F(x) \geq 0 \quad \text{and} \quad \langle F(x), x \rangle = 0.$$

This is a particular case of the variational problem $V(f, g, K)$ that we have studied in the previous section and in which the set K is the positive octant of $\mathbb{R}^n$; the function g is the identity map and $f = F$.

The complementarity problem is often used as a general model for studying important problems that arise in economic equilibrium, engineering mechanics, and optimization. The aim of this section is to present a solution point analysis and a global convergence analysis of a descent algorithm for the complementarity problem (CP) in the case where F is a continuous nonsmooth function.

Nonsmooth Merit Functions

As we have seen, under certain conditions, a local solution to a programming problem satisfies the Kuhn–Tucker condition. This rule can in its turn be expressed as a complementarity problem. To see this, let us consider the following minimization problem (P),

$$\begin{aligned} &\text{minimize} \quad f(y) \\ &\text{subject to } A(y) \geq b \\ &\qquad\qquad y \geq 0, \end{aligned}$$

where $f : \mathbb{R}^n \to \mathbb{R}$ is a differentiable function, and A is an $m \times n$ matrix, whose rows are $a_1, \ldots, a_m$ and $b = (b_1, \ldots, b_m)$ is a vector of $\mathbb{R}^m$. If y_0 is a local solution of this problem and the matrix A is of maximal rank, then there exist nonnegative numbers $\lambda_1, \ldots, \lambda_m$ and $\mu_1, \ldots, \mu_n$ satisfying:

$$\nabla f(y_0) - \sum_{i=1}^{m} \lambda_i a_i - \sum_{i=1}^{n} \mu_i = 0$$
$$\lambda_i (\langle a_i, x_0 \rangle - b_i) = 0$$
$$\mu_i x_i = 0.$$

By defining the new variable $x = (y, \lambda) \in \mathbb{R}^n \times \mathbb{R}^m$ and the function $F : \mathbb{R}^n \times \mathbb{R}^m \to \mathbb{R}^n \times \mathbb{R}^m$ by $F(x) = (\nabla f(y) - A^{tr}(\lambda), A(y) - b)$, one deduces that the system above is equivalent to the complementarity problem (CP).

It turns out that the converse is also true, that is, the complementarity problem can be formulated as a minimization problem by means of the so-called merit functions. Generally, a nonnegative function $\theta : K \to \mathbb{R}_+$ is called a merit function for the problem (CP) provided that a point x_0 is a solution to the problem (CP) if and only if the value of θ at this point is zero, or equivalently x_0 is a global solution to the problem

$$\begin{aligned} &\text{minimize} \quad \theta(x) \\ &\text{subject to } x \in K \end{aligned}$$

whose optimal value is zero.

There exist several merit functions for a given complementarity problem. Here is a quite simple one, for instance,

$$\theta(x) := \sum_{i=1}^{n}(\min\{x_i, f_i(x)\})^2.$$

Another merit function that we use is based on the Fischer-Burmeister function $\phi : \mathbb{R}^2 \to \mathbb{R}$ which is defined by

$$\phi(a, b) := \sqrt{a^2 + b^2} - a - b.$$

The associated merit function $\Psi : \mathbb{R}^n \to [0, \infty)$ is given by

$$\Psi(x) := \frac{1}{2}\sum_{i=1}^{n}\phi(x_i, F_i(x))^2.$$

To see that, in fact, it is a merit function for the problem (CP), we observe that if $x \in \mathbb{R}^n$ is a solution of the problem (CP), then $\Psi(x) = 0$ which means that x is a global minimizer of the function Ψ on $\mathbb{R}^n$. Conversely, if x is a global minimizer of the function Ψ, then because this function is separable, each component x_i of x is a global minimizer of the function $\phi(x_i, F_i(x))$ on $\mathbb{R}^n$. Consequently, $x_i \geq 0, F_i(x) \geq 0$, and $x_i F_i(x) = 0$. By this x is a solution of the complementarity problem (CP).

Let us now obtain a composite expression for the merit function Ψ. Define $\varphi : \mathbb{R}^2 \to [0, \infty)$ and $g : \mathbb{R}^{2n} \to [0, \infty)$ by

$$\varphi(a, b) := \frac{1}{2}\phi(a, b)^2,$$

$$g(x, y) := \frac{1}{2}\sum_{i=1}^{n}\phi(x_i, y_i)^2 = \sum_{i=1}^{n}\varphi(x_i, y_i). \tag{5.41}$$

For $\mathbf{F} : \mathbb{R}^n \to \mathbb{R}^{2n}$ given by

$$\mathbf{F}(x) := \begin{pmatrix} x \\ F(x) \end{pmatrix}, \tag{5.42}$$

the merit function $\Psi : \mathbb{R}^n \to [0, \infty)$ can now be written as $\Psi = g \circ \mathbf{F}$ or

$$\Psi(x) = g(x, F(x)) = \frac{1}{2}\sum_{i=1}^{n}\phi(x_i, F_i(x))^2. \tag{5.43}$$

Here are some basic properties of the functions φ and g defined in (5.41).

The notations $\nabla_1\varphi$ and $\nabla_2\varphi$ stand for the partial derivatives of φ with respect to the first and to the second variables.

Lemma 5.4.1 *The functions φ and g are continuously differentiable on $\mathbb{R}^n$. Moreover, the following properties are valid for all $a, b \in \mathbb{R}$.*

(i) $\nabla_1\varphi(a,b) = \nabla_2\varphi(a,b) = 0$ if and only if $\varphi(a,b) = 0$.
(ii) $\nabla_1\varphi(a,b) = \nabla_2\varphi(a,b) = 0$ if and only if $\nabla_1\varphi(a,b)\nabla_2\varphi(a,b) = 0$.
(iii) $\nabla_1\varphi(a,b)\nabla_2\varphi(a,b) \geq 0$.

Proof. The first part of the lemma is evident. For the second part, let us compute the partial derivatives of the function φ:

$$\nabla_1\varphi(a,b) = \varphi(a,b)\left(\frac{a}{\sqrt{a^2+b^2}} - 1\right)$$

$$\nabla_2\varphi(a,b) = \varphi(a,b)\left(\frac{b}{\sqrt{a^2+b^2}} - 1\right).$$

It follows that if $\varphi(a,b) = 0$, then $\nabla_1\varphi(a,b) = \nabla_2\varphi(a,b) = 0$. If $\frac{a}{\sqrt{a^2+b^2}} - 1 = 0$ and $\frac{b}{\sqrt{a^2+b^2}} - 1 = 0$, then both a and b are zero, which imply that $\varphi(a,b) = 0$. Thus (i) holds. The second assertion is deduced from the first one. For the last assertion, it suffices to notice that $a \leq \sqrt{a^2+b^2}$ and $b \leq \sqrt{a^2+b^2}$ so that the product $\left(\frac{a}{\sqrt{a^2+b^2}} - 1\right)\left(\frac{b}{\sqrt{a^2+b^2}} - 1\right)$ is nonnegative. $\square$

The merit function Ψ is a composite function, therefore we need the following optimality condition for composite functions.

Lemma 5.4.2 Let $x \in \mathbb{R}^n$, let $F\colon \mathbb{R}^n \to \mathbb{R}^m$ be a continuous map, and let $g\colon \mathbb{R}^m \to \mathbb{R}$ be a continuously differentiable function. Assume that F admits a pseudo-Jacobian map ∂F which is upper semicontinuous at x. If $x \in \mathbb{R}^n$ is a local minimum of $g \circ F$, then

$$0 \in \nabla g(F(x)) \circ [\overline{co}(\partial F(x)) \cup co((\partial F(x))_\infty\backslash\{0\})].$$

Proof. Because x is a local minimizer, it follows from the chain rule (Corollary 2.3.4) and the optimality condition (Theorem 2.1.13) that for every $\epsilon > 0$

$$0 \in \overline{co}[(\nabla g(F(x)) + \epsilon B_n) \circ \partial F(x)].$$

Take $\epsilon = 1/k, k = 1, 2, \ldots$. Then there exist $a_{jk} \in B_n$, $b_{jk} \in \partial F(x)$, $c_k \in B_n$, $\lambda_{jk} \in [0,1]$, $j = 1, 2, \ldots, n+1$ with $\sum_{j=1}^{n+1}\lambda_{jk} = 1$ such that

$$0 = \sum_{j=1}^{n+1}\lambda_{jk}(\nabla g(F(x)) + \frac{1}{k}a_{jk}) \circ b_{jk} + \frac{1}{k}c_k.$$

Define

$$J_1 := \{j \mid \{b_{jk}\}_k \text{ is bounded }\};$$
$$J_2 := \{j \mid \{b_{jk}\}_k \text{ is unbounded }\}.$$

Then the above sum can be rewritten as

$$0 = \sum_{j \in J_1} \lambda_{jk}(\nabla g(F(x)) + \frac{1}{k}a_{jk}) \circ b_{jk} + \sum_{j \in J_2} \lambda_{jk}(\nabla g(F(x)) + \frac{1}{k}a_{jk}) \circ b_{jk} + \frac{1}{k}c_k.$$

Now we may assume, without loss of generality, that $\lambda_{jk} \to \lambda_j$ for some $\lambda_j \in [0,1]$, $j = 1, \ldots, n+1$ with $\sum_{j=1}^{n+1} \lambda_j = 1$. Then one of the following two cases holds.

Case (i). $J_2 = \emptyset$. In this case we may assume that $b_{jk} \to b_j$ for some $b_j \in \partial F(x)$, for $j = 1, 2, \ldots, n+1$. As $k \to \infty$, the previous sum gives us

$$0 = \nabla g(F(x)) \circ \sum_{j=1}^{n+1} \lambda_j b_j \in \nabla g(F(x)) \circ \text{co}(\partial F(x)).$$

Case(ii). $J_2 \neq \emptyset$. If $\{\lambda_{jk}b_{jk}\}_k$ is bounded for each $j \in J_2$, then $\lambda_j = 0$ for each $j \in J_2$ and so $\sum_{j \in J_1} \lambda_j = 1$. We may now assume that $\lambda_{jk}b_{jk} \to b_j^\infty \in (\partial F(x))_\infty$, for $j \in J_2$, and $b_{jk} \to b_j \in \partial F(x)$, for $j \in J_1$. By passing to the limit, we get

$$0 = \nabla g(F(x)) \circ \left(\sum_{j \in J_1} \lambda_j b_j + \sum_{j \in J_2} b_j^\infty \right)$$
$$\in \nabla g(F(x)) \circ (\text{co}(\partial F(x)) + \text{co}((\partial F(x))_\infty))$$
$$\subset \nabla g(F(x)) \circ \overline{\text{co}}(\partial F(x)).$$

This follows from the fact that $\text{co}((\partial F(x))_\infty) \subset (\text{co}(\partial F(x)))_\infty$ because $\partial F(x) \subset \text{co}(\partial F(x))$ and $(\text{co}(\partial F(x)))_\infty$ is a closed convex cone, and that

$$\text{co}(\partial F(x)) + \text{co}((\partial F(x))_\infty) \subset \text{co}(\partial F(x)) + (\text{co}(\partial F(x)))_\infty$$
$$\subset \overline{\text{co}}(\partial F(x)) + (\overline{\text{co}}(\partial F(x)))_\infty$$
$$= \overline{\text{co}}(\partial F(x)).$$

If there exists $l \in J_2$ such that $\{\lambda_{lk}b_{lk}\}_k$ is unbounded, then, by taking subsequences instead, we may assume that there exists $l_0 \in J_2$ such that

$$||\lambda_{l_0 k}b_{l_0 k}|| \geq ||\lambda_{jk}b_{jk}||, \ \forall j \in J_2, \ \forall k \in N.$$

So,

$$\frac{\lambda_{jk}b_{jk}}{||\lambda_{l_0 k}b_{l_0 k}||} \to b_j^\infty \in (\partial F(x))_\infty, \ j \in J_2.$$

Let $J_3 := \{j \in J_2 \mid b_j \neq 0\}$. Then $J_3 \neq \emptyset$ as $b_{l_0}^\infty \neq 0$. Dividing the sum by $||\lambda_{l_0 k}b_{l_0 k}||$ and passing to the limit with k, we get

$$0 = \nabla g(F(x)) \circ \sum_{j \in J_3} b_j^\infty \in \nabla g(F(x)) \circ \text{co}((\partial F(x))_\infty \backslash \{0\}).$$

Thus

$$0 \in \nabla g(F(x)) \circ [\overline{\text{co}}(\partial F(x)) \cup \text{co}((\partial F(x))_\infty \backslash \{0\})]$$

and the conclusion holds. $\square$

The following example shows that the necessary condition in the lemma above is, in general, not valid without a recession cone condition.

Example 5.4.3 Let $F : \mathbb{R}^2 \to \mathbb{R}^2$ and $g : \mathbb{R}^2 \to \mathbb{R}$ be defined by

$$F(x, y) = \left(x^{2/3}\text{sign}(x) + \frac{y^4}{2}, \sqrt{2}x^{1/3} + \frac{y^2}{\sqrt{2}}\right)$$

$$g(u, v) = u + v^2.$$

Then F is continuous, but not Lipschitz, g is continuously differentiable, and the composite function $g \circ F$ is given by

$$(g \circ F)(x, y) = x^{2/3}(\text{sign}(x) + 2) + y^4 + 2x^{1/3}y^2.$$

The function $g \circ F$ attains its local minimum at $(0, 0)$. A pseudo-Jacobian of F at $(0, 0)$ and its recession cone are given, respectively, by

$$\partial F(0, 0) = \left\{\begin{pmatrix} \alpha & 0 \\ \alpha^2 & 0 \end{pmatrix} : \alpha \geq 1\right\},$$

$$\partial F(0, 0)_\infty = \left\{\begin{pmatrix} 0 & 0 \\ \beta & 0 \end{pmatrix} : \beta \geq 0\right\}.$$

Clearly, $0 \notin \nabla g(F(0, 0)) \circ \overline{\text{co}}(\partial F(0, 0))$. However,

$$0 \in \nabla g(F(0, 0)) \circ \text{co}((\partial F(0, 0))_\infty \backslash \{0\}).$$

We now see how Lemma 5.4.2 can be used for characterizing optimality of the merit function in terms of pseudo-Jacobian matrices.

We say that an $n \times n$-matrix M is a P_0-*matrix* if for each $x \neq 0$ there exists an index $i \in \{1, 2, \ldots, n\}$ such that $x_i \neq 0$ and $x_i(Mx)_i \geq 0$. A useful characterization of P_0-matrices is that a matrix is P_0 if and only if its principal minors are all nonnegative. In particular, positive semidefinite matrices are P_0-matrices, but the converse is not true in general.

Theorem 5.4.4 *Let F be a continuous map on $\mathbb{R}^n$. Suppose that F admits a pseudo-Jacobian map ∂F which is upper semicontinuous at $x \in \mathbb{R}^n$. If all elements of $\text{co}(\partial F(x))$ are P_0-matrices, then the following assertions are equivalent:*

(i) $\Psi(x) = 0$.

(ii) $0 \in \nabla_1 g(x, F(x)) + \nabla_2 g(x, F(x)) \circ [\overline{\mathrm{co}}(\partial F(x)) \cup \mathrm{co}((\partial F(x))_\infty \backslash \{0\})]$.

Proof. For $\mathbf{F} : \mathbb{R}^n \to \mathbb{R}^{2n}$ as defined by (5.42),

$$\partial \mathbf{F}(x) := \begin{pmatrix} I \\ \partial F(x) \end{pmatrix}$$

is a pseudo-Jacobian of $\mathbf{F}$ at x, where $I \in \mathbb{R}^{n \times n}$ denotes the identity matrix. If $\Psi(x) = 0$, then x is a local minimum of $\Psi = g \circ \mathbf{F}$ and so,

$$0 \in \nabla_1 g(x, F(x)) + \nabla_2 g(x, F(x)) \circ [\overline{\mathrm{co}}(\partial F(x)) \cup \mathrm{co}((\partial F(x))_\infty \backslash \{0\})]$$

follows from Lemma 5.4.2.

Conversely, if we assume the latter, we deduce the existence of $D \in [\overline{\mathrm{co}}(\partial F(x)) \cup \mathrm{co}((\partial F(x))_\infty \backslash \{0\})]$ such that

$$0 = \nabla_1 g(x, F(x)) + \nabla_2 g(x, F(x)) \circ D. \tag{5.44}$$

If all the matrices in $\mathrm{co}(\partial F(x))$ are P_0-matrices, then all the matrices in $\overline{\mathrm{co}}(\partial F(x))$ and in $\mathrm{co}((\partial F(x))_\infty)$ are also P_0-matrices. The latter follows from the fact that $a \in (\partial F(x))_\infty$ if and only if there exist sequences $\{a_j\} \subset \partial F(x)$ and $\{t_j\} \subset (0, \infty)$ with $\lim_{j \to \infty} t_j = 0$ so that $a = \lim_{j \to \infty} t_j a_j$. Because a_j is P_0-matrix $t_j a_j$ is also a P_0-matrix as $t_j > 0$. Hence D is a P_0-matrix.

By Lemma 5.4.1 (ii) and (iii), it is known that for each i and all $x \in \mathbb{R}^n$,

$$\nabla_1 \varphi(x_i, F_i(x)) \nabla_2 \varphi(x_i, F_i(x)) \geq 0, \ \nabla_1 \varphi(x_i, F_i(x)) \nabla_2 \varphi(x_i, F_i(x)) = 0$$

$$\Rightarrow \quad \nabla_1 \varphi(x_i, F_i(x)) = \nabla_2 \varphi(x_i, F_i(x)) = 0.$$

Therefore, (5.44) together with the fact that D is a P_0-matrix yields

$$\nabla_1 \varphi(x_i, F_i(x)) = \nabla_2 \varphi(x_i, F_i(x)) = 0$$

for each i. This together with Lemma 5.4.1 (i) gives $\varphi(x_i, F_i(x)) = 0$ for each i. Thus $g(x, F(x)) = \Psi(x) = 0$ follows. $\qquad\square$

When the function F is locally Lipschitz and the Clarke generalized Jacobian is used, condition (ii) of Theorem 5.4.4 is simplified as follows.

Corollary 5.4.5 *Let F be Lipschitz continuous. If all elements of $\partial^C F(x)$ are P_0-matrices, then the following are equivalent.*

(i) $\Psi(x) = 0$.
(ii) $0 \in \nabla_1 g(x, F(x)) + \nabla_2 g(x, F(x)) \circ \partial^C F(x)$.

Proof. This follows from the previous theorem by choosing $\partial^C F(x)$ as a pseudo-Jacobian of F at x. In this case $\mathrm{co}\,((\partial^C F(x))_\infty \backslash \{0\}) = \emptyset$.   □

A Derivative-Free Descent Method

In this part we present conditions under which a line search method possesses global convergence properties. This method has the particularity that it works with the values of F instead of additionally using derivate information.

Now to formulate the derivative-free line search algorithm let g, $\mathbf{F}$, and Ψ be given as in (5.41), (5.42), and (5.43). We make use of the search direction

$$s(x) := -\nabla_2 g(x, F(x))$$

for all $x \in \mathbb{R}^n$. Then we define the function $\theta : \mathbb{R}^n \to \mathbb{R}$ by

$$\theta(x) = \nabla_1 g(x, F(x)) \circ \nabla_2 g(x, F(x)).$$

By Lemma 5.4.2 the function $\theta(x)$ is always nonnegative and it is 0 if and only if $\Psi(x) = 0$ (i.e., if and only if x solves (CP)). The next lemma shows that $s(x)$ is a descent direction for Ψ at x and that the local descent can be measured by means of $\theta(x)$.

Lemma 5.4.6 *Let $F : \mathbb{R}^n \to \mathbb{R}^n$ be a monotone continuous map. Assume that F is comonotone at each $x \in \mathbb{R}^n$ in each direction $u \in \mathbb{R}^n$ for which*

$$\limsup_{t \downarrow 0} \frac{\|F(x + tu) - F(x)\|}{t} = +\infty.$$

is satisfied. Moreover, let $\bar{\sigma} \in (0, 1)$ be given. If $\Psi(x) > 0$, then there exists a number $t(x) > 0$ such that

$$\Psi(x + ts(x)) \leq \Psi(x) - \bar{\sigma} t \theta(x) \; \forall t \in [0, t(x)]. \tag{5.45}$$

Proof. Let $x, y \in \mathbb{R}^n$ and $\bar{\sigma} \in (0, 1)$ be arbitrary but fixed. Because g is continuously differentiable, there is some function $\varepsilon : (0, \infty) \to \mathbb{R}$ so that, for all $p, q \in \mathbb{R}^n$,

$$g(x + p, y + q) - g(x, y) \leq \nabla g(x, y) \circ \begin{pmatrix} p \\ q \end{pmatrix} + \varepsilon(\|p\| + \|q\|)$$

and

$$\lim_{\tau \downarrow 0} \frac{\varepsilon(\tau)}{\tau} = 0. \tag{5.46}$$

Letting

$$y := F(x), \quad p := p(t) := ts(x), \quad q := q(t) := F(x + ts(x)) - F(x)$$

and
$$\tau(t) := \|p(t)\| + \|q(t)\|,$$

we obtain

$$
\begin{aligned}
\Psi(x + ts(x)) - \Psi(x) &= g(x + ts(x), F(x + ts(x))) - g(x, F(x)) \\
&\le t\nabla_1 g(x, F(x)) \circ s(x) \\
&\quad + \nabla_2 g(x, F(x)) \circ q(t) + \varepsilon(\tau(t)).
\end{aligned}
$$

Thus, using the definitions of $\theta(x)$, $s(x)$, $\tau(t)$, and $p(t)$, it follows that

$$\Psi(x + ts(x)) - \Psi(x) \le -t\theta(x) - q(t) \circ s(x) + \varepsilon(\tau(t)) \tag{5.47}$$

and

$$
\begin{aligned}
\Psi(x + ts(x)) - \Psi(x) &\le -t\theta(x) - q(t) \circ s(x) \\
&\quad + \left(\|ts(x)\| + \|q(t)\| \right) \frac{\varepsilon(\tau(t))}{\tau(t)}.
\end{aligned}
\tag{5.48}
$$

We now distinguish two cases, namely whether (5.6) is satisfied for the direction $u := s(x)$.

(a) If

$$\limsup_{t \downarrow 0} \frac{\|q(t)\|}{t} = \limsup_{t \downarrow 0} \frac{\|F(x + ts(x)) - F(x)\|}{t} = +\infty \tag{5.49}$$

then the comonotonicity assumption on F yields

$$-q(t) \circ s(x) = -(F(x + ts(x)) - F(x)) \circ s(x) \le -\gamma(x, s(x))\|q(t)\|$$

for all $t > 0$ sufficiently small. Hence we obtain from (5.48) that

$$
\begin{aligned}
\Psi(x + ts(x)) - \Psi(x) &\le -t\Big\{ \theta(x) + \frac{\|q(t)\|}{t} \left(\gamma(x, s(x)) - \frac{\varepsilon(\tau(t))}{\tau(t)} \right) \\
&\quad - \|s(x)\| \frac{\varepsilon(\tau(t))}{\tau(t)} \Big\}.
\end{aligned}
$$

Therefore, the desired inequality (5.45) follows for all $t > 0$ sufficiently small.

(b) If, otherwise,

$$\limsup_{t \downarrow 0} \frac{\|q(t)\|}{t} = \limsup_{t \downarrow 0} \frac{\|F(x + ts(x)) - F(x)\|}{t} < +\infty, \tag{5.50}$$

we first note that the monotonicity of F implies that, for all $t \in [0, \infty)$,

$$-q(t) \circ s(x) = -(F(x + ts(x)) - F(x)) \circ s(x) \le 0.$$

This and (5.47) yield

$$\Psi(x + ts(x)) - \Psi(x) \leq -t\left\{\theta(x) - \frac{\tau(t)}{t}\frac{\varepsilon(\tau(t))}{\tau(t)}\right\}$$

and furthermore,

$$\Psi(x + ts(x)) - \Psi(x) \leq -t\left\{\theta(x) - \left(\|s(x)\| + \frac{\|q(t)\|}{t}\right)\frac{\varepsilon(\tau(t))}{\tau(t)}\right\}.$$

Taking into account (5.50) and (5.46), we see that (5.45) is satisfied for all $t > 0$ sufficiently small. Thus a positive number $t(x)$ exists so that (5.45) is satisfied.   □

Based on (5.45) the descent direction $s(x)$ is now exploited by means of the following standard line search algorithm. Moreover, note that in Lemma 5.1 and in the subsequent theorem the comonotonicity of F at x is required only for those directions u which satisfy condition (5.6). Therefore, no comonotonicity assumption is necessary for locally Lipschitz or directionally differentiable maps.

The Algorithm

Let us describe an algorithm for solving the complementarity problem: Given $x^0 \in \mathbb{R}^n$, $\rho, \sigma \in (0,1)$, for $k = 0, 1, 2, \ldots$, repeat the following steps:

(i)   Calculate $\Psi(x^k)$. If $\Psi(x^k) = 0$, stop.
(ii)   If $\Psi(x^k) \neq 0$, set $s^k = s(x^k)$ and choose $t_k \in \{\rho^j \,|\, j \in \mathbb{N}\}$ as large as possible such that

$$\Psi(x^k + t_k s^k) \leq \Psi(x^k) - \sigma t_k \theta(x^k).$$

(iii)   Set $x^{k+1} = x^k + t_k s^k$. Set $k = k + 1$ and go to (i).

The convergence of the algorithm is seen in the next result.

Theorem 5.4.7 *Let* $F : \mathbb{R}^n \to \mathbb{R}^n$ *be a monotone continuous map. If* F *is comonotone at each* $x \in \mathbb{R}^n$ *in each direction* $u \in \mathbb{R}^n$ *for which*

$$\limsup_{t\downarrow 0} \frac{\|F(x + tu) - F(x)\|}{t} = +\infty$$

is satisfied, then the algorithm is well defined and any accumulation point of the sequence $\{x^k\}$ *generated by the algorithm solves the complementarity problem (CP).*

Proof. First note that $s(x)$ and $\theta(x)$ are well defined for all $x \in \mathbb{R}^n$. Furthermore, for any x^k generated by the algorithm, Lemma 5.4.2 ensures $t_k > 0$. Thus the algorithm is well defined. Because $\{\Psi(x^k)\}$ is monotone, decreasing and bounded below, the limit

$$\bar{\Psi} := \lim_{k \to \infty} \Psi(x^k)$$

exists. Suppose that $\bar{\Psi} > 0$. Furthermore, let $\bar{x}$ denote an accumulation point of the sequence $\{x^k\}$. Then, there is an infinite set $N \subseteq \mathbb{N}$ such that $\lim_{k \in N} x^k = \bar{x}$. For $\bar{\sigma} := (\sigma + 1)/2$, Lemma 5.4.2 provides $t(\bar{x}) > 0$. Due to the fact that $\theta(\bar{x}) > 0$ (as explained at the beginning of this section) and due to the continuity of $F, g, \nabla g, \Psi, s$, and θ, a number $\delta > 0$ exists so that, for all $x \in \bar{x} + \delta B(0, 1)$,

$$|\theta(x) - \theta(\bar{x})| \le \frac{1}{4}(1 - \sigma)\theta(\bar{x}) \tag{5.51}$$

and, for all $x \in \bar{x} + \delta B(0, 1)$ and all $t \in [0, t(\bar{x})]$,

$$|\Delta\Psi(x, t)| \le \frac{1}{4}\rho t(\bar{x})(1 - \sigma)\theta(\bar{x}), \tag{5.52}$$

where

$$\Delta\Psi(x, t) := \Psi(x + ts(x)) - \Psi(\bar{x} + ts(\bar{x})) + \Psi(\bar{x}) - \Psi(x).$$

Taking into account (5.52) and the fact that (5.45) holds for $x := \bar{x}$ and all $t \in [0, t(\bar{x})]$, we get

$$\begin{aligned}
\Psi(x + ts(x)) - \Psi(x) &= \Psi(\bar{x} + ts(\bar{x})) - \Psi(\bar{x}) + \Delta\Psi(x, t) \\
&\le -t\bar{\sigma}\theta(\bar{x}) + |\Delta\Psi(x, t)| \\
&\le -t\sigma\theta(\bar{x}) - \tfrac{1}{2}t(1 - \sigma)\theta(\bar{x}) + \tfrac{1}{4}\rho t(\bar{x})(1 - \sigma)\theta(\bar{x})
\end{aligned}$$

for all $x \in \bar{x} + \delta B_n$ and all $t \in [0, t(\bar{x})]$. If we now consider $t \in [\rho t(\bar{x}), t(\bar{x})]$, we have $\rho t(\bar{x}) \le t$. Thus, using (5.51), it follows that

$$\begin{aligned}
\Psi(x + ts(x)) - \Psi(x) &\le -t\sigma\theta(\bar{x}) - \tfrac{1}{4}t(1 - \sigma)\theta(\bar{x}) \\
&\le -t\sigma\theta(x) + t\sigma(\theta(x) - \theta(\bar{x})) - \tfrac{1}{4}t(1 - \sigma)\theta(\bar{x}) \\
&\le -t\sigma\theta(x) + \tfrac{1}{4}t\sigma(1 - \sigma)\theta(\bar{x}) - \tfrac{1}{4}t(1 - \sigma)\theta(\bar{x}) \\
&\le -t\sigma\theta(x)
\end{aligned}$$

is valid for all $x \in \bar{x} + \delta B_n$ and all $t \in [\rho t(\bar{x}), t(\bar{x})]$. Therefore, because $x^k \in \bar{x} + \delta B(0, 1)$ for all $k \in N$ large enough, the step length procedure used in the algorithm provides $t_k \ge \rho t(\bar{x})$ and, thus,

$$\Psi(x^{k+1}) \le \Psi(x^k) - \sigma t_k \theta(x^k) \le \Psi(x^k) - \sigma\rho t(\bar{x})\theta(x^k)$$

for all $k \in N$ sufficiently large. Using (5.51), we obtain

$$\Psi(x^{k+1}) \le \Psi(x^k) - \frac{3}{4}\sigma\rho t(\bar{x})\theta(\bar{x})$$

for infinitely many $k \in N$. Moreover $\Psi(x^{k+1}) < \Psi(x^k)$ is valid for all $k \in \mathbb{N}$. Thus, because $\theta(\bar{x}) > 0$, we have $\lim_{k\to\infty} \Psi(x^k) = -\infty$. This contradicts $\bar{\Psi} > 0$. Hence, by the continuity of Ψ, $0 = \bar{\Psi} = \Psi(\bar{x})$ must be valid.   □

We complete this section by observing that the boundedness of the level set

$$\Omega := \{x \in \mathbb{R}^n \,|\, \Psi(x) \le \Psi(x^0)\}$$

obviously guarantees the existence of an accumulation point of the sequence $\{x^k\}$ generated by the algorithm.

Bibliographical Notes

Chapter 1

Basic references on nonsmooth analysis are Clarke [11], Mordukhovich [91], [94], and Rockafellar and Wets [107] in which several definitions of generalized derivatives, their calculus, and applications can be found. The concept of pseudo-Jacobian was first introduced in [50]. It should be noted that this concept was termed as an approximate Jacobian in [50] and in other related papers of Jeyakumar and Luc [50, 52, 53, 55]. The notions of Gâteaux and Fréchet pseudo-Jacobians were introduced in Luc [79]. The Gâteaux derivative, Fréchet derivative, and strict derivative as well as the Clarke generalized gradients are discussed in Clarke [11]. Mordukhovich's coderivative was given in [91, 92, 93]; its relationship to pseudo-Jacobians was analyzed in [96]. The connections to Warga's derivative containers [118, 117] and pseudo-Jacobians were established in [52]. The notions of prederivatives were introduced and extensively studied in Ioffe [41, 42, 43, 44], whereas H-differentials were given in Gowda and Ravendran [28].

For real-valued functions various definitions of subdifferentials can be found in books dealing with nonsmooth analysis as well as convex analysis: Aubin and Frankowska [1], Borwein and Lewis [4], Hiriart-Urruty and Lemarechal [39], Rockafellar [106], Rockafellar and Wets [107], and Zalinescu [123]. Some recent improvements of convex subdifferential calculus and analysis can be found in [7, 8, 9, 10, 48, 49, 57]. A survey of subdifferential calculus can also be found in Borwein and Zhu [5]. See also [90] for Michel and Penot's subdifferentials, [114] and [115] for Treiman's linear generalized gradients, and [122] for Zagrodny's mean value theorem. A treatment of quasidifferentials can be found in Demyanov and Rubinov [17]. An equivalent notion of pseudo-differentials was first given in Studniarski and Jeyakumar [111] in terms of a two-sided convex approximation and then was refined and discussed in Demyanov and Jeyakumar [16] as a small subdifferential.

Pseudo-Hessian matrices were first introduced in [50] and [58]. Other notions of generalized Hessians can be found in [12, 40, 93]. The concept of a partial pseudo-Jacobian was investigated in Jeyakumar and Luc [53]. Properties of recession cones were given in [2, 70, 72, 85, 106]. For absolutely continuous functions see [97]. The independence of the Clarke generalized Jacobian upon the set of null measure that contains all nondifferentiable points of a locally Lipschitz function (Section 1.5) is given in [22].

Chapter 2

The elementary calculus rules of pseudo-differentials can be found in [51]. Rules for max-functions and min-functions, given in [51], are improved in Section 2.1. A mean value theorem for continuous maps and a characterization of locally Lipschitz functions were given in [50]. Mean value theorems for locally Lipschitz vector functions were given in [38]. The results on sup-functions and inf-functions of Section 2.2 are new. Generalizations of Taylor's expansion in terms of pseudo-Jacobians were given in Jeyakumar and Luc [50] and Jeyakumar and Wang [58]. Other extensions were given in [40, 59, 76, 121]. The fuzzy chain rule was proven in [81]. Other chain rules of Section 2.3 are based on the papers [24, 51, 52, 53, 55].

Chapter 3

The open mapping theorem and implicit function theorem for continuously differentiable functions are well known and can be found in any advanced calculus books. The first extension of the open mapping theorem to locally Lipschitz functions is due to Clarke [11]. Related extensions using set-valued derivatives can be found in [6, 19, 25, 33, 34, 65, 67, 69, 102, 103]. A complete characterization of openness and metric regularity of set-valued maps was given in Mordukovich [92] by means of coderivatives (see also [45] for the case of general metric spaces). Several sufficient conditions in terms of pseudo-Jacobians for openness of nonsmooth continuous maps were given in [52, 53, 61]. Inverse and implicit function theorems for locally Lipschitz functions can be found in [11, 67]. These theorems for nonsmooth functions using quasidifferentials and derivative containers were, respectively, given in [17] and [117, 118]. Interior mapping as well as implicit function theorems using pseudo-Jacobians were given in [53]. The convex interior mapping theorem using Fréchet pseudo-Jacobians is new.

Following the work of Robinson [105], various conditions for stability, metric regularity, and the pseudo-Lipschitz property of the solution maps of parametric inequality systems involving nonsmooth functions and sets can be found in [3, 64, 92, 99, 107]. These results for (not necessarily locally Lipschitz) continuous systems using pseudo-Jacobian maps were given in [61], [82]. The proof of Ekeland's variational principle [21] is taken from [4]. Proposition 3.5.1 is a consequence of Robinson–Ursescu's theorem on metric regularity given in [107].

Chapter 4

First-order necessary optimality conditions for constrained nonsmooth optimization problems involving locally Lipschitz functions using Clarke generalized subdifferentials can be found in [11]. Improved forms of such optimality conditions were given in [5, 14, 37, 95, 104]. Sharp optimality conditions for locally Lipschitz optimization problems using pseudo-differentials were given in [116]. Optimality conditions for locally Lipschitz optimization problems using other generalized subdifferentials can be found in [44, 115]. First order necessary optimality conditions for problems involving nonsmooth continuous functions were given in [53, 61], whereas for problems involving composite functions were given in [46, 47, 54, 55]. First-order optimality conditions for cone-constrained continuous problems were given in [61].

Second-order optimality conditions for optimization problems involving continuously differentiable functions were given in [50, 58, 80]. Second-order conditions for $C^{1,1}$-optimization problems can be found in [12, 40]. Second-order conditions for composite optimization problems involving continuously differentiable functions were given in [55].

First-order optimality conditions for multiobjective programming problems with (not necessarily locally Lipschitz) functions were given in [70, 78]. Second-order conditions for such problems were given in [29, 30, 31], see also [20] and [26]. Second-order conditions for multiobjective convex composite problems can be found in [60, 120].

Further applications of pseudo-Jacobians in dynamic optimization were recently developed in [15] and not included in this book.

Chapter 5

Characterizations of (strong) monotone and generalized monotone operators in terms of pseudo-Jacobians were given in [56]. Similar characterizations by means of Clarke generalized Jacobians can be found in [86]. Characterizations of generalized convexity in terms of pseudo-differentials are new. Comonotonicity was introduced in [24]. For more on generalized convex functions and generalized monotone maps, see [13, 32, 62, 63, 71, 73, 74, 75, 83, 86, 87, 88, 89, 98, 101]. Basic results on variational inequalities and complementarity problems with applications can be found in [23, 27, 35, 36, 66, 68, 110, 112]. Conditions for existence and uniqueness of solutions of variational inequalities by way of pseudo-Jacobians were given in [77, 79, 84]. Solution point characterizations of complementarity problems involving nonsmooth continuous maps were examined in [24] by means of a nonsmooth merit function. A derivative-free descent method for complementarity problems was developed in [24].

References

1. J.-P. Aubin and H. Frankowska, *Set-Valued Analysis*, Wiley, New York, 1984.
2. A. Auslender and M. Teboulle, *Asymptotic Cones and Functions in Optimization and Variational Inequalities*, Springer, New York, 2002.
3. J. M. Borwein, *Stability and regular points of inequality systems*, J. Optim. Theory Appl. 48, (1986), 9–52.
4. J. M. Borwein and A. S. Lewis, *Convex Analysis and Nonlinear Optimization*, Springer, New York, 2000.
5. J. M. Borwein and Q. J. Zhu, *A survey of subdifferential calculus with applications,* Nonlinear Anal. 35 (1999), pp. 687–773.
6. J. M. Borwein and D. M. Zhuang, *Verifiable necessary and sufficient conditions for regularity of set-valued and single-valued maps*, J. Math. Anal. Appl. 134(1988), pp. 441–459.
7. R. S. Burachik and V. Jeyakumar, *A new geometric condition for Fenchel duality in infinite dimensions*, Math. Program., Ser. B, 104(2005), pp. 229–233.
8. R. S. Burachik and V. Jeyakumar, *A dual condition for the convex subdifferential sum formula with applications*, J. Convex Anal. 12(2005), pp. 279–290.
9. R. S. Burachik and V. Jeyakumar, *A simple closure condition for the normal cone intersection formula*, Proc. Amer. Math. Soc. 133(2005), pp. 1741–1748.
10. R.S. Burachik, V. Jeyakumar, and Z. Y. Wu, *Necessary and sufficient conditions for stable conjugate duality*, J. Nonlinear Analysis, Ser. A, 64(2006), pp. 1998–2006.
11. F. H. Clarke, *Optimization and Nonsmooth Analysis*, Wiley, New York, 1983.
12. R. Cominetti and R. Correa, *A generalized second-order derivative in nonsmooth optimization*, SIAM J. Control Optim. 28 (1990), pp. 789–809.
13. R. Correa, A. Jofre, and L. Thibault, *Characterization of lower semicontinuous convex functions*, Proc. Amer. Math. Soc. 116 (1992), pp. 67–72.
14. B. D. Craven, *Mathematical Programming and Control Theory*, Chapman and Hall, London, 1978.
15. G. Crespi, D. T. Luc, and N. B. Minh, *Pseudo-Jacobians and a necessary condition in dynamic optimization*, Prepublication, Laboratoire d'Analyse Non Linéaire et de Géométrie, Université d'Avignon, May 2006.
16. V. F. Demyanov and V. Jeyakumar, *Hunting for a smaller convex subdifferential*, J. Global Optim. 10 (1997), pp. 305–326.
17. V. F. Demyanov and A. M. Rubinov, *Constructive Nonsmooth Analysis*, Verlag Peter Lang, 1995.
18. P. H. Dien, *Some results on locally Lipschitzian mappings*, Acta Math. Vietnamica, 6 (1981), pp. 97–105.
19. A. L. Donchev and W. W. Hager, *Implicit functions, Lipschitz maps and stability in optimization*, Math. Oper. Res. 19 (1994), pp. 753–768.

20. J. Dutta and S. Chandra, *Convexifactors, generalized convexity, and optimality conditions*, J. Optim. Theory Appl. 113(2002), pp. 41–64.
21. I. Ekeland, *On the variational principle*, J. Math. Anal. Appl. 47(1974), pp. 324–358.
22. M. Fabian and D. Preiss, *On the Clarke generalized Jacobian*, Rend. Circ. Mat. Palermo 52 Suppli. N. 14(1987), pp. 305–307.
23. F. Facchinei and J. S. Pang, *Finite-Dimensional Variational Inequalities and Complementarity Problems*, Vol. 1, Springer, New York, 2003.
24. A. Fischer, V. Jeyakumar, and D. T. Luc, *Solution point characterizations and convergence analysis of a descent algorithm for nonsmooth continuous complementarity problems*, J. Optim. Theory Appl. 110(2001), pp. 493–513.
25. H. Frankowska, *An open mapping principle for set-valued maps*, J. Math. Anal. Appl. 127(1987), pp. 172–180.
26. N. Gadhi, *Sufficient second order optimality conditions for C^1 multiobjective optimization problems*, Serdica. Math. J. 29(2003), pp. 225–238.
27. F. Giannessi and A. Maugeri, *Variational Inequalities and Network Equilibrium Problems,* Plenum Press, New York, 1995.
28. M. S. Gowda and G. Ravindran, *Algebraic univalence theorems for nonsmooth functions*, J. Math. Anal. Appl. 252(2000), pp. 917–935.
29. A. Guerraggio and D. T. Luc, *Optimality conditions for $C^{1,1}$ vector optimization problems*, J. Optim. Theory Appl. 109(2001), pp. 615–629.
30. A. Guerraggio and D. T. Luc, *Optimality conditions for $C^{1,1}$ constrained multiobjective problems*, J. Optim. Theory Appl. 116(2003), pp. 117–129.
31. A. Guerraggio, D.T. Luc, and N.B. Minh, *Second-order optimality conditions for C^1 multiobjective programming problems*, Acta Math. Vietnam. 26(2002), pp. 257–268.
32. N. Hadjisavvas, S. Komlosi and S. Schaible, eds., *Handbook of Generalized Convexity and Generalized Monotonicity*, Springer, New York, 2005.
33. H. Halkin, *Interior mapping theorem with set-valued derivatives*, J. Anal. Math. 30(1976), pp. 200–207
34. H. Halkin, *Mathematical programming without differentiability*, in D. Russel. ed., *Calculus of Variations and Control Theory*, Academic Press, New York, 1976, pp. 279–297.
35. J. P. T. Harker and J. S. Pang, *Finite dimension variational inequality and nonlinear complementary problems: a survey of theory, algorithms and applications*, Math. Program. 48(1990), pp. 161–220.
36. P. Hartman and G. Stampacchia, *On some nonlinear elliptic differential functional equations*, Acta Math. 115(1966), pp. 153–188.
37. J.-B. Hiriart-Urruty, *Refinements of necessary optimality conditions in nondifferentiable programming*, Appl. Math. Optim. 5(1979), pp. 63–82.
38. J.-B. Hiriart-Urruty, *Mean value theorems for vector valued mappings in nonsmooth optimization*, Numer. Funct. Anal. Optim. 2(1980), pp. 1–30.
39. J.-B. Hiriart-Urruty and C. Lemarechal, *Convex Analysis and Minimization Algorithms*, Volumes I and II, Springer-Verlag, Berlin, 1993.
40. J.-B. Hiriart-Urruty, J. J. Strodiot, and V. Hien Nguyen, *Generalized Hessian matrix and second-order optimality conditions for problems with $C^{1,1}$ data*, Appl. Math. Optim. 11(1984), pp. 43–56.
41. A. D. Ioffe, *Nonsmooth analysis: Differential calculus of nondifferentiable mapping*, Trans. Amer. Math. Soc. 266(1981), pp. 1–56.
42. A. D. Ioffe, *Approximate subdifferentials and applications I: The finite dimensional theory*, Trans. Amer. Math. Soc. 281(1984), pp. 389–416.
43. A.D. Ioffe, *On the local surjection property*, Nonlinear Anal. 11(1987), pp. 565–592.
44. A.D. Ioffe, *A Lagrange multiplier rule with small convex-valued subdifferentials for nonsmooth problems of mathematical programming involving equality and nonfunctional constraints*, Math. Program. 588(1993), pp. 137–145.

45. A.D. Ioffe, *Metric regularity and subdifferential calculus*, Russian Math. Surveys, 55(2000), 501–558.

46. V. Jeyakumar, *Composite nonsmooth optimization*, Encyclopedia of Optimization, Kluwer Academic, Dordrecht, I (2001), pp. 307–310.

47. V. Jeyakumar, *Composite nonsmooth programming with Gâteaux differentiability*, SIAM J. Optim. 1(1991), pp. 30–41.

48. V. Jeyakumar, *The conical hull intersection property for convex programming*, Math. Program. Ser. A, 106(2006), pp. 81–92.

49. V. Jeyakumar, G. M. Lee, and N. Dinh, *New sequential Lagrange multiplier conditions characterizing optimality without constraint qualifications for convex programs*, SIAM J. Optim. 14(2003), pp. 534–547.

50. V. Jeyakumar and D. T. Luc, *Approximate Jacobian matrices for nonsmooth continuous maps and C^1-Optimization*, SIAM J. Control Optim. 36(1998), pp. 1815–1832.

51. V. Jeyakumar and D. T. Luc, *Nonsmooth calculus, minimality and monotonicity of convexificators*, J. Optim. Theory Appl. 101(1999), pp. 599–621.

52. V. Jeyakumar and D. T. Luc, *An open mapping theorem using unbounded generalized Jacobians*, Nonlinear Anal. 50(2002), pp. 647–663.

53. V. Jeyakumar and D. T. Luc, *Convex interior mapping theorems for continuous nonsmooth functions and optimization*, J. Nonlinear Convex Anal. 3(2002), pp. 251–266.

54. V. Jeyakumar and D. T. Luc, *Sharp variational conditions for convex composite nonsmooth functions*, SIAM J. Optim. 13(2003), pp. 904–920.

55. V. Jeyakumar, D. T. Luc, and P. N. Tinh, *Convex composite non-Lipschitz programming*, Math. Program. Ser. A, 25(2002), pp. 177–195.

56. V. Jeyakumar, D. T. Luc, and S. Schaible, *Characterizations of generalized monotone nonsmooth continuous maps using approximate Jacobians*, J. Convex Anal. 5(1998), pp. 119–132.

57. V. Jeyakumar and H. Mohebi, *Limiting ϵ-subgradient characterizations of constrained best approximation*, J. Approx. Theory. 135(2005), pp. 145–159.

58. V. Jeyakumar and Y. Wang, *Approximate Hessian matrices and second order optimality conditions for nonlinear programming problems with C^1 data*, J. Aust. Math. Soc. Ser. B, 40(1999), pp. 403–420.

59. V. Jeyakumar and X. Q. Yang, *Approximate generalized Hessians and Taylor's expansions for continuously Gateaux differentiable functions*, Nonlinear Anal. 36(1999), pp. 353–368.

60. V. Jeyakumar and X. Q. Yang, *Convex composite multi-objective nonsmooth programming*, Math. Program., Ser. A, 59 (1993), pp. 325–343.

61. V. Jeyakumar and N. D. Yen, *Solution stability of nonsmooth continuous systems with applications to cone-constrained optimization*, SIAM J. Optim. 14(2004), pp. 1106–1127.

62. A. Jofre, D. T. Luc, and M. Thera, *ϵ-Subdifferential calculus for nonconvex functions and ϵ-monotonicity*, C. R. A. S. Paris, Ser. I Math. 323(1996), pp. 735–740.

63. A. Jofre, D. T. Luc and M. Thera, *ϵ-Subdifferential and ϵ-monotonicity*, Nonlinear Anal. 33(1998), pp. 71–90.

64. A. Jourani and L. Thibault, *Metric regularity for strongly compactly Lipschitzian mappings*, Nonlinear Anal. 24 (1995), pp. 229–240.

65. D. Klatte and R. Henrion, *Regularity and stability in nonlinear semi-infinite optimization. Semi-infinite programming*, 69–102, Nonconvex Optim. Appl., 25 (1998), pp. 68–102.

66. D. Kinderlehrer and G. Stampacchia, *An Introduction to Variational Inequalities and Their Application*, Academic Press, New York, 1980.

67. B. Kummer, *An implicit function theorem for $C^{0,1}$-equations and parametric $C^{1,1}$-optimization*, J. Math. Anal. Appl. 158(1991), pp. 35–46.

68. J. Kyparisis, *Uniqueness and differentiability of solutions of parametric nonlinear complementarity problems*, Math. Program. 36(1986), pp.105–113.

69. Y.S. Ledyaev and Q.J. Zhu, *Implicit multifunction theorems*, Set-valued Analysis, 7(1999), 209–238.

70. D. T. Luc, *Theory of Vector Optimization*, LNEMS 319, Springer-Verlag, Berlin, 1989.

71. D. T. Luc, *On the maximal monotonicity of subdifferentials*, Acta Math. Vietnam. 18(1993), 99–106.

72. D. T. Luc, *Recession maps and applications*, Optimization 27(1993), pp. 1–15.

73. D. T. Luc, *Characterizations of quasiconvex functions*, Bull. Aust. Math. Soc. 48(1993), pp. 393–405.

74. D.T. Luc, *On generalized convex nonsmooth functions*, Bull. Aust. Math. Soc. 49(1994), pp. 139–149.

75. D. T. Luc, *Generalized monotone maps and bifunctions*, Acta Math. Vietnam. 21(1996), pp. 213–253.

76. D. T. Luc, *Taylor's formula for $C^{k,1}$ functions*, SIAM J. Optim. 5(1995), pp. 659–669.

77. D. T. Luc, *Existence results for densely pseudomonotone variational inequalities*, J. Math. Anal. Appl. 254(2001), pp. 291–308.

78. D. T. Luc, *A multiplier rule for multiobjective programming problems with continuous data*, SIAM J. Optim. 13(2002), pp. 168–178.

79. D. T. Luc, *Frechet approximate Jacobian and local uniqueness of solutions in variational inequalities*, J. Math. Anal. Appl. 268(2002), pp. 629–646.

80. D. T. Luc, *Second-order optimality conditions for problems with continuously differentiable data*, Optimization 51(2002), pp. 497–510.

81. D. T. Luc, *Chain rules for approximate Jacobians of continuous functions*, Nonlinear Anal. 61(2005), pp. 97–114.

82. D. T. Luc and N. B. Minh, *Equi-surjective systems of linear operators and applications*, Prepublication N.50, Laboratoire d'Analyse Non Linéaire et de Géométrie, Université d'Avignon, June 2005.

83. D. T. Luc, H. V. Ngai, and M. Thera, *On ϵ-monotonicity and ϵ-convexity*, in *Calculus of Variations and Differential Equations* (Haifa, 1998), pp. 82–100, Chapman and Hall/CRC, Boca Raton, FL, 2000.

84. D. T. Luc and M. A. Noor, *Local uniqueness of solutions of general variational inequalities*, J. Optim. Theory Appl. 117(2003), pp. 149–154.

85. D. T. Luc and J.-P. Penot, *Convergence of asymptotic directions*, Trans. Amer. Math. Soc. 353(2001), pp. 4095–4121.

86. D. T. Luc and S. Schaible, *On generalized monotone nonsmooth maps*, J. Convex Anal. 3(1996), pp. 195–205.

87. D. T. Luc and S. Schaible, *Efficiency and generalized concavity*, J. Optim. Theory Appl. 94(1997), pp. 147–153.

88. D. T. Luc and S. Swaminathan, *A characterization of convex functions*, Nonlinear Anal. 20(1993), pp. 697–701.

89. D. T. Luc and M. Volle, *Level sets, infimal convolution and level addition*, J. Optim. Theory Appl. 94(1997), pp. 695–714.

90. P. Michel and J.-P. Penot, *Calcul sous-différentiel pour des fonctions Lipschitziennes et non-Lipschitziennes*, C. R. A. S. Paris, Ser. I Math. 298(1985), pp. 269–272.

91. B. Mordukhovich, *Approximation Methods in Problems of Optimization and Control*, Nauka, Moscow, Russian, 1988.

92. B. Mordukhovich, *Complete characterizations of openness, metric regularity, and Lipschitzian properties of multifunctions*, Trans. Amer. Math. Soc. 340(1993), pp. 1–35.

93. B.S. Mordukhovich, *Generalized differential calculus for nonsmooth and set-valued mappings*, J. Math. Anal. Appl. 183(1994), pp. 250–288.

94. B. Mordukhovich, *Variational Analysis and Generalized Differentiation*, Vols 1 and 2, Springer, New York, 2006.

95. B. Mordukhovich, J. S. Treiman, Q. J. Zhu, *An extended extremal principle with applications to multiobjective optimization*, SIAM J. Optim. 14(2003), pp. 359–379.

96. N. M. Nam and N. D. Yen, *Relationship between approximate Jacobians and coderivatives*, J. Nonlinear Convex Anal., 2006(to appear).

97. I. P. Natason, *Theory of Functions of a Real Variable*, Frederick Ungar, New York, 1964.

98. H. V. Ngai, D. T. Luc, and M. Thera, *Approximate convex functions*, J. Nonlinear Convex Anal. 1(2000), pp. 155–176.

99. J.-P. Penot, *Metric regularity, openness and Lipschitzian behavior of multifunctions*, Nonlinear Anal. 13(1989), pp. 629–643.

100. J.-P. Penot, *Sub-Hessians, super-Hessians and conjugation*, Nonlinear Anal. 23 (1994), pp. 689–702.

101. R. R. Phelps, *Convex Functions, Monotone Operators, and Differentiability*, Lecture Notes in Math. 1364, Springer, New York, 1989.

102. B. H. Pourciau, *Analysis and optimization of Lipschitz continuous mappings*, J. Optim. Theory Appl. 22(1977), pp. 311–351.

103. B. H. Pourciau, *Modern multiplier rules*, Amer. Math. Monthly 87(1980), pp. 433–452.

104. B. Pschenichnii, *Necessary Conditions for an Extremum*, Marcel Dekker, New York, 1971.

105. S. M. Robinson, *Stability theory for sytems of inequalities, part II: differentiable nonlinear systems* SIAM J. Numer. Anal., 13(1976), pp. 497–513.

106. R. T. Rockafellar, *Convex Analysis*, Princeton University Press, Princeton, NJ, 1970.

107. R. T. Rockafellar and R. J. Wets, *Variational Analysis*, Springer, New York, 1997.

108. A. Rubinov and X. Q. Yang, *Lagrange-type Functions in Constrained Nonconvex Optimization*, Kluwer Academic, Boston, 2003.

109. M. A. Tawhid, *On the Local uniqueness of solutions of variational inequalities under H-differentiability*, J. Optim. Theory Appl. 113(2002), pp.149–154.

110. G. Stampacchia, *Formes bilinéaires coercives sur les ensembles convexes*, C. R. A. S. Paris, Ser. I Math. 258(1964), pp.4413–4416.

111. M. Studniarski and V. Jeyakumar, *A generalized mean-value theorem and optimality conditions in composite nonsmooth minimization*, Nonlinear Anal. 24(1995), pp. 883–894.

112. M. Thera, *A note on the Hartman-Stampacchia theorem*, Nonlinear Anal. Appl., V. Lakshamikantham (ed.), Dekker, New York (1987), pp. 573–577.

113. L. Thibault, *On generalized differentials and subdifferentials of Lipschitz vector valued functions*, Nonlinear Anal. 6(1982), 1037–1053.

114. J. S. Treiman, *The linear nonconvex generalized gradient and Lagrange multipliers*, SIAM J. Optim. 5(1995), pp. 670–680.

115. J. S. Treiman, *Lagrange multipliers for nonconvex generalized gradients with equality, inequality and set constraints*, SIAM J. Control Optim. 37(1999), pp. 1313–1329.

116. X. Wang and V. Jeyakumar, *A sharp Lagrange multiplier rule for nonsmooth mathematical programming problems involving equality constraints*, SIAM J. Optim. 10(1999), pp. 1136–1148.

117. J. Warga, *An implicit function theorem without differentiability*, Proc. Amer. Math. Soc. 69(1978), pp. 65–69.

118. J. Warga, *Fat homeomorphisms and unbounded derivate containers*, J. Math. Anal. Appl. 81(1981), pp. 545–560.

119. X. Q. Yang, *Second-order global optimality conditions for convex composite optimization*, Math. Programming, 81 (1998), pp. 327–347.

120. X. Q. Yang and V. Jeyakumar, *First and second order optimality conditions for convex composite multiobjective optimization*, J. Optim. Theory Appl. 95(1997), pp. 209–224.

121. X. Q. Yang and V. Jeyakumar, *Generalized second-order directional derivatives and optimization with $C^{1,1}$ functions*, Optimization, 26 (1992), pp. 165–185.

122. D. Zagrodny, *Approximate mean value theorem for upper subderivatives,* Nonlinear Anal. 12(1988), pp. 1413–1428.

123. C. Zalinescu, *Convex Analysis in General Vector Spaces,* World Scientific, London, 2003.

Notations

IN: the natural numbers
IR: the real numbers
IR^n: Euclidean n-dimensional space
$L(IR^n, IR^m)$: space of $m \times n$ matrices
B_n: closed unit ball in IR^n
$B_{m \times n}$: closed unit ball in $L(IR^n, IR^m)$
$\|x\|$: Euclidean norm
$\langle x, y \rangle$: canonical scalar product
$\mathcal{O}_n$: origin of IR^n
$cl(A)$, $\overline{A}$: closure
$int(A)$: interior
$co(A)$: convex hull
$\overline{co}(A)$: closed convex hull
$cone(A)$: conic hull
K^*: positive polar cone
K^δ: conic δ-neighborhood
A_∞: recession/asymptotic cone
$N(A, x)$: normal cone
$T(A, x)$: Bouligant contingent cone
$T_0(A, x)$: cone of feasible directions
$T_1(S, x)$: first-order tangent cone
$T_2(S, x)$: second-order tangent cone
$C_{(f,g)}(K, x)$: critical cone
$\leq_K$: partial order generated by K
$dom(f)$: effective domain
$epi(f)$: epigraph
$d(x, C)$: distance function
σ_C: support function

$\phi^+(x; u)$: upper Dini directional derivative
$\phi^-(x; u)$: lower Dini directional derivative
$\phi'(x; u)$: directional derivative
$\nabla f(x)$: Jacobian matrix
$\partial f(x)$: pseudo-Jacobian
$\partial_x f(x, y)$: partial pseudo-Jacobian
$\phi^0(x; u)$: Clarke's directional derivative
$\partial^C f(x)$: Clarke's subdifferential
$D^M f(x)$: Mordukhovich's coderivative
$\partial^M f(x)$: basic subdifferential
$\partial^{ca} f(x)$: convex analysis subdifferential
$\partial_\epsilon f(x)$: ϵ-subdifferential
$\phi^\uparrow(x; u)$: Clarke–Rockafellar's directional derivative
$\partial^{CR} f(x)$: Clarke–Rockafellar's subdifferential
$\partial^B f(x)$: B-subdifferential
$\partial^{IA} f(x)$: Ioffe's approximate subdifferential
$\partial^{MP} f(x)$: Michel–Penot's subdifferential
$\partial^l f(x)$: Treiman's linear generalized gradient
$\partial^2 f(x)$: pseudo-Hessian
$\partial_H^2 f(x)$: Hiriart-Urruty, Strodiot, and Hien's generalized Hessian
$\partial^{00} f(x)$: Cominetti and Correa's generalized Hessian
$\hat{F}(x)$: Kuratowski–Painleve's upper limit
$F^\infty(x)$: recession (upper horizon) limit

Index